Springer INdAM Series

Volume 35

Springer INdAM Series

This series will publish textbooks, multi-authors books, thesis and monographs in English language resulting from workshops, conferences, courses, schools, seminars, doctoral thesis, and research activities carried out at INDAM - Istituto Nazionale di Alta Matematica, http://www.altamatematica.it/en. The books in the series will discuss recent results and analyze new trends in mathematics and its applications.

THE SERIES IS INDEXED IN SCOPUS

More information about this series at http://www.springer.com/series/10283

Carlotta Giannelli · Hendrik Speleers
Editors

Advanced Methods for Geometric Modeling and Numerical Simulation

Springer

Editors
Carlotta Giannelli
Department of Mathematics and Computer Science "U. Dini"
University of Florence
Florence, Italy

Hendrik Speleers
Department of Mathematics
University of Rome Tor Vergata
Rome, Italy

ISSN 2281-518X ISSN 2281-5198 (electronic)
Springer INdAM Series
ISBN 978-3-030-27333-0 ISBN 978-3-030-27331-6 (eBook)
https://doi.org/10.1007/978-3-030-27331-6

This Springer imprint is published by the registered company Springer Nature Switzerland AG
The registered company address is: Gewerbestrasse 11, 6330 Cham, Switzerland

Preface

This book brings together nine contributions selected from among those presented at the INdAM workshop "DREAMS," which was held in Rome (Italy) from January 22 to 26, 2018. This international workshop was the closing event of the "Futuro in Ricerca" project entitled "*Design of Reliable, Exact, and Application-oriented technologies for geometric Modeling and numerical Simulation* (DREAMS)," funded by the Ministero dell'Istruzione, dell'Università e della Ricerca (MIUR, Italy). The project was organized in two research units, located in Florence and Rome, and coordinated by ourselves. We are very grateful for this support.

The DREAMS project focused on advanced computer aided geometric design (CAGD) methods and related application areas. Its research plan explored the identification of novel geometric representations suitable for automatic control and numerical simulation. The following research topics were covered by the project: algebraic-geometric structures in CAGD and related algorithms; adaptive splines and splines on unstructured meshes in isogeometric analysis (IgA); and interconnections between applied geometry, CAGD, approximation theory, linear algebra, and IgA.

The INdAM workshop provided an opportunity for interaction among researchers interested in CAGD and related application areas, particular attention being paid to the topics of the DREAMS project. The scientific activities of the DREAMS week promoted stimulating discussion and collaboration among the participants.

The first contribution in the book, written by Alessandra Aimi, is entitled "An Isogeometric Approach to Energetic BEM: Preliminary Results." It presents an isogeometric method for the solution of a space-time variational formulation of the two-dimensional wave equation.

The second contribution, "Approximate Reconstructions of Perturbed Rational Planar Cubics," by Michal Bizzarri, Miroslav Lávička, and Jan Vršek, addresses the problem of approximating a non-rational planar cubic curve by a rational one, employing hybrid (symbolic-numeric) computations.

The contribution, "Quadrature Rules in the Isogeometric Galerkin Method: State of the Art and an Introduction to Weighted Quadrature," by Francesco Calabrò, Gabriele Loli, Giancarlo Sangalli, and Mattia Tani, reviews some quadrature rules used for efficient matrix assembly in isogeometric methods, with a special focus on weighted quadrature.

In the contribution "Eigenvalue Isogeometric Approximations Based on B-Splines: Tools and Results," Sven-Erik Ekström and Stefano Serra-Capizzano focus on the spectral analysis of discretization matrices resulting from the isogeometric B-spline discretization of second-order differential equations.

The contribution, "Lofting with Patchwork B-Splines," by Nora Engleitner and Bert Jüttler, presents a new method to construct lofting surfaces, one of the fundamental operations for creating free-form surfaces in computer aided design, using patchwork B-splines.

In the contribution "A Study on Spline Quasi-interpolation Based Quadrature Rules for the Isogeometric Galerkin BEM," Antonella Falini and Tadej Kanduč discuss the use of quadrature rules based on quasi-interpolation schemes for weakly singular integrals in the context of isogeometric boundary element methods.

The contribution, "New Developments in Theory, Algorithms, and Applications for Pythagorean–Hodograph Curves," by Rida T. Farouki, Carlotta Giannelli, and Alessandra Sestini, provides a survey of the recent results in the field of Pythagorean–hodograph curves and related topics.

The contribution, "Tchebycheffian B-Splines Revisited: An Introductory Exposition," by Tom Lyche, Carla Manni, and Hendrik Speleers, gives a refreshing review on the construction of B-splines for smooth piecewise function spaces, where the pieces are taken from extended Tchebycheff spaces, and their properties.

The last contribution, "Template Mapping Using Adaptive Splines and Optimization of the Parameterization," by Svajūnas Sajavičius, Bert Jüttler, and Jaka Špeh, presents a method to construct an accurate spline map that transforms a template domain into a target domain, by applying an iterative procedure based on least-squares fitting.

We wish to thank all the speakers of the workshop, as well as all the participants, for the interesting talks, the scientific discussions, and the pleasant atmosphere during the week in Rome. This book would not have been possible without the authors of the nine papers, who deserve our deepest gratitude for their interesting contributions. Finally, we would like to thank INdAM for their support for the workshop organization, which also included the nice facilities in Rome. Special thanks are due to Mauro Petrucci, who was always available to help us with any organizational issues and practical arrangements.

Florence, Italy
Rome, Italy
April 2019

Carlotta Giannelli
Hendrik Speleers

Contents

About the Editors

Carlotta Giannelli is an Associate Professor of Numerical Analysis at the Department of Mathematics and Computer Science, University of Florence, Italy. She received her Ph.D. in Computer Science and Applications from the University of Florence in 2010. Her primary research interest is in computer aided geometric design and related application areas. She is the author of more than 40 peer-reviewed research publications.

Hendrik Speleers received his Ph.D. in Engineering (Numerical Analysis and Applied Mathematics) from the University of Leuven, Belgium in 2008. He is currently an Associate Professor of Numerical Analysis at the Department of Mathematics, University of Rome Tor Vergata, Italy. His main research interest is in the construction, analysis, and application of multivariate splines. He is the author of more than 60 peer-reviewed scientific papers.

An Isogeometric Approach to Energetic BEM: Preliminary Results

Alessandra Aimi

Abstract The Energetic Boundary Element Method (BEM) is a discretization technique for the numerical solution of wave propagation problems, introduced in [2] and applied in the last decade for the simulation of wave propagation inside bounded domains or outside bounded obstacles, in several space dimensions. The differential initial-boundary value problem at hand is converted into a space-time Boundary Integral Equation (BIE), which is then written in weak form through energy considerations and discretized by a Galerkin approach. Taking into account the model problem of 2D soft scattering by open arcs, the aim of this paper is to explore the introduction of the powerful IgA approach, firstly introduced by Hughes and collaborators [18] in the context of Finite Element Methods, into Energetic BEM for what concerns space discretization, in order to take the same benefits already observed in IgA-SGBEM (Symmetric Galerkin BEM) applied to BIEs related to elliptic problems [8]. Numerical challenges to be faced for an efficient integration of weakly singular kernels related to the fundamental solution of the wave operator will be outlined and preliminary numerical results will be given and discussed.

1 Introduction

A variety of engineering and physical applications, such as the propagation or the scattering of acoustic or elastic waves, leads to the problem of solving linear hyperbolic Partial Differential Equations (PDEs) in two or three dimensional space. These problems are usually considered in an unbounded homogeneous domain and advanced numerical techniques to solve them, such as Finite Element Methods (FEMs) and Finite Difference Methods (FDMs), even if well established, suffer from improper domain truncation and from the choice of boundary conditions to be suitably imposed on the artificial boundary. A nowadays widely used method to tackle

A. Aimi (✉)
Department of Mathematical, Physical and Computer Sciences, University of Parma, Parma, Italy
e-mail: alessandra.aimi@unipr.it

C. Giannelli and H. Speleers (eds.), *Advanced Methods for Geometric Modeling and Numerical Simulation*, Springer INdAM Series 35,
https://doi.org/10.1007/978-3-030-27331-6_1

this situation is to reformulate the PDE as a Boundary Integral Equation (BIE) on the usually bounded boundary of the problem's original domain; the BIE can then be numerically solved by the Boundary Element Method (BEM) [12, 17]. In some applications, the physically relevant data are given not by the solution in the interior of the domain but rather by the boundary values of the solution or its derivatives. These data can be obtained directly from the solution of BIEs, whereas it is well known that boundary values obtained from FEM solutions are in general not so accurate. Furthermore, since wave propagation phenomena are often observed in semi-infinite media (domain) where Sommerfeld radiation condition holds, a suitable numerical method has to ensure that this condition is not violated. For example, FEMs need the application of special techniques to fulfill this condition that, on the contrary, is implicitly fulfilled by BEM. Anyway, when a local non-linearity appears, a suitable BEM-FEM coupling, when applicable, gives undoubted advantages.

For the numerical solution of wave propagation problems, one needs consistent approximations and accurate simulations even on large time intervals. In principle, both frequency-domain [14] and time-domain [10, 11] BEM can be used for hyperbolic initial-boundary value problems. Space-time BEM has the advantage that it directly yields the unknown time-dependent quantities, thus avoiding the use of the Laplace transform and its inversion. In the time-domain framework, the construction of the BIEs, via representation formula in terms of single and double layer potentials, uses the fundamental solution of the hyperbolic PDE and jump relations [16]. The mathematical background of time-dependent BIEs is summarized by M. Costabel in [15].

For the numerical solution of the wave equation inside bounded domains or outside bounded obstacles, the so-called space-time Energetic BEM has been recently introduced in the literature, for problems defined in several space dimensions [2, 3, 6]. Energetic BEM comes from a Galerkin type discretization of a weak formulation based on energy arguments, directly expressed in the space-time domain.

On the other side, the isogeometric approach applied to BEMs is an emerging research area in different fields of applications (see e.g. [8, 9, 20, 22] for problems modeled by elliptic PDEs). The Isogeometric Analysis (IgA) paradigm, introduced by Hughes and collaborators in the seminal paper [18] in the context of FEMs, uses the same B-spline or NURBS basis, as adopted in CAD systems, to describe both the boundary of the domain and the approximate solution of the problem at hand, giving surprising computational advantages. Since then, IgA has brought a renewed interest in BEMs, because these methods need to work only on the boundary avoiding the domain meshing and drastically reducing the number of Degrees of Freedom (DoF).

Taking into account the model problem of 2D soft scattering by open arcs, the aim of this paper is to explore the introduction of the powerful B-spline based IgA paradigm into Energetic BEM for what concerns space discretization, in order to take the same benefits already observed in IgA-SGBEM (Symmetric Galerkin BEM) applied to BIEs related to elliptic problems [8].

The paper is structured as follows. First, the differential model problem on an unbounded 2D domain and its energetic boundary integral weak formulation are presented. Then, the consequent BEM discretization is illustrated, highlighting numerical aspects of the quadrature of weakly singular space-time integrals. Finally, preliminary numerical results will be given and discussed, showing the superiority of the IgA approach versus the standard Energetic BEM; the latter based on the classical local Lagrangian basis for the approximation of the problem solution.

2 Model Problem and Its Energetic Boundary Integral Weak Formulation

Let us consider a Dirichlet 2D problem for the wave equation exterior to an open arc $\Gamma \subset \mathbf{R}^2$, which models a *soft scattering* phenomenon:

$$\left[\Delta u - \frac{1}{c^2} u_{tt}\right](\mathbf{x}, t) = 0, \qquad \mathbf{x} \in \mathbf{R}^2 \setminus \Gamma,\ t \in [0, T], \tag{1}$$

$$u(\mathbf{x}, 0) = u_t(\mathbf{x}, 0) = 0, \qquad \mathbf{x} \in \mathbf{R}^2 \setminus \Gamma, \tag{2}$$

$$u(\mathbf{x}, t) = g(\mathbf{x}, t), \qquad \mathbf{x} \in \Gamma,\ t \in [0, T], \tag{3}$$

where c is the propagation velocity of a perturbation inside the domain, T is the final time instant of analysis and the boundary datum $g(\mathbf{x}, t)$ represents the time history of the excitation field over Γ.

Since the above problem will be discretized using a boundary element technique, it must be rewritten in a boundary integral form. This can be done using classical arguments and the knowledge of the fundamental solution of the 2D wave operator. The starting point is the single-layer representation of the solution of (1)–(3):

$$u(\mathbf{x}, t) = \int_\Gamma \int_0^t G(\mathbf{r}, t - \tau)\psi(\mathbf{y}, \tau)\, d\tau\, d\gamma_{\mathbf{y}}, \qquad \mathbf{x} \in \mathbf{R}^2 \setminus \Gamma,\ t \in (0, T], \tag{4}$$

where $\mathbf{r} := \mathbf{x} - \mathbf{y}$, the unknown density $\psi = \left[\frac{\partial u}{\partial \mathbf{n}}\right]_\Gamma$ represents the time history of the jump of the normal derivative of u along Γ and

$$G(\mathbf{x}, t) = \frac{c}{2\pi} \frac{H(ct - \|\mathbf{x}\|)}{\sqrt{c^2t^2 - \|\mathbf{x}\|^2}} \tag{5}$$

is the forward fundamental solution of the 2D wave operator, with $H(\cdot)$ the Heaviside function and $\|\cdot\|$ the Euclidean norm.

From (4), it is clear that in order to recover the solution of the differential problem at any point outside the obstacle and at any time instant, one has to proceed with a post-processing phase, provided that the density function $\psi(\mathbf{x}, t)$ is known. To this aim, performing a limit process for $\mathbf{x}$ tending to Γ in (4) and using the assigned Dirichlet boundary condition (3), the weakly singular space-time BIE

$$\int_{\Gamma}\int_{0}^{t} G(\mathbf{r}, t-\tau)\psi(\mathbf{y}, \tau)\, d\tau\, d\gamma_{\mathbf{y}} = g(\mathbf{x}, t), \qquad \mathbf{x} \in \Gamma,\ t \in [0, T], \tag{6}$$

in the unknown $\psi(\mathbf{x}, t)$ can be obtained and written in the compact notation

$$V\psi = g\,. \tag{7}$$

Problem (7) is then set in weak form. The so-called energetic weak formulation of (7) can be deduced observing that, multiplying the PDE (1) by u_t, integrating over $[0, T] \times (\mathbf{R}^2 \setminus \Gamma)$ and using integration by parts in space, one obtains that the energy of the solution u at the final time of analysis T, defined by

$$\mathcal{E}(u, T) := \frac{1}{2}\int_{\mathbf{R}^2\setminus\Gamma} \Big[\|\nabla_{\mathbf{x}} u(\mathbf{x}, T)\|^2 + \frac{1}{c^2}u_t^2(\mathbf{x}, T)\Big] d\gamma_{\mathbf{x}}, \tag{8}$$

can be rewritten as

$$\mathcal{E}(u, T) = \int_{\Gamma}\int_{0}^{T} u_t(\mathbf{x}, t)\Big[\frac{\partial u}{\partial \mathbf{n_x}}\Big]_{\Gamma}(\mathbf{x}, t)\, dt\, d\gamma_{\mathbf{x}} = \int_{\Gamma}\int_{0}^{T} (V\,\psi)_t(\mathbf{x}, t)\, \psi(\mathbf{x}, t)\, dt\, d\gamma_{\mathbf{x}}\,. \tag{9}$$

Hence, after projecting (7), derived w.r.t. time, by means of test functions ϕ belonging to the same functional space of the unknown density ψ, the energetic weak problem finally reads as follows: *find* $\psi \in L^2([0, T]; H^{-1/2}(\Gamma))$ *such that*

$$\int_{\Gamma}\int_{0}^{T} (V\,\psi)_t(\mathbf{x}, t)\phi(\mathbf{x}, t)\, dt\, d\gamma_{\mathbf{x}} = \int_{\Gamma}\int_{0}^{T} g_t\ (\mathbf{x}, t)\phi(\mathbf{x}, t)\, dt\, d\gamma_{\mathbf{x}}\,, \qquad \forall\phi \in L^2([0, T]; H^{-1/2}(\Gamma))\,. \tag{10}$$

Remark 1 The theoretical analysis of the quadratic form coming from the left-hand side of (10) was carried out in [6] where, under suitable assumptions, coercivity was proved with some technicalities. This property allowed to deduce stability and convergence of the related Galerkin approximate solution.

3 Energetic BEM Discretization

At the beginning of this section, the definition of the B-spline basis used in the IgA approach is briefly recalled, while the definition of the local Lagrangian basis used below to compare IgA versus standard Energetic BEM is not written, because it is really basic.

Given a partition $\Delta := \{a = \sigma_0 < \cdots < \sigma_n = b\}$ of an interval $[a, b] \subset \mathbf{R}$, a general polynomial spline space $\mathcal{B}$ of order k on such partition is composed by piecewise polynomial functions of degree $d = k - 1$ which are required to have an assigned regularity C^{k-1-m_i} at the breakpoints σ_i, $i = 1, \ldots, n-1$, with m_i denoting an integer between 1 and k.[1] For example when all the m_i are fixed equal to 1 or to $k-1$ or to k, respectively, $\mathcal{B}$ is a subset of $C^{k-2}[a,b]$, it is included in $C[a,b]$ or it is just a subset of $L^2(a,b)$. It is quite easy to verify that the dimension of such space is $dim(\mathcal{B}) = k + \sum_{i=1}^{n-1} m_i$. The easiest way to define in $\mathcal{B}$ a B-spline basis $B_{i,k}(t), i = 0, \dots, N$, with $N+1 = dim(\mathcal{B})$, is based on the usage of a recursion formula and can be described through two easy steps [13]. The first step consists in associating to $\mathcal{B}$ an *extended knot vector* $Z = \{\zeta_0, \dots, \zeta_{N+k}\}$ whose elements constitute a non-decreasing sequence of abscissas, where $\{\zeta_{k-1}, \dots, \zeta_{N+1}\}$ are the *internal knots* with $\zeta_{k-1} = \sigma_0$, $\zeta_{N+1} = \sigma_n$ and $\{\zeta_k, \dots, \zeta_N\} = \{\sigma_1, \dots, \sigma_1, \dots, \sigma_{n-1}, \dots, \sigma_{n-1}\}$, where each σ_i has m_i occurrences and it is said *multiple* if $m_i > 1$. The remaining knots in Z, $\{\zeta_0, \dots, \zeta_{k-2}\}$ and $\{\zeta_{N+2}, \dots, \zeta_{N+k}\}$, form two sets of $k-1$ knots called *auxiliary left* and *right knots* which are only required to verify the following inequalities, $\zeta_0 \le \dots \le \zeta_{k-2} \le \zeta_{k-1} = a$ and $b = \zeta_{N+1} \le \zeta_{N+2} \le \dots \le \zeta_{N+k}$. Note that the standard assumption of selecting an *open* extended knot vector, that is $\zeta_0 = \dots = \zeta_{k-2} = \zeta_{k-1} = a$ and $b = \zeta_{N+1} = \zeta_{N+2} = \dots = \zeta_{N+k}$, will be always considered in the numerical simulations. In the second step, the basis is defined by using the following recursion [13]:

$$B_{i,1}(s) := \begin{cases} 1, \text{ if } \zeta_i \le s < \zeta_{i+1}, \\ 0, \text{ otherwise;} \end{cases}$$

$$B_{i,j}(s) := \omega_{i,j}(s)\, B_{i,j-1}(s) + (1 - \omega_{i+1,j}(s))\, B_{i+1,j-1}(s), \quad 1 < j \le k,$$

with

$$\omega_{i,j}(s) := \begin{cases} \frac{s-\zeta_i}{\zeta_{i+j-1}-\zeta_i} & \text{if } \zeta_i < \zeta_{i+j-1}, \\ 0 & \text{otherwise;} \end{cases}$$

and fractions with zero denominator are considered to be zero. Note that from the above recursive definition it is easy to verify that the B-splines are non-negative and that the support of $B_{i,k}$ is the subinterval $[\zeta_i\,,\ \zeta_{i+k}]$.

Now, let us suppose that the obstacle Γ is given by a B-spline parametric representation of order k over the interval $[a,b]$, i.e.

$$\Gamma := \{\mathbf{x} \in \mathbf{R}^2 \,|\, \mathbf{x} = \mathbf{C}(s) = \sum_{i=0}^{N} \mathbf{P}_i\, B_{i,k}(s),\ s \in [a,b]\}, \tag{11}$$

where $\mathbf{P}_i$, $i = 0, \dots, N$ are suitably given control points. For the approximation of the unknown boundary density ψ in space variable, the functional background compels to choose space basis functions belonging to $L^2(\Gamma)$, but for smooth obstacles one can require at least $C^0(\Gamma)$ regularity.

In the IgA paradigm, an Energetic BEM space discretization involves boundary element functions $w_j(\mathbf{x}) := B_{j,k}(\mathbf{C}^{-1}(\mathbf{x}))$, $j = 0, \dots, N$ defined by means of the B-spline basis related to the partition Δ as described above.

[1] When $m_i = k$, this means that the function has a finite jump at σ_i.

On the other side, the partition Δ induces on the obstacle Γ a boundary mesh constituted by n non-overlapping curvilinear elements $\{e_1, \ldots, e_n\}$, such that $\bigcup_{i=1}^{n} \overline{e}_i = \overline{\Gamma}$. Then, for the standard Energetic BEM, a local Lagrangian basis of degree $d = k - 1$ can be lifted from Δ to Γ, giving rise to classical piecewise polynomial boundary element functions $w_j^L(\mathbf{x}),\ j = 0, \ldots, N_L$ suitably defined on the boundary mesh.

In the following M will denote N or N_L in relation to the use of IgA or standard Energetic BEM, and the total number of DoF in space variable will be $N_{DoF} = M + 1$. Further, in order to unify the notation, in the sequel the upper index L will be dropped from w_j^L, being clear from the context which kind of space basis functions will be employed for the simulations.

For the time discretization, a uniform decomposition of the time interval $[0, T]$ with time step $\Delta t = T/N_{\Delta t}\,,\ N_{\Delta t} \in \mathbf{N}^+$, generated by the $N_{\Delta t} + 1$ instants $t_k = k\,\Delta t,\ k = 0, \ldots, N_{\Delta t}\,,$ is considered and the time mesh is equipped by piecewise constant time shape functions. Note that, for this particular choice which will allow analytical double integration in time variables, shape functions, denoted by $v_k(t),\ k = 0, \ldots, N_{\Delta t} - 1$, will be defined as

$$v_k(t) = H(t - t_k) - H(t - t_{k+1})\,. \tag{12}$$

Hence, the approximate solution of the problem at hand will be expressed as

$$\psi(\mathbf{x}, t) \simeq \tilde{\psi}(\mathbf{x}, t) := \sum_{k=0}^{N_{\Delta t}-1} \sum_{j=0}^{M} \alpha_j^{(k)}\, w_j(\mathbf{x})\, v_k(t). \tag{13}$$

The Galerkin BEM discretization coming from the energetic weak formulation (10) produces the linear system

$$A\,\alpha = \beta\,, \tag{14}$$

of order $N_{DoF} \cdot N_{\Delta t}$, where the matrix A has a block lower triangular Toeplitz structure, due to respectively the choice (12) and the fact that kernel G in (4) depends on the difference between t and τ. Note that higher order time basis functions would produce, instead, a block Hessenberg type structure.

Each block has dimension N_{DoF}. If we indicate with $A^{(\ell)}$ the block obtained when $t_h - t_k = \ell\,\Delta t,\ \ell = 0, \ldots, N_{\Delta t} - 1$, the linear system can be written as

$$\begin{pmatrix} A^{(0)} & 0 & 0 & \cdots & 0 \\ A^{(1)} & A^{(0)} & 0 & \cdots & 0 \\ A^{(2)} & A^{(1)} & A^{(0)} & \cdots & 0 \\ \cdots & \cdots & \cdots & \cdots & 0 \\ A^{(N_{\Delta t}-1)} & A^{(N_{\Delta t}-2)} & \cdots & A^{(1)} & A^{(0)} \end{pmatrix} \begin{pmatrix} \alpha^{(0)} \\ \alpha^{(1)} \\ \alpha^{(2)} \\ \vdots \\ \alpha^{(N_{\Delta t}-1)} \end{pmatrix} = \begin{pmatrix} \beta^{(0)} \\ \beta^{(1)} \\ \beta^{(2)} \\ \vdots \\ \beta^{(N_{\Delta t}-1)} \end{pmatrix}, \tag{15}$$

where $\alpha^{(\ell)} = \left(\alpha_j^{(\ell)}\right)$ and $\beta^{(\ell)} = \left(\beta_j^{(\ell)}\right),\ \ell = 0, \ldots, N_{\Delta t} - 1\,,\ j = 0, \ldots, M.$

The solution of (15) is obtained by a block forward substitution, i.e. at every time instant t_ℓ, one solves a reduced linear system of the type

$$A^{(0)}\alpha^{(\ell)} = \beta^{(\ell)} - (A^{(1)}\alpha^{(\ell-1)} + \cdots + A^{(\ell)}\alpha^{(0)}). \tag{16}$$

Procedure (16) is a time marching technique, where the only matrix to be inverted once for all is the symmetric, even if dense, non-singular $A^{(0)}$ diagonal block, while all the other blocks are used to update at every time step the right-hand side. Owing to this procedure, one can construct and store only the blocks $A^{(0)}, \ldots, A^{(N_{\Delta t}-1)}$ with a considerable reduction of computational cost and memory requirement.

On the other side, the only drawback is the necessity of calculating, as discretization matrix elements, double integrals, involving weakly singular kernel in the space variables, as it happens for SGBEM applied in the context of Dirichlet elliptic exterior problems. In fact, setting $\Delta_{hk} = t_h - t_k$, the matrix elements in blocks $A^{(\ell)}$, after a double analytic integration in the time variables, are of the form

$$\sum_{\mu,\nu=0}^{1} (-1)^{\mu+\nu} \int_\Gamma w_l(\mathbf{x}) \int_\Gamma H(c\,\Delta_{h+\mu,k+\nu} - r)\,\mathcal{V}(r, t_{h+\mu}, t_{k+\nu}) w_m(\mathbf{y})\, d\gamma_{\mathbf{y}}\, d\gamma_{\mathbf{x}}, \tag{17}$$

where

$$\mathcal{V}(r, t_h, t_k) = \frac{1}{2\pi}\left[\log\left(c\,\Delta_{hk} + \sqrt{c^2\,\Delta_{hk}^2 - r^2}\right) - \log r\right], \tag{18}$$

and $H(c\Delta_{h+\mu,k+\nu} - r)$ models the wavefront propagation.

Using the standard element-by-element technique, the evaluation of every double integral in (17) is reduced to the assembling of local contributions of the type

$$\int_{e_i} w_l(\mathbf{x}) \int_{e_j} H(c\,\Delta_{hk} - r)\,\mathcal{V}(r, t_h, t_k) w_m(\mathbf{y})\, d\gamma_{\mathbf{y}}\, d\gamma_{\mathbf{x}}, \tag{19}$$

where e_i, e_j belong to (or coincide with) the support of w_l, w_m, respectively.

Numerical challenges for an efficient integration of the whole kernel in (17) were faced in [4] for standard Energetic BEM, but for the readers' convenience they will be outlined in the next section.

4 Technical Issues in Matrix Element Evaluation

Looking at (18), a space singularity of type $\log r$ as $r \to 0$, which is typical of $2D$ weakly singular kernels related to elliptic problems, is observed. Hence, efficient evaluation of double integrals (19) is particularly required when $e_i \equiv e_j$ or $e_j = e_{i+1}$. Further, the Heaviside function $H(c\,\Delta_{hk} - r)$ in (19) and the square root function $\sqrt{c^2\,\Delta_{hk}^2 - r^2}$ in (18) give rise to other different type of troubles, which have to be

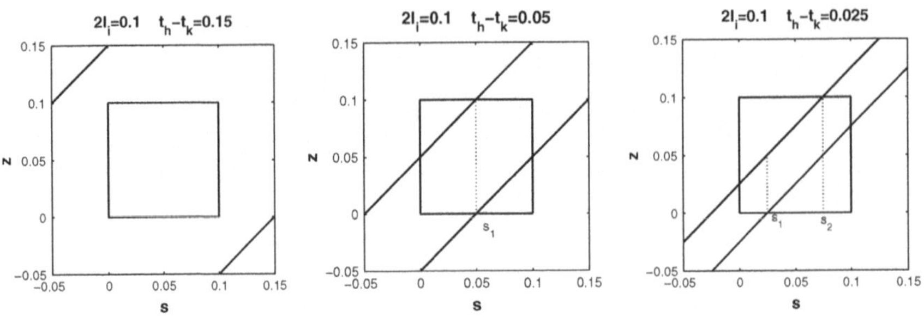

Fig. 1 Double integration domain (coincident elements) for different values of Δ_{hk} (increasing time from right to left)

properly managed. For the sake of clearness, let us describe the above issues in a geometrical setting where $\mathbf{C} = Id$, fixing the velocity $c = 1$.

When the double integration has to be done over $e_i \equiv e_j$, which represent in this case two coincident straight boundary elements, double integral (19) simplifies in

$$\int_{e_i} w_l(s) \int_{e_i} H(\Delta_{hk} - r)\, \mathcal{V}(r, t_h, t_k) w_m(z)\, dz\, ds\,, \tag{20}$$

with $r = |s - z|$ in the local variables of integration $s, z \in (0, 2l_i)$, where $2l_i$ represents the length of the element e_i.

Let us note that the double integration domain is constituted by the intersection between the square $e_i \times e_i$ and strip $|s - z| < \Delta_{hk}$ where the Heaviside function is not trivial. The singularities arise for $r = 0$ along a diagonal of the square, precisely along the line $z = s$. Figure 1 shows these intersections for different values of Δ_{hk} and fixed length $2l_i = 0.1$.

Setting

$$M_s = \max(0, s - \Delta_{hk}), \qquad m_s = \min(2l_i, s + \Delta_{hk})\,, \tag{21}$$

the double integral (20) can be rewritten as

$$\int_0^{2l_i} w_l(s) \int_{M_s}^{m_s} \mathcal{V}(r, t_h, t_k)\, w_m(z)\, dz\, ds\,. \tag{22}$$

The numerical quadrature in the outer variable of integration s has been performed subdividing, when necessary, the outer interval of integration. Without this subdivision, one should use a lot of quadrature nodes for the outer numerical integration in order to achieve the single precision accuracy. In fact, the derivative with respect to s of the outer integrand function, after the inner integration, presents jumps in correspondence to possible subdivision points given by

$$s_1 = \Delta_{hk}, \qquad s_2 = 2l_i - \Delta_{hk}\,. \tag{23}$$

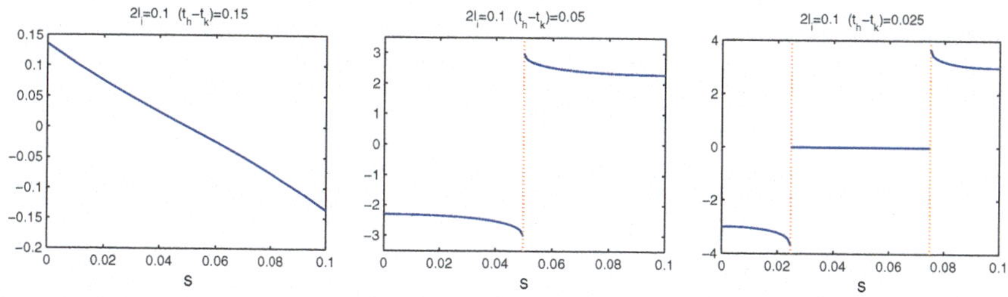

Fig. 2 Behavior of the outer integrand function derivative for different values of Δ_{hk}

Note that in this simple geometrical case, if $\Delta_{hk} > 2l_i$ these points do not belong to the integration interval $[0, 2l_i]$; if $\Delta_{hk} = 2l_i$ these points coincide with the endpoints of the integration interval $[0, 2l_i]$; when $0 < \Delta_{hk} < 2l_i$, $\Delta_{hk} \neq l_i$ both points belong to $[0, 2l_i]$; at last, when $\Delta_{hk} = l_i$ only one point belongs to the integration interval ($s_1 = s_2$). Almost all these geometrical situations are shown in Fig. 1. Further, as an example, Fig. 2 presents the behavior of the derivative

$$\frac{d}{d\,s}\left[\int_{M_s}^{m_s} \log(\Delta_{hk} + \sqrt{\Delta_{hk}^2 - |s - z|^2})\,dz\right],$$

referred to the domains of the previous figure and to an integrand function related to the non-singular part (for $r \to 0$) of the kernel (18). Hence, (22) will be eventually decomposed into the sum of double integrals of the form

$$\int_{\tilde{a}}^{\tilde{b}} w_l(s) \int_{M_s}^{m_s} \mathcal{V}(r, t_h, t_k)\, w_m(z)\, dz\, ds\,, \tag{24}$$

where $[\tilde{a}, \tilde{b}] \subset [0, 2l_i]$. Of course when no subdivision is needed, one will have to deal with only one double integral (24) where $[\tilde{a}, \tilde{b}] \equiv [0, 2l_i]$.

In Table 1 the computational gain obtained to achieve single precision accuracy when using the suggested splitting is displayed in the numerical evaluation of the outer integral in

$$\int_{0}^{2l_i} \int_{M_s}^{m_s} \log(\Delta_{hk} + \sqrt{\Delta_{hk}^2 - |s - z|^2})\; dz\, ds\,, \tag{25}$$

for $2l_i = 0.1$ and $\Delta_{hk} = 0.05,\ 0.025$, using a classical Gauss–Legendre rule. Here, the inner integral has been evaluated analytically.

In the case of double integration over consecutive elements e_i and e_j, with $e_j \equiv e_{i+1}$, the distance between the source and the field point can be written as $r = s + z$, with the origin of a local reference system fixed in the common endpoint, $s \in [0, 2l_i]$ and $z \in [0, 2l_j]$. Hence the double integration domain is represented by the intersection between the rectangle $e_i \times e_j$, where the only singular point lies in

Table 1 Relative errors in the outer numerical integration of (25) when $\Delta_{hk} = 0.05, 0.025$, with and without the proposed splitting. The symbol $--$ means that the single precision accuracy has been achieved

	$\Delta_{hk} = 0.05$		$\Delta_{hk} = 0.025$	
n.nodes	Without splitting	With splitting	Without splitting	With splitting
4	1.5871×10^{-2}	1.2663×10^{-5}	4.8220×10^{-4}	4.7584×10^{-6}
8	4.4374×10^{-3}	4.8539×10^{-7}	5.2265×10^{-2}	7.3393×10^{-7}
16	1.1928×10^{-3}	$--$	2.1787×10^{-4}	$--$
32	3.1162×10^{-4}		3.9416×10^{-4}	
64	7.9969×10^{-5}		2.1043×10^{-5}	
128	2.0301×10^{-5}		2.5823×10^{-5}	
256	5.1162×10^{-6}		2.0254×10^{-6}	
512	1.2786×10^{-6}		1.1237×10^{-6}	
1024	3.1285×10^{-7}		6.6507×10^{-7}	
2048	$--$		$--$	

the vertex of coordinates (0, 0), and the half plane: $z < \Delta_{hk} - s$, where the Heaviside function is not trivial. This intersection will not be empty if $0 < s < \Delta_{hk}$. Therefore, setting

$$m_0 = \min(\Delta_{hk}, 2l_i)\,, \qquad m_s = \min(2l_j, \Delta_{hk} - s)\,, \tag{26}$$

the double integral (19) becomes

$$\int_0^{m_0} w_l(s) \int_0^{m_s} \mathcal{V}(r, t_h, t_k)\, w_m(z)\, dz\, ds\,. \tag{27}$$

Also in this case, the numerical quadrature in the outer variable of integration s has been performed by subdividing, when necessary, the outer interval of integration. In fact, the derivative with respect to s of the outer integrand function, after the inner integration, presents a jump in correspondence to a possible subdivision point given by

$$s_1 = \Delta_{hk} - 2l_j\,. \tag{28}$$

Note that in this geometrical case, it is easy to verify that if $\Delta_{hk} < 2l_j$ or $\Delta_{hk} > 2l_i + 2l_j$ this point does not belong to the integration interval $[0, m_0]$, while if $2l_j < \Delta_{hk} < 2l_j + 2l_i$, s_1 breaks the outer integration interval in two subintervals. Hence, (27) will be eventually decomposed into the sum of double integrals of the form

$$\int_{\tilde{a}}^{\tilde{b}} w_l(s) \int_0^{m_s} \mathcal{V}(r, t_h, t_k)\, w_m(z)\, dz\, ds\,, \tag{29}$$

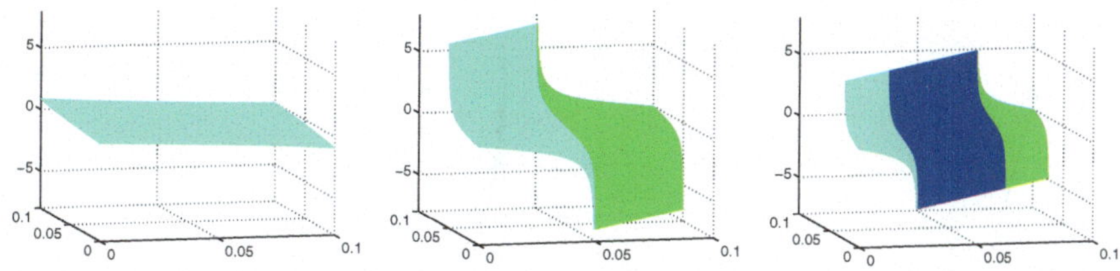

Fig. 3 Behavior of $\frac{\partial}{\partial z}\sqrt{\Delta_{hk}^2 - |s-z|^2}$ for $\Delta_{hk} = 0.15, 0.05, 0.025$

where $[\tilde{a}, \tilde{b}] \subset [0, m_0]$. Of course when no subdivision is needed, we will have to deal with only one double integral (29) where $[\tilde{a}, \tilde{b}] \equiv [0, m_0]$.

Another numerical issue arises in relation to the function $\log(c\,\Delta_{hk}+\sqrt{c^2\,\Delta_{hk}^2-r^2})$ in (18), even if it is, of course, not singular for $r \to 0$. Nevertheless the inner numerical integration of this function has to be performed carefully when the Heaviside function reduces the 2D integration domain: the problem one has to deal with is due to the presence of the square root function $\sqrt{c^2\,\Delta_{hk}^2 - r^2}$, as mentioned at the beginning of this section.

This issue will be illustrated for the same geometrical setting as before, always fixing $c = 1$ and considering now only the case of coincident elements. The argument of $\sqrt{\Delta_{hk}^2 - |s-z|^2}$ is always positive but it can assume very small values and in the limit for the argument tending to zero the derivative of the square root with respect to the inner variable of integration z becomes unbounded. This behavior, shown in Fig. 3, happens along the oblique boundary of the double integration domain, as shown in Fig. 1 for $\Delta_{hk} = 0.05,\ 0.025$, and produces a bad performance of the classical Gauss–Legendre quadrature formula, for instance in the inner numerical integration of (25), in the sense that one should use a lot of quadrature nodes to achieve the single precision accuracy.

To overcome this difficulty, the inner integration has been performed considering the regularization procedure in [19], which suitably pushes the Gaussian nodes towards the endpoints of the interval $[M_s, m_s]$ and modifies the Gaussian weights in order to regularize integrand functions with mild boundary singularities (the higher certain positive integer parameters p, q, the smoother the regularization). The outer integral in (25) is numerically evaluated by a classical Gauss–Legendre rule.

In Fig. 4, the computational cost of the Gaussian quadrature formula and of the regularization procedure just explained is shown, in relation to the achievement of the single precision accuracy (horizontal line) in the evaluation of the double integral (25) for $2l_i = 0.1$ and $\Delta_{hk} = 0.15, 0.05$ (results for $\Delta_{hk} = 0.025$ are similar to those for $\Delta_{hk} = 0.05$ and they have not been reported). The outer integral has been numerically evaluated with the 8-nodes classical Gauss–Legendre formula, recalling that for $\Delta_{hk} = 0.15$ the outer integration interval does not need a subdivision while for $\Delta_{hk} = 0.05$ the outer interval has been divided in two subintervals, and taking into account the numerical analysis presented in Table 1.

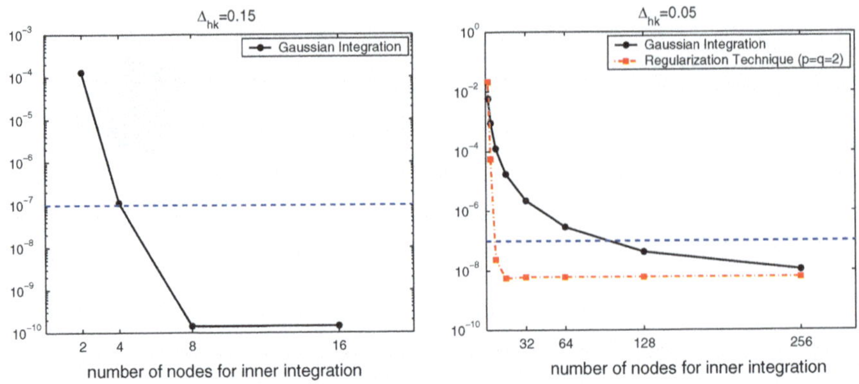

Fig. 4 Comparison between computational costs of Gaussian quadrature and regularization procedure [19] for the inner numerical integration of (25)

To conclude, in order to obtain preliminary numerical results of IgA-Energetic BEM, presented in the next section, the numerical evaluation of discretization matrix elements has been operated through quadrature schemes widely used for weakly singular kernels in the context of SGBEM coming from elliptic problems [7], coupled with a suitable regularization technique [19], after a careful subdivision of the integration domain due to the presence of the Heaviside function. The interested reader is referred to [4] for a wider and deeper treatment of the above described technical issues, usable when $\mathbf{C} = Id$.

The extension of the above explained decomposition technique, when $\mathbf{C}$ is not the identity, is the core of a general implementation of IgA-Energetic BEM, which is currently under study. Working in the framework of an element-by-element technique on the partition Δ introduced in the parametrization interval $[a, b]$, the key point for an efficient and accurate evaluation of (19) will be the automatic detection of possible subdivision points, both for the inner and the outer variables of integration. While in the case $\mathbf{C} = Id$ these points are expressed in closed form (see e.g. (21), (23)), in the general case they have to be found as roots of the non-linear equation

$$c^2\,\Delta_{hk}^2 - \|\mathbf{C}(s) - \mathbf{C}(z)\|^2 = 0 \tag{30}$$

either in z or in s variable, fixing, respectively, the outer integration variable s to find the eventually reduced inner integration interval, or z as the endpoints of $\mathbf{C}^{-1}(e_j)$ to find possible subdivision points in $\mathbf{C}^{-1}(e_i)$. Once these points are found, the double integral (19) can be evaluated, for instance in the case of coincident elements, as

$$\int_{\tilde{a}}^{\tilde{b}} w_l(s)\,J(s) \int_{M_s}^{m_s} \mathcal{V}(r, t_h, t_k)\,w_m(z)\,J(z)\,dz\,ds\,, \tag{31}$$

where $J(\cdot) := \|\mathbf{C}'(\cdot)\|$ is the parametric speed of the boundary Γ, with quadrature schemes for singular integrals on curvilinear boundaries described in [7], together with the already cited regularization technique [19].

Remark 2 Recently, in [1], efficient B-spline based weighted quadrature rules for IgA-SGBEM applied to elliptic problems have been introduced, abandoning the element-by-element framework to work directly on the B-spline's support. In the context of IgA-Energetic BEM, these formulas should be revised in order to take into account the possible truncation of the integration domain due to the presence of the Heaviside function in the kernel.

5 Preliminary Numerical Results

In the following, some preliminary numerical results obtained by the B-spline based Energetic BEM applied to the analysis of soft scattering of 2D damped waves are presented. A Fortran code has been implemented modifying an existing standard Energetic BEM sofware,[2] principally removing Lagrangian basis routines and inserting B-spline basis routines. Therefore, here and in the sequel, reported results obtained by the new code using first order B-splines have been checked and perfectly match those obtained by the old one using Lagrangian basis of degree 0 and published in [5].

Let us consider the model problem (1)–(3) fixing $\Gamma = \{\mathbf{x} = (x, 0) \in \mathbf{R}^2 \mid x \in [-0.5, 0.5]\}$, $c = 1$ and Dirichlet boundary datum, taken from [5], given by

$$g(x,t) = -H(t)f(t)x, \quad \text{where} \quad f(t) = \begin{cases} \sin^2(4\pi t), \text{ if } & 0 \leq t \leq \dfrac{1}{8}, \\ 1, & \text{if } \; t \geq \dfrac{1}{8}. \end{cases} \tag{32}$$

For an analysis on the time interval $[0, T] = [0, 2]$, the time step $\Delta t = 0.025$ and a uniform decomposition of Γ constituted by 40 elements, equipped at first by piecewise constant space basis functions, have been chosen.

Figure 5 presents the time history of the BIE numerical solution on the tenth element of the boundary mesh, i.e. for $x \in e_{10}$. In fact, all the points of e_{10} behave in the same way, since we are considering first order B-splines for the approximation of the BIE solution in space variable. At the beginning of the simulation points in the neighborhood of $x = 1/4$ are excited till the time instant $t = 1/8$, owing to the nature of Dirichlet datum. Further, they receive a solicitation at $t = 1/4$ by the wave coming from the right endpoint of Γ and traveling with unitary velocity. Finally the Energetic BEM solution stabilizes at a constant value. This is due to the fact that the boundary datum becomes independent of time, hence one can expect a convergence to a stationary solution, i.e. the solution ψ_∞ of the BIE on Γ related to the simplified

[2]This software is part of the BEM Fortran library developed by the Numerical Analysis Research Group of the University of Parma.

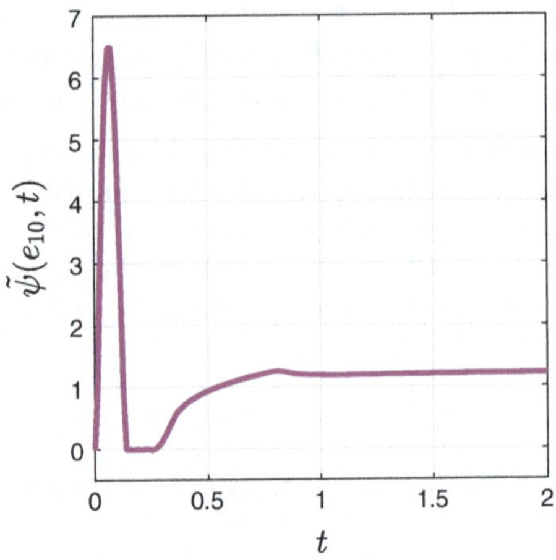

Fig. 5 Approximated density $\tilde{\psi}(x,t)$ for $x \in e_{10}$

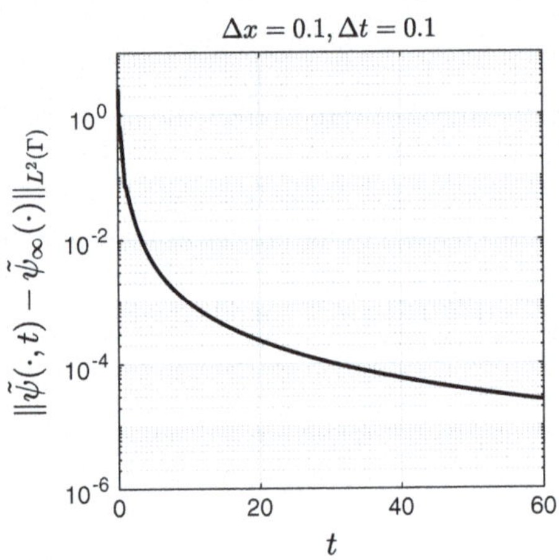

Fig. 6 Convergence of the approximate transient BIE solution to the approximate stationary one, using B-splines of order 1

problem

$$\Delta u_\infty(\mathbf{x}) = 0, \qquad \mathbf{x} \in \mathbf{R}^2 \setminus \Gamma, \tag{33}$$

$$u_\infty(\mathbf{x}) = -x, \qquad \mathbf{x} \in \Gamma, \tag{34}$$

$$u_\infty(\mathbf{x}) = O(1), \qquad \|\mathbf{x}\| \to \infty\,. \tag{35}$$

To analyze the convergence of the transient solution to the stationary one, the time interval of analysis has been extended to $[0, T] = [0, 60]$, fixing $\Delta t = 0.1$. In Fig. 6, the time history of the difference, evaluated in $L^2(\Gamma)$ norm, between the Energetic BEM approximate transient solution $\tilde{\psi}$ and the SGBEM approximate stationary solution $\tilde{\psi}_\infty$, fixing the same discretization parameter in space, $\Delta x = 0.1$ and piecewise constant basis functions, is plotted.

Table 2 Difference in $L^2(\Gamma)$ norm between IgA-Energetic BEM approximate transient solution, at different time instants, and IgA-SGBEM stationary solution

d	DoF	$t = 2$	$t = 5$	$t = 10$
1	11	4.95×10^{-2}	7.49×10^{-3}	1.77×10^{-3}
2	12	5.00×10^{-2}	7.46×10^{-3}	1.76×10^{-3}
3	13	4.74×10^{-2}	7.29×10^{-3}	1.73×10^{-3}
4	14	5.00×10^{-2}	7.16×10^{-3}	1.68×10^{-3}
5	15	4.61×10^{-2}	6.95×10^{-3}	1.64×10^{-3}
6	16	4.77×10^{-2}	6.85×10^{-3}	1.60×10^{-3}

Table 3 Difference in $L^2(\Gamma)$ norm between standard Energetic BEM approximate transient solution, at different time instants, and standard SGBEM stationary solution

d	DoF	$t = 2$	$t = 5$	$t = 10$
1	11	4.95×10^{-2}	7.49×10^{-3}	1.77×10^{-3}
2	21	5.23×10^{-2}	7.47×10^{-3}	1.75×10^{-3}
3	31	5.01×10^{-2}	7.12×10^{-3}	1.66×10^{-3}
4	41	4.82×10^{-2}	6.85×10^{-3}	1.60×10^{-3}
5	51	4.67×10^{-2}	6.58×10^{-3}	1.54×10^{-3}
6	61	4.61×10^{-2}	6.54×10^{-3}	1.53×10^{-3}

In the following, a comparison between IgA and standard Energetic BEMs is given, fixing $\Delta t = 0.1$ on the time interval $[0, T] = [0, 10]$ for all the simulations.

Table 2 shows, for different values of degree d of the employed B-spline basis functions, the DoF in space, i.e. the dimension of the matrix blocks $A^{(\ell)}$, and the difference in $L^2(\Gamma)$ norm between the IgA-Energetic BEM approximate transient solution and the IgA-SGBEM approximate stationary solutions, fixing the same discretization parameter in space, i.e. $\Delta x = 0.1$, at time instants $t = 2, 5, 10$. As one can observe, for increasing time, there is a convergence of the approximate transient solution $\tilde{\psi}(\cdot, t)$ to the corresponding approximate stationary one $\tilde{\psi}_\infty(\cdot)$.

The same behavior happens if standard Energetic BEM and SGBEM are taken into account, as it is visible in Table 3. The only difference is that the number of DoF increases much more rapidly for increasing Lagrangian basis degree.

In Fig. 7, on the left, the time history of the difference, evaluated in $L^2(\Gamma)$ norm, between the IgA-Energetic BEM transient solution and the corresponding Energetic SGBEM stationary solution, i.e. $\|\tilde{\psi}(\cdot, t) - \tilde{\psi}_\infty(\cdot)\|_{L^2(\Gamma)}$, fixing the same discretization parameter $\Delta x = 0.1$ in space and cubic B-splines basis functions, is shown, enlarging the time interval of analysis to $[0, 60]$. In particular, $\|\tilde{\psi}(\cdot, 60) - \tilde{\psi}_\infty(\cdot)\|_{L^2(\Gamma)} = 4.63 \times 10^{-5}$. For cubic Lagrangian basis functions, a similar difference (i.e. between the standard Energetic BEM transient solution and the corresponding standard SGBEM stationary solution) is plotted in the same Figure on the right and $\|\tilde{\psi}(\cdot, 60) - \tilde{\psi}_\infty(\cdot)\|_{L^2(\Gamma)} = 4.46 \times 10^{-5}$.

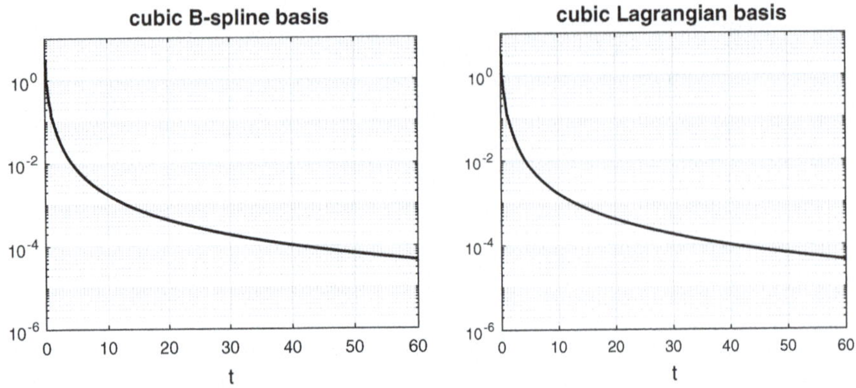

Fig. 7 Convergence of the approximate transient BIE solution to the related approximate stationary one, in case of cubic B-spline (left) and Lagrangian (right) basis functions

Fig. 8 Analytical solution of the stationary BIE related to (33)–(35)

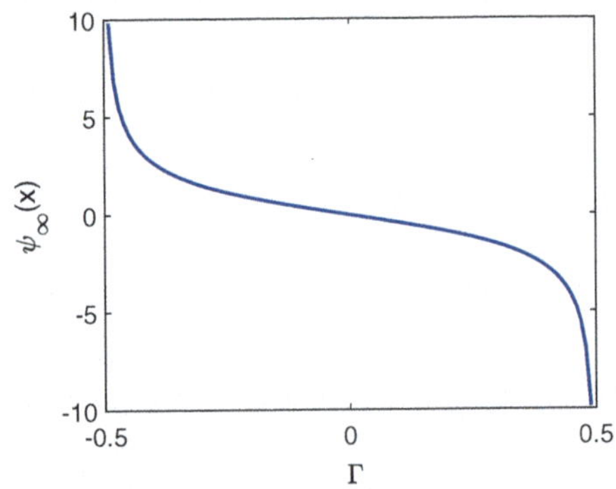

Now, let us analyze the convergence to the analytical solution of the stationary BIE related to the limit problem (33)–(35), which is known in closed form and reads

$$\psi_\infty(x) = \frac{-2x}{\sqrt{\frac{1}{4} - x^2}}. \tag{36}$$

This analytical solution is shown in Fig. 8.

In this context, it is useful to recall a theoretical result in [21] given for elliptic problems, stating that, using the Lagrangian basis of degree d, the Galerkin BEM absolute error $E_d = \|\tilde{\psi} - \psi\|_{H^{-1/2}(\Gamma)}$ decays as $O(d^{-1})$. Hence, one can derive the following formula for estimating the rate of convergence, both for the stationary approximate solution and for the transient approximate solution evaluated in a fixed, sufficiently high time instant:

Table 4 Convergence to the analytical stationary solution, using Lagrangian basis functions in space

d	DoF	E_d stat.	E_d trans.	rate$_d$	rate$_{N_{DoF,d}}$
1	11	3.16×10^{-1}	3.15×10^{-1}	−−	−−
2	21	2.50×10^{-1}	2.49×10^{-1}	0.34	0.36
3	31	1.67×10^{-1}	1.66×10^{-1}	1.00	1.04
4	41	1.29×10^{-1}	1.27×10^{-1}	0.93	0.96
5	51	1.02×10^{-1}	1.03×10^{-1}	0.94	0.96
6	61	8.94×10^{-2}	8.65×10^{-2}	0.96	0.98

Table 5 Convergence to the analytical stationary solution, using B-spline basis functions in space

d	DoF	E_d stat.	E_d trans.	rate$_d$	rate$_{N_{DoF,d}}$
1	11	3.16×10^{-1}	3.15×10^{-1}	−−	−−
2	12	2.74×10^{-1}	2.73×10^{-1}	0.21	1.64
3	13	2.32×10^{-1}	2.31×10^{-1}	0.41	2.09
4	14	1.92×10^{-1}	1.91×10^{-1}	0.66	2.57
5	15	1.58×10^{-1}	1.56×10^{-1}	0.91	2.93
6	16	1.25×10^{-1}	1.23×10^{-1}	1.30	3.68

$$\text{rate}_d := \log\left(\frac{E_d}{E_{d+1}}\right) / \log\left(\frac{d+1}{d}\right). \tag{37}$$

Further, since for $C^0(\Gamma)$ Lagrangian basis functions it holds $N_{DoF,d} \simeq n \cdot d$, the following formula, obtained from (37), can be considered too:

$$\text{rate}_{N_{DoF,d}} := \log\left(\frac{E_d}{E_{d+1}}\right) / \log\left(\frac{N_{DoF,d+1}}{N_{DoF,d}}\right). \tag{38}$$

Table 4 presents, for increasing values of degree d of the Lagrangian basis, the absolute errors E_d for standard SGBEM and standard Energetic BEM solutions, the latter evaluated at $t = 10$, together with the evaluated rates of convergence (37) and (38), which confirm the theoretical expectations, i.e. an error decay as $O(d^{-1})$ or equivalently as $O(N_{DoF,d}^{-1})$.

Table 5 presents, instead, for increasing values of degree d of the B-spline basis, the absolute errors E_d for IgA-SGBEM and IgA-Energetic BEM solutions, the latter evaluated at $t = 10$, together with the rates of convergence still evaluated by (37) and (38), which show that this approach has an error decay similar to the Lagrangian approach w.r.t. the basis degree (as expected due to the nature of the analytical stationary solution having mild singularity at the endpoints of Γ and to the uniform space mesh employed), but a higher speed of convergence w.r.t. DoF.

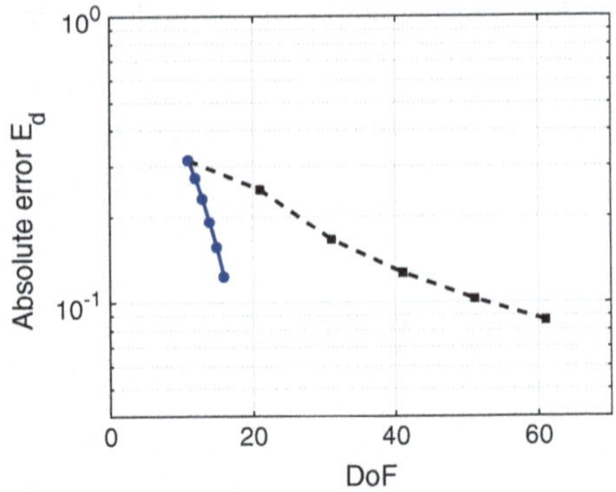

Fig. 9 Errors E_d versus DoF, for IgA (continuous line) and standard (dotted line) simulations

Table 6 Condition number of the matrix block $A^{(0)}$, together with its dimension

d	B-spline basis	N_{DoF}	Lagrangian basis	N_{DoF}
0	2.62×10^0	10	2.62×10^0	10
1	1.03×10^1	11	1.03×10^1	11
2	4.06×10^1	12	3.76×10^1	21
3	1.30×10^2	13	5.33×10^1	31
4	3.60×10^2	14	1.16×10^2	41
5	9.60×10^2	15	1.49×10^2	51
6	2.66×10^3	16	3.44×10^2	61

This feature is perfectly visible in Fig. 9 where absolute errors E_d of the last two Tables, concerning IgA (continuous line) and standard (dotted line) simulations, are plotted versus DoF. Hence, these preliminary results show that a B-spline based IgA approach to Energetic BEM seems to give the same computational benefits as observed for IgA-SGBEM applied to elliptic problems in [8].

Finally, in Table 6, the condition number of the matrix block $A^{(0)}$, together with its dimension, related to the B-spline and the Lagrangian basis is shown, being the latter only slightly better than the former one.

Remark 3 The convergence of the transient solution to the stationary one can be observed also outside the obstacle. In fact, once the unknown density over Γ is obtained, in the post-processing phase the representation formula (4) allows to evaluate the approximate solution of (1)–(3) in any area around the obstacle. In Figs. 10, 11 and 12, snapshots of the transient solution $u(\mathbf{x}, t)$ in a square S around Γ, evaluated at $t = 2, 4, 6, 8, 10$ after recovering the IgA-Energetic BEM solution on Γ using first order B-splines, together with the plot of the corresponding stationary solution $u_\infty(\mathbf{x})$ are shown. After the wavefront has left the considered area, the transient approximate solution is going to overlap the stationary one. In Fig. 13, time

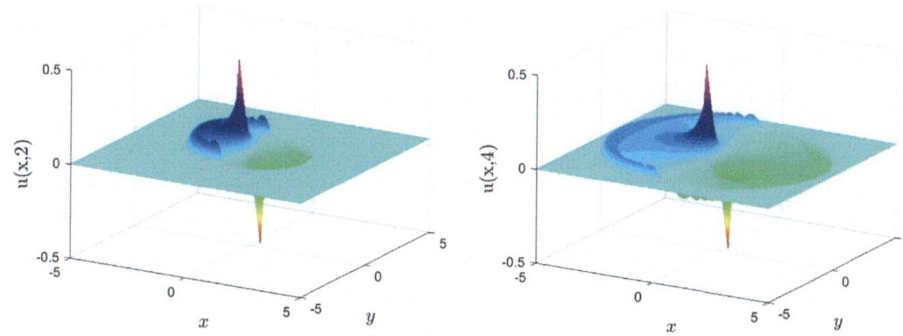

Fig. 10 Snapshots of the transient solution around Γ, evaluated at $t = 2, 4$

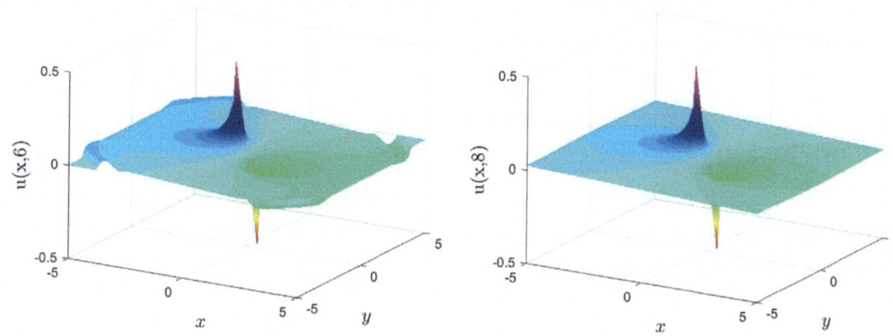

Fig. 11 Snapshots of the transient solution around Γ, evaluated at $t = 6, 8$

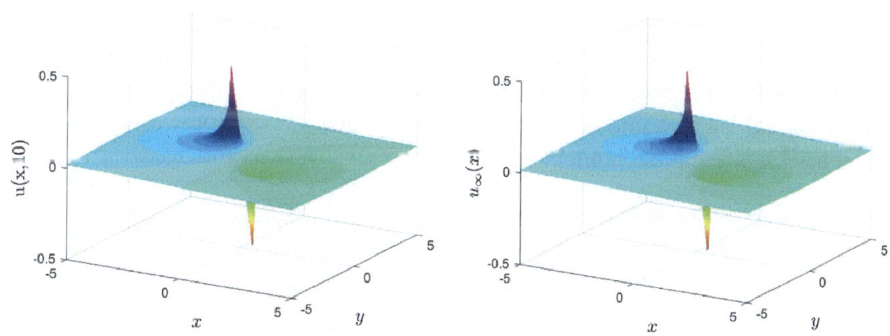

Fig. 12 Transient solution around Γ evaluated at $t = 10$ and stationary one

history of the difference in $L^2(S)$ norm between transient and stationary solutions around the obstacle is presented. These results perfectly match those in [5] obtained by applying standard Energetic BEM with Lagrangian basis functions of degree 0 and the subsequent post-processing phase.

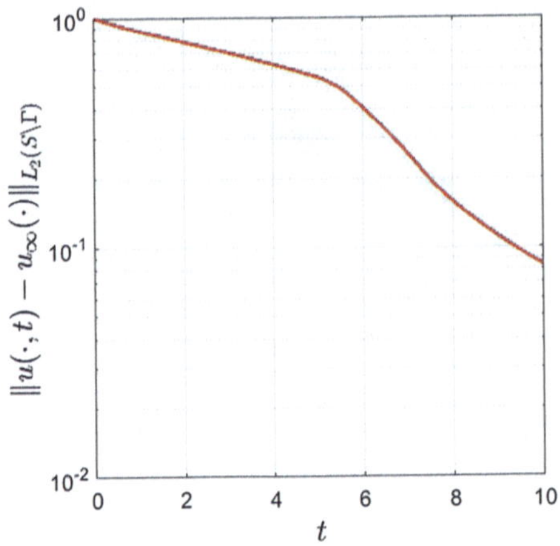

Fig. 13 Time history of the difference in L^2 norm between the transient and the stationary solution around Γ

6 Conclusions

In this paper, a numerical investigation related to a B-spline based IgA approach to the so-called Energetic BEM, applied to the numerical solution of 2D wave propagation exterior problems equipped by Dirichlet boundary conditions, has been started. Preliminary numerical results on straight obstacles indicate the superiority of the proposed approach w.r.t. standard Energetic BEM, based on local Lagrangian basis functions for the approximation of the boundary unknown in space variable, as already observed for IgA-SGBEM applied to elliptic problems [8]. Current work is devoted to extend the numerical analysis of IgA-Energetic BEM, performing simulations involving different types of curvilinear boundaries having a B-spline parametric representation, taking also into account Neumann boundary conditions, which allow to model hard scattering phenomena.

Acknowledgements The author is a member of the INdAM-GNCS Research Group and has been financially supported through the INdAM-GNCS Research Projects funding program.

References

1. Aimi, A., Calabrò, F., Diligenti, M., Sampoli, M., Sangalli, G., Sestini, A.: Efficient assembly based on B-spline tailored quadrature rules for the IgA-SGBEM. Comput. Methods Appl. Mech. Eng. **331**, 327–342 (2018)
2. Aimi, A., Diligenti, M.: A new space-time energetic formulation for wave propagation analysis in layered media by BEMs. Int. J. Numer. Methods Eng. **75**, 1102–1132 (2008)
3. Aimi, A., Diligenti, M., Frangi, A., Guardasoni, C.: Neumann exterior wave propagation problems: computational aspects of 3D energetic Galerkin BEM. Comput. Mech. **51**, 475–493 (2013)

4. Aimi, A., Diligenti, M., Guardasoni, C.: Numerical integration schemes for application of energetic Galerkin BEM to wave propagation problems. Riv. Mat. Univ. Parma **2**, 147–187 (2011)
5. Aimi, A., Diligenti, M., Guardasoni, C.: Energetic BEM for the numerical analysis of 2D Dirichlet damped wave propagation exterior problems. Commun. Appl. Ind. Math. **8**(1), 103–127 (2017)
6. Aimi, A., Diligenti, M., Guardasoni, C., Mazzieri, I., Panizzi, S.: An energy approach to space-time Galerkin BEM for wave propagation problems. Int. J. Numer. Methods Eng. **80**, 1196–1240 (2009)
7. Aimi, A., Diligenti, M., Monegato, G.: Numerical integration schemes for the BEM solution of hypersingular integral equations. Int. J. Numer. Methods Eng. **45**, 1807–1830 (1999)
8. Aimi, A., Diligenti, M., Sampoli, M., Sestini, A.: Isogeometric analysis and symmetric Galerkin BEM: a 2D numerical study. Appl. Math. Comput. **272**, 173–186 (2016)
9. Aimi, A., Diligenti, M., Sampoli, M., Sestini, A.: Non-polynomial spline alternatives in isogeometric symmetric Galerkin BEM. Appl. Numer. Math. **116**, 10–23 (2017)
10. Bamberger, A., Ha Duong, T.: Formulation variationelle espace-temps pour le calcul par potential retardé de la diffraction d'une onde acoustique (I). Math. Methods Appl. Sci. **8**, 405–435 (1986)
11. Bamberger, A., Ha Duong, T.: Formulation variationelle pour le calcul de la diffraction d'une onde acoustique par une surface rigide. Math. Methods Appl. Sci. **8**, 598–608 (1986)
12. Banerjee, P., Butterfield, P.: Boundary Element Methods in Engineering. McGraw-Hill U.K, Ltd, London (1981)
13. de Boor, C.: A Practical Guide to Splines, Revised edition. Springer, Berlin (2001)
14. Chaillat, S., Desiderio, L., Ciarlet, P.: Theory and implementation of $\mathcal{H}$-matrix based iterative and direct solvers for Helmholtz and elastodynamic oscillatory kernels. J. Comput. Phys. **341**, 429–446 (2017)
15. Costabel, M.: Time-dependent problems with the boundary integral equation method. In: Stein, E., et al. (eds.) Encyclopedia of Computational Mechanics, pp. 1–28. Wiley, Hoboken (2004)
16. Ha Duong, T.: On retarded potential boundary integral equations and their discretization. In: Davies, P., et al. (eds.) Topics in Computational Wave Propagation. Direct and Inverse Problems, pp. 301–336. Springer, Berlin (2003)
17. Hartmann, F.: Introduction to Boundary Element Theory Method in Engineering. McGraw-Hill U.K. Ltd., London (1981)
18. Hughes, T.J., Cottrel, J., Bazilevs, Y.: Isogeometric analysis: CAD, finite elements, NURBS, exact geometry and mesh refinement. Comput. Methods Appl. Mech. Eng. **194**, 4135–4195 (2005)
19. Monegato, G., Scuderi, L.: Numerical integration of functions with boundary singularities. J. Comput. Appl. Math. **112**, 201–214 (1999)
20. Nguyen, B., Tran, H., Anitescu, C., Zhuang, X., Rabczuk, T.: An isogeometric symmetric Galerkin boundary element method for two-dimensional crack problems. Comput. Methods Appl. Mech. Eng. **306**, 252–275 (2016)
21. Postell, F., Stephan, E.: On the h-, p- and h-p versions of the boundary element method - numerical results. Comput. Methods Appl. Mech. Eng. **83**, 69–89 (1990)
22. Simpson, R., Bordas, S., Trevelyan, J., Rabczuk, T.: A two-dimensional isogeometric boundary element method for elastostatic analysis. Comput. Methods Appl. Mech. Eng. **209–212**, 87–100 (2012)

Approximate Reconstructions of Perturbed Rational Planar Cubics

Michal Bizzarri, Miroslav Lávička and Jan Vršek

Abstract This paper is devoted to a problem from geometric modelling and related applications when exact symbolic computations are sometimes used also on objects given inexactly, i.e., when it is not adequately respected that numerical or input errors may significantly influence fundamental properties of considered algebraic varieties, including e.g. their rationality. We formulate a simple algorithm for an approximation of a non-rational planar cubic which is assumed to be a perturbation of some unknown rational planar cubic. The input curve is given by a perturbed polynomial or by perturbed points sampled from the original curve. The algorithm consists of two main parts. First, we suggest geometric methods for the estimation of a singular point of the original curve. Then we select from the six-dimensional subspace of all rational cubics with a given singular point a suitable one that may also satisfy some further criteria. The designed method is presented on several commented examples.

1 Introduction and Motivation

Symbolic-numeric (or *hybrid*) *computation* is a powerful technique nowadays very often used to improve the stability and robustness of numerical computations. Its main idea consists of identifying and employing always the most suitable framework for specific computations. Symbolic computations tend to be slow but are always exact,

---

M. Bizzarri
NTIS – New Technologies for the Information Society, University of West Bohemia, Pilsen, Czech Republic
e-mail: bizzarri@ntis.zcu.cz

M. Lávička (✉) · J. Vršek
Department of Mathematics and NTIS, University of West Bohemia, Pilsen, Czech Republic
e-mail: lavicka@kma.zcu.cz

J. Vršek
e-mail: vrsekjan@kma.zcu.cz

C. Giannelli and H. Speleers (eds.), *Advanced Methods for Geometric Modeling and Numerical Simulation*, Springer INdAM Series 35,
https://doi.org/10.1007/978-3-030-27331-6_2

while numerical computations are fast but their accuracy may not always be sufficient. The combination of both allows us to replace numerically critical and repeating computations by symbolic ones, thereby improving the quality and reliability of the results, in particular in geometric computing, for more details see e.g. [1–3].

Nonetheless, one has to be aware of a certain inexactness noticeable in the approaches from recent years. Especially in applications, but also in some fields of applied mathematics, standard methods of *exact* symbolic computation (based on computer algebra/algebraic geometry) are often used, without any further discussion, also on mathematical objects which are assumed to be given by the description with perturbed float coefficients (obtained e.g. as results from some computer computations), see also [4].

Let us demonstrate it on one simple (illustrative) example. In particular, we will deal with a situation when some input error (or some error caused by numerical computations) occurs. Consider nine particular points, among them also the point $\mathbf{q} = (-2, 0)$, which uniquely determine a planar cubic curve C_0,

$$f_0(x, y) = -x^3 - 2x^2 + 2y^2, \tag{1}$$

see Fig. 1 (left). C_0 possesses a double point $\mathbf{p} = (0, 0)$, hence its genus is 0 and it has a polynomial parameterization

$$\mathbf{x}_0(t) = \big(2(t^2 - 1), 2(t^3 - t)\big), \qquad t \in \mathbb{R}, \tag{2}$$

see [5]. However when perturbing coordinates of only one point, e.g. consider the point $\overline{\mathbf{q}} = (-1.99, 0)$ instead of the original one $\mathbf{q} = (-2, 0)$, we obtain a new cubic curve C, see Fig. 1 (right), with almost the same implicit equation

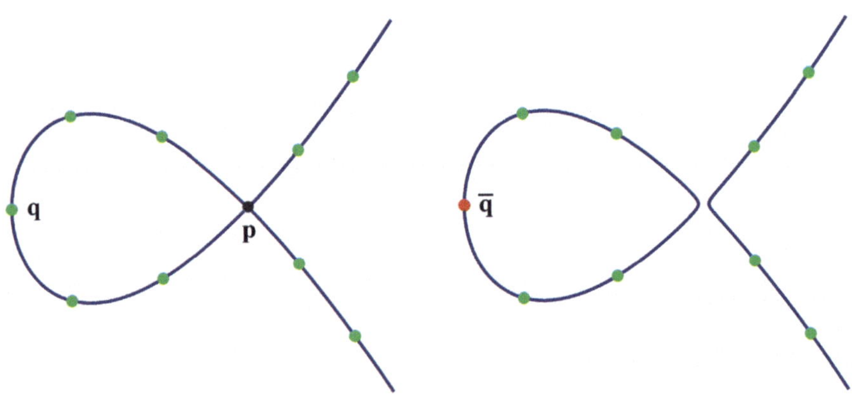

Fig. 1 Cubic curves determined by nine points. Left: a rational cubic C_0 with a singular point $\mathbf{p}$. Right: an elliptic cubic C without singularities (obtained by a perturbation of C_0)

$$f(x, y) = -x^3 - 1.99353x^2 + 0.0179xy^2 - 0.00522x + \\ + 1.98067y^2 + 0.00357 \approx -x^3 - 2x^2 + 2y^2 = f_0(x, y), \tag{3}$$

but with completely different geometric properties. The new perturbed curve *is not rational* anymore—it is an elliptic curve without singularities and hence it does not possess a description analogous to (2). This means that all subsequent exact algorithms and techniques formulated for rational algebraic varieties automatically fail. Thus a natural question reads: How can we find a rational (cubic) curve “close” to C? By “close” one can mean the minimum distance of the coefficients of their implicit equations or another suitable criterion. Of course, not only numerical but also geometric properties must be considered. And the problem even escalates when considering more complicated varieties as e.g. curves of higher degree, curves with more complicated singularities, spatial curves or surfaces.

To sum up, this paper aims at studying important aspects of geometric modelling stemming from numerical or input errors and their influence on algebraic varieties, which is a problem that falls within the scope of approximate algebraic geometry, see [4]. This problem is relatively new in geometric modelling and many challenges must be still solved and answered, see [6, 7]. Since there is no hope that the problem could be solved in full generality, special algorithms must be formulated for distinguished classes of algebraic varieties.

We will focus on the first interesting example and present some methods for computing approximate reconstructions of perturbed rational planar cubics, given implicitly by polynomials of degree three. We recall that cubic curves are popular in geometric modelling because curves of lower degree commonly have too little flexibility, while curves of higher degree are in some situations considered unnecessarily complex and make it easy to introduce an undesired behavior. Low degree curves are also preferred when they shall be used for generating other shapes, e.g. surfaces.

2 Preliminaries

Let us consider a *planar cubic curve* C defined by a homogeneous equation

$$F(X, Y, Z) = \sum_{\substack{i, j, k \geq 0 \\ i + j + k = 3}}^{3} a_{ijk} X^i Y^j Z^k = 0, \tag{4}$$

where $X : Y : Z$ are the homogeneous coordinates for the projective plane $\mathbb{P}^2_{\mathbb{C}}$. The inhomogeneous version $f(x, y) = 0$ for the affine plane $\mathbb{A}^2_{\mathbb{C}}$ is obtained by setting $x = X/Z$, $y = Y/Z$. In what follows we assume that a_{ijk} are real and moreover the polynomial $F(X, Y, Z)$ is irreducible over $\mathbb{C}$.

Let us write

$$F(X, Y, Z) = B(X, Y, Z)\mathbf{c}^{\top}, \tag{5}$$

where $B = (X^3 : X^2Y : \cdots : Z^3)$ and $\mathbf{c} = (a_{300} : a_{210} : \cdots : a_{003})$. Hence the space of all planar cubic curves can be identified with the projective space $\mathbb{P}^9_{\mathbb{R}}$. In what follows, we will not distinguish between C and $\mathbf{c}$ when there is no danger of confusion. Then all cubics passing through a given point form a hyperplane in $\mathbb{P}^9_{\mathbb{R}}$ and thus one can find a cubic going through any 9 prescribed points, but it may be reducible and not unique. In addition, this cubic is unique and irreducible if the given 9 points are in general position, see [8, 9].

A cubic curve is rational, i.e., it has a rational parameterization iff it possesses a singular point. The singular points of irreducible cubics are either one double point (a node), or one cusp. A smooth non-singular cubic is an elliptic curve with its genus equal to one. See e.g. [8] for more detail about cubics and further algebraic curves.

Furthermore, any irreducible (rational or elliptic) cubic curve $f(x, y) = 0$ is birationally equivalent to a planar curve in the *Weierstrass form*

$$y^2 = x^3 + ax + b, \tag{6}$$

where a and b are real numbers. A sequence of transformations yielding the Weierstrass form of cubics is described e.g. in [10]. The cubic in the Weierstrass form is non-singular iff the right-hand side of (6) has three distinct roots $\alpha_1, \alpha_2, \alpha_3$, i.e., iff the discriminant

$$\Delta = -16(4a^3 + 27b^2) \tag{7}$$

does not vanish. The real part of a non-singular cubic has two components if Δ is positive, and one component if it is negative. Next, if two of the α_i coincide and the third one is different then the cubic is nodal, if all three roots α_i are identical then the cubic is cuspidal.

Given a cubic polynomial $x^3 + a_2x^2 + a_1x + a_0$ then the linear transformation $x \mapsto x - \frac{a_2}{3}$ removes the quadratic term, see (6). However this special form does not play any role in our algorithm, so we will understand under the Weierstrass form of a cubic curve any expression

$$y^2 = p(x), \tag{8}$$

where $p(x)$ is a cubic polynomial.

The first illustrative example, given in Sect. 1, demonstrates clearly that perturbations of the input (e.g. the coordinates of the points that determine a cubic) can cause significant instabilities in geometric features (in our case the rationality) of the result. This instability brings fatal consequences for further steps based on symbolic computations and thus for all algorithms based on hybrid computing. Hence it is an interesting question how to find a rational cubic $\widetilde{C}$ close to a given perturbed cubic.

Let us emphasize that the problem is complicated even for curves of low degree which cubics are. In particular, all singular cubics form an algebraic hypersurface in $\mathbb{P}^9_{\mathbb{R}}$ given by the equation

$$D\,(a_{300}, a_{210}, \ldots, a_{003}) = 0, \tag{9}$$

where D is the discriminant of $f(x, y)$, see [11] for more details about computing discriminants. Then finding a close rational cubic $\widetilde{C}$ is, in this setup, transformed to the problem of projecting a point in $\mathbb{P}^9_{\mathbb{R}}$ to this hypersurface. Unfortunately the discriminant D is of degree 12 and it contains 2040 monomials. This makes the direct approach useless (and the situation is much worse for varieties of higher degree). So from the practical point of view, it is essential to find other alternatives and computationally accessible approaches.

Let us recall that the situation is simpler when the potential singular point is known. Then all rational cubics with a given singular point $\mathbf{p}$ form a 6-dimensional subspace $\mathbf{L_p}$ of the 9-dimensional space of all cubics which is given by the linear equations with unknowns a_{ijk}

$$F_X(\mathbf{p}) = F_Y(\mathbf{p}) = F_Z(\mathbf{p}) = 0, \tag{10}$$

where $F_X = \frac{\partial F}{\partial X}$, $F_Y = \frac{\partial F}{\partial Y}$, $F_Z = \frac{\partial F}{\partial Z}$, or in the matrix representation

$$\mathbf{M_p}\,(a_{300} : a_{210} : \cdots : a_{003})^\top = \mathbf{0}, \qquad \mathbf{M_p} = \begin{pmatrix} B_X(\mathbf{p}) \\ B_Y(\mathbf{p}) \\ B_Z(\mathbf{p}) \end{pmatrix}. \tag{11}$$

3 Formulation of the Problem and Algorithm

First, we would like to emphasize that in real situations the original entity C_0 is not known. This makes it impossible to measure the exact error of a constructed approximate shape $\widetilde{C}$ from this unknown object. Hence, it is difficult to state which approximate reconstruction is the best one as in the considered neighbourhood one can find more potential candidates, some of them possibly closer to the perturbed object than to the original one. This significantly depends on the considered inexactness.

The input to our algorithm is a non-rational cubic C which is a perturbation of some unknown rational planar cubic C_0. This perturbed curve is described by a cubic polynomial $F(X, Y, Z)$ or $f(x, y)$ obtained either by some computations yielding directly implicit curve equations, or after solving a system of linear equations (when exactly 9 points in general position are given), or from some optimization when more perturbed points are prescribed.

As the cubic C is non-rational, it does not possess a singular point. Nonetheless, the original curve C_0 was by assumption singular—but due to a certain perturbation, its singularity blew up into two (for the nodal cubic) or three (for the cuspidal cubic) close smooth points.

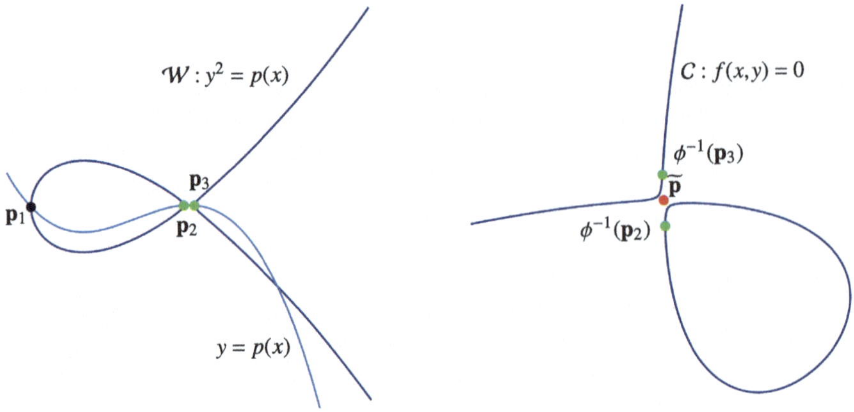

Fig. 2 Left: Choosing two "close" roots of $p(x)$ yielding two points $\mathbf{p}_2 = (\alpha_2, 0)$ and $\mathbf{p}_3 = (\alpha_3, 0)$ (green) on the Weierstrass curve $\mathcal{W}$. Right: The preimages $\phi^{-1}(\mathbf{p}_2)$ and $\phi^{-1}(\mathbf{p}_3)$ (green) on C and the computed approximation $\widetilde{\mathbf{p}}$ of the prospective singular point (red)

(I) The initial and crucial step of the reconstruction algorithm is to find a suitable approximate singular point of $\widetilde{C}$. In what follows we present three possible simple methods how to proceed.

(I.1) As a first method, we can apply an efficient approach based on properties of the Weierstrass form, i.e., we construct an isomorphism $\phi : C \to \mathcal{W}$, where $\mathcal{W}$ is a Weierstrass cubic (8). Then we compute the roots α_1, α_2 (both real, or both imaginary) and α_3 (real) of $p(x)$. If all α_i are real, then we consider α_i, α_j as a pair of the closest roots. Otherwise, we consider the imaginary roots α_1, α_2. A candidate for the singular point of $\widetilde{C}$ is consequently obtained by averaging the suitable preimages as

$$\widetilde{\mathbf{p}} = \frac{\phi^{-1}(\alpha_i, 0) + \phi^{-1}(\alpha_j, 0)}{2}, \tag{12}$$

see Fig. 2. It is true that the coefficients of the polynomial $F(X, Y, Z)$ or $f(x, y)$ are given as floating point numbers, in general. Anyway, it is useful to convert them by rounding to approximate rational numbers, which enables us to apply the symbolic computation approach (for instance for computing the mappings ϕ and ϕ^{-1} we use the CAS `Maple` and the package `algcurves`). In addition, as we want to avoid working with large numbers, it is also convenient to round the coordinates of the approximate singular point with respect to a prescribed height—we mean by the *height of a rational number* $x = a/b$, $\gcd(a, b) = 1$, the value $h(x) = \lceil \log_{10} \max\{|a|, |b|\} \rceil$.

Remark 1 In some situations more pairs of real roots α_i, α_j may be considered as equivalently close because all the distances $|\alpha_i - \alpha_j|$ differ less than some prescribed ϵ. Then we have to consider all the obtained approximations as equivalently feasible. Moreover, it is possible to join by averaging all $\phi^{-1}(\alpha_i, 0)$. This case is also to

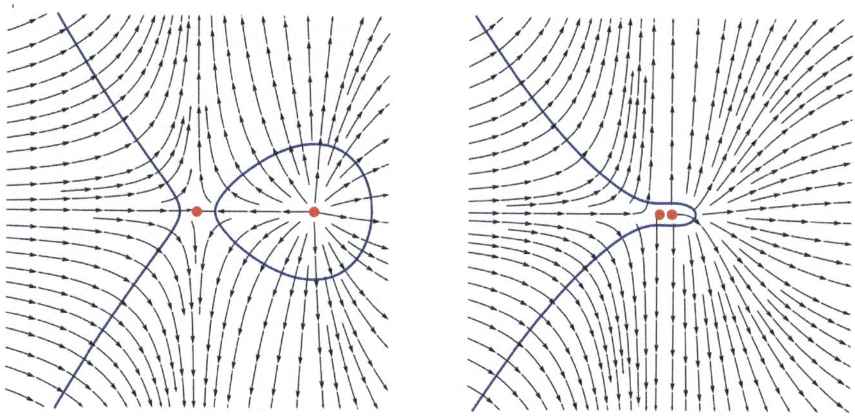

Fig. 3 The gradient vector fields $\nabla f(x, y)$ (black) of cubic curves (blue) defined by $f(x, y) = 0$ and the real solutions of (13) (red)

consider when the real root α_1 and the value $\alpha = (\alpha_2 + \alpha_3)/2$, obtained from two conjugated imaginary roots α_2, α_3, are closer than the prescribed ϵ.

(I.2) A different approach for computing an approximate singular point $\widetilde{\mathbf{p}}$ can exploit the *gradient vector field* $\nabla f(x, y) = (\partial f/\partial x, \partial f/\partial y)$ associated to the curve C defined by the polynomial $f(x, y)$. The singular point $\mathbf{p}$ on the (unknown) singular cubic C_0 simultaneously satisfies $f_0(\mathbf{p}) = 0$ and $\nabla f_0(\mathbf{p}) = 0$ and hence we can seek an approximate singular point $\widetilde{\mathbf{p}}$ on $\widetilde{C}$ by solving

$$\nabla f(x, y) \equiv \mathbf{0}, \tag{13}$$

where $f(x, y) = 0$ describes again a known perturbed cubic C. Equation (13) yields 4 solutions $\widetilde{\mathbf{p}}_i$, $i = 1, \dots, 4$ (counted with multiplicities) obtained as intersections of two conic sections $\partial f/\partial x = 0$, $\partial f/\partial y = 0$. When there is some complex conjugate couple, we join it together by averaging to obtain one real point (i.e., we consider their real coordinates only). Now between all the real points we select the one with the minimal value of $f^2(\widetilde{\mathbf{p}}_i)$, see Fig. 3.

Remark 2 As there are only few potential singular points (max. 4) it is not a computational problem to consider all of them, continue with part (II) and postpone a final choice to the end with respect to a suitable criterion measuring the quality of the approximation, see (III).

(I.3) The last presented approach for determining an approximate singularity from the perturbed cubic is based on finding and identifying "right" *critical points*. Recall that a point $\mathbf{r}$ on the affine curve C defined by the polynomial $f(x, y)$ is called *critical* w.r.t. the direction $\mathbf{d}$ iff

$$\nabla f(\mathbf{r}) \cdot \mathbf{d} = 0, \tag{14}$$

see also [12, 13]. When a rational cubic C_0 being perturbed, its node decomposes to two near critical points (w.r.t. a generic direction). Hence we can choose a direction $\mathbf{d}$ and by solving $f(x, y) = \nabla f(x, y) \cdot \mathbf{d} = 0$ compute all 6 critical points $\mathbf{r}_i$ w.r.t. $\mathbf{d}$ (intersection points of a conic section and a cubic). First, we choose from all couples $\mathbf{r}_i, \mathbf{r}_j, i \neq j$, only those which might have arisen from one (singular) point, i.e., the corresponding coordinates are either both real or complex conjugate. Secondly, we select such a couple with the minimal absolute value $|\mathbf{r}_i - \mathbf{r}_j|$ and compute the midpoint $\widetilde{\mathbf{p}} = (\mathbf{r}_i + \mathbf{r}_j)/2$, which serves as a candidate for the sought-after singular point.

Of course, this process depends on the choice of the direction $\mathbf{d}$. So we perform the estimation for more directions

$$\mathbf{d}_i = \big(\cos(2\pi i/k), \sin(2\pi i/k)\big), \quad i = 0, \ldots, k-1, \tag{15}$$

and for each $\mathbf{d}_i$ we find the corresponding candidate $\widetilde{\mathbf{p}}_i$. Finally, we choose as a resulting approximate singular point $\widetilde{\mathbf{p}}$ the trimmed mean of all $\widetilde{\mathbf{p}}_i$, see Fig. 4. We recall that the *trimmed* (or *truncated*) *mean* is a statistical measure of central tendency, less sensitive to outliers than the classical mean. It involves the calculation of the mean after discarding given parts of a probability distribution at the high and low end. In most statistical applications, 5–25% of the ends are discarded. In our computations we will use the *interquartile mean*, i.e., 25% trimmed mean (the lowest 25% and the highest 25% are discarded). Our numerical experiments have shown that this value is sufficient for the presented approach and yields good results.

Remark 3 Let us note that for a generic cubic, the value of k has an insignificant influence for the computed singular point. However, when the input cubic is in some special position then employing only one direction may lead to the incorrect result. This is why we use more directions. According to our experimental results setting $k = 10$ eliminates this inconvenience and the method works correctly.

`(II)` The second step of the designed reconstruction algorithm is to find a suitable cubic $\widetilde{C}$ sufficiently "close" to the given perturbed cubic C when the singular point $\widetilde{\mathbf{p}}$ of $\widetilde{C}$ is prescribed. For this purpose we have to introduce a metric on the projective space $\mathbb{P}^9_{\mathbb{R}}$. In particular, we consider the standard metric originated in the ray model and we will measure the angle between the lines through the origin which represent the points in the projective space, i.e., we have

$$\delta(\mathbf{x}, \mathbf{y}) = \arccos\left(\frac{|\mathbf{x} \cdot \mathbf{y}|}{\|\mathbf{x}\| \, \|\mathbf{y}\|}\right), \tag{16}$$

where '$\cdot$' and $\| \ \|$ denote the standard inner product and the standard norm in the corresponding vector space. The angle δ is real-valued, and runs from 0 to $\frac{\pi}{2}$.

`(II.1)` The (standard) inner product enables us to work with orthogonal vectors and orthogonal subspaces. Thus, an orthogonal projection is a projection for which the range and the null space are orthogonal subspaces. Furthermore, let $\mathbf{K}$ be the matrix

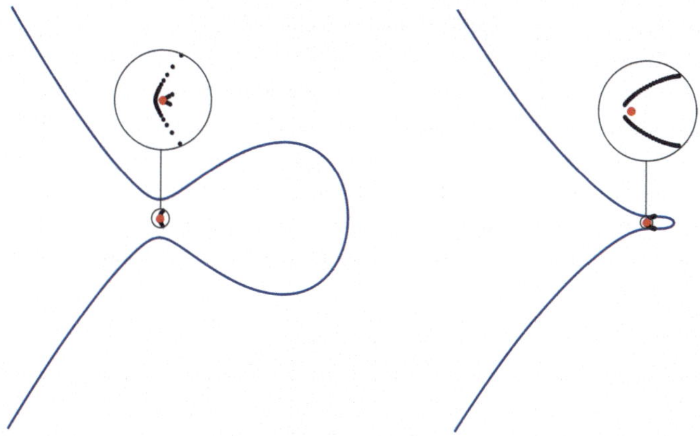

Fig. 4 Candidates $\widetilde{\mathbf{p}}_i$ corresponding to different directions $\mathbf{d}_i$ (black) and their trimmed mean $\widetilde{\mathbf{p}}$ (red)

whose rows are the basis vectors of a subspace U then the orthogonal projection onto the subspace U is given by the matrix

$$\mathbf{T} = \mathbf{K}^\top \left(\mathbf{K}\mathbf{K}^\top\right)^{-1} \mathbf{K}. \tag{17}$$

This approach induces a projection in the corresponding projective space. Moreover, it holds

Lemma 1 *Let H be a subspace of $\mathbb{P}^n_{\mathbb{R}}$. Consider a point $\mathbf{p} \in \mathbb{P}^n_{\mathbb{R}}$ and denote by $\mathbf{p}^\perp$ the orthogonal projection of $\mathbf{p}$ to H. Then for all $\mathbf{q} \in H$ it holds $\delta(\mathbf{p}, \mathbf{q}) \geq \delta(\mathbf{p}, \mathbf{p}^\perp)$ and the equality occurs iff $\mathbf{q} = \mathbf{p}^\perp$.*

Proof Consider the vector space $\mathbb{R}^{n+1}$ inducing the projective space $\mathbb{P}^n_{\mathbb{R}}$ and its subspace $U \subset \mathbb{R}^{n+1}$ inducing $H \subset \mathbb{P}^n_{\mathbb{R}}$. Let $\langle \mathbf{p} \rangle$ be the direction given by a representative of the point $\mathbf{p}$. The angle between the direction $\langle \mathbf{p} \rangle$ and the subspace U is defined as the angle between $\mathbf{p}$ and its orthogonal projection $\mathbf{p}^\perp$ onto U, and it is a minimum of the angles between $\langle \mathbf{p} \rangle$ and any direction given by vectors from U. □

Hence, we can compute the orthogonal projection of the point $\mathbf{c}$ representing the cubic C in $\mathbb{P}^9_{\mathbb{R}}$ to the space $\mathbf{L}_{\widetilde{\mathbf{p}}}$ of all rational cubics with the singular point $\widetilde{\mathbf{p}}$. The solution of homogeneous system (11), when considering the coordinates of $\widetilde{\mathbf{p}}$, is the same as the kernel of the matrix $\mathbf{M}_{\widetilde{\mathbf{p}}}$, i.e.,

$$\mathbf{L}_{\widetilde{\mathbf{p}}} = \ker(\mathbf{M}_{\widetilde{\mathbf{p}}}). \tag{18}$$

Considering the basis of this kernel we can describe the projection by the matrix (17) and using this matrix we arrive at the point

$$\widetilde{\mathbf{c}} = \mathbf{c}\mathbf{T}, \tag{19}$$

which represents the sought approximate rational cubic $\widetilde{C}$.

(II.2) Alternatively, when C was given at the beginning by the set of perturbed points $\mathbf{p}_i$, $i = 1, \ldots, n$ and computed by some approximation method (e.g. least squares minimization), one can once more use the least squares method to determine $\widetilde{f}(x, y) = 0$. In particular, we consider again (18) and find the (homogeneous) implicit equation of all rational cubics with the singular point $\widetilde{\mathbf{p}}$ (which depends on 7 parameters, see the last paragraph of Sect. 2). We denote it by $\widetilde{F}_{\boldsymbol{\lambda}}(X, Y, Z) = 0$, where $\boldsymbol{\lambda} = (\lambda_0 : \cdots : \lambda_6)$. In this 7-parametric system we find the cubic whose implicit equation minimizes

$$\sum_{i=1}^{n} \left(\widetilde{F}_{\boldsymbol{\lambda}}(\mathbf{p}_i)\right)^2, \tag{20}$$

subject to the constrain $\|\boldsymbol{\lambda}\| = 1$. The constraint is considered to exclude the trivial solution $\boldsymbol{\lambda} = \mathbf{0}$.

In particular, the sought solution $\boldsymbol{\lambda}$ is then the eigenvector of the matrix $\mathbf{A}^\top \mathbf{A}$ corresponding to the smallest eigenvalue, where $\mathbf{A}$ is the $n \times 7$ matrix with rows composed of the coefficients of $\widetilde{F}_{\boldsymbol{\lambda}}(\mathbf{p}_i)$ w.r.t. $\boldsymbol{\lambda}$, i.e.,

$$\mathbf{A} = \left(\mathbf{L}_{\widetilde{\mathbf{p}}} B(\mathbf{p}_i)^\top\right)^\top. \tag{21}$$

And then

$$\widetilde{\mathbf{c}} = \boldsymbol{\lambda} \mathbf{L}_{\widetilde{\mathbf{p}}}. \tag{22}$$

(III.) Finally, to measure suitably a quality of the approximation (i.e., the deviation δ between the perturbed and the constructed cubic) we will consider the standard metric, see (16), applied on determining the distance between the points $\widetilde{\mathbf{c}}$, $\mathbf{c}$ in the projective space of cubics $\mathbb{P}^9_{\mathbb{R}}$, in particular

$$\delta(\widetilde{\mathbf{c}}, \mathbf{c}) = \arccos\left(\frac{|\widetilde{\mathbf{c}} \cdot \mathbf{c}|}{\|\widetilde{\mathbf{c}}\| \, \|\mathbf{c}\|}\right). \tag{23}$$

In addition, when $\widetilde{f}(x, y) = 0$ was computed using the least squares method it is also natural to quantify the deviation via measuring the approximation of the squared Euclidean distance of points $\mathbf{p}_i$ to the curve $\widetilde{C}$, see [14],

$$\Delta(\widetilde{C}, \mathbf{p}_i) = \sum_{i=1}^{n} \frac{\left(\widetilde{f}(\mathbf{p}_i)\right)^2}{\|\nabla \widetilde{f}(\mathbf{p}_i)\|^2}. \tag{24}$$

Remark 4 When starting with an exact rational cubic, all the presented methods for determining the singular point yield the actual correct singularity. Then since we simply choose between all cubics with this (correct) singularity the one with the minimal distance from the input rational one, we arrive at the input cubic.

Table 1 The 25% trimmed mean of the distances between ${}^i\widetilde{\mathbf{p}}$ and $\mathbf{p}$ in each row for 1000 different inputs

ϵ	mean $(\Vert\mathbf{p} - {}^1\widetilde{\mathbf{p}}\Vert, 0.25)$	mean $(\Vert\mathbf{p} - {}^2\widetilde{\mathbf{p}}\Vert, 0.25)$	mean $(\Vert\mathbf{p} - {}^3\widetilde{\mathbf{p}}\Vert, 0.25)$
10^{-1}	6.92154×10^{-3}	2.53897×10^{-3}	1.87996×10^{-3}
10^{-2}	6.81900×10^{-4}	2.36461×10^{-4}	1.62944×10^{-4}
10^{-3}	4.97281×10^{-5}	1.78523×10^{-5}	1.45099×10^{-5}
10^{-4}	5.90521×10^{-6}	1.97876×10^{-6}	1.55910×10^{-6}
10^{-5}	5.96431×10^{-7}	1.73101×10^{-7}	1.25627×10^{-7}
10^{-6}	6.79475×10^{-8}	1.86060×10^{-8}	1.38855×10^{-8}
10^{-7}	1.30508×10^{-8}	1.72472×10^{-9}	1.37039×10^{-9}

Table 2 The 25% trimmed mean of the angles between $\widetilde{C}_i$ and C in each row for 1000 different inputs

ϵ	mean $(\delta({}^1\widetilde{\mathbf{c}}, \mathbf{c})), 0.25)$	mean $(\delta({}^2\widetilde{\mathbf{c}}, \mathbf{c})), 0.25)$	mean $(\delta({}^3\widetilde{\mathbf{c}}, \mathbf{c})), 0.25)$
10^{-1}	5.36635×10^{-3}	2.75295×10^{-3}	9.56900×10^{-4}
10^{-2}	3.78011×10^{-4}	1.74204×10^{-4}	7.28357×10^{-5}
10^{-3}	4.38528×10^{-5}	2.01407×10^{-5}	8.54439×10^{-6}
10^{-4}	4.79020×10^{-6}	1.69576×10^{-6}	6.38216×10^{-7}
10^{-5}	4.44271×10^{-7}	6.18986×10^{-8}	5.43790×10^{-8}
10^{-6}	3.28066×10^{-8}	7.87332×10^{-9}	5.91876×10^{-9}
10^{-7}	1.01274×10^{-9}	6.78943×10^{-10}	5.82319×10^{-10}

In what follows, we demonstrate the behavior of the proposed method on randomly chosen and perturbed rational cubics. Hence, we generate a random cubic parameterization

$$\mathbf{x}_0(t) = (a_1(t)/b(t), a_2(t)/b(t)), \tag{25}$$

where $a_1(t)$, $a_2(t)$ and $b(t)$ are cubic polynomials from $\mathbb{Q}[t]$ and by a suitable implicitization method, see e.g. [15–18], we arrive at the defining polynomial $f_0(x, y)$ of C_0. Of course, (10) yields the exact singular point $\mathbf{p}$.

Next, we employ the normal distribution with the mean 0 and the standard deviation $\epsilon = 10^{-\ell}$ to perturb C_0 which leads to a non-rational cubic C. By performing all three presented methods for computing the approximate singular point of C we obtain three candidates ${}^i\widetilde{\mathbf{p}}$, $i = 1, 2, 3$. In Table 1, the 25% trimmed mean of the distances between ${}^i\widetilde{\mathbf{p}}$ and $\mathbf{p}$ for 1000 different random inputs (depending on ϵ) is measured. Our numerical experiments show that when perturbing a generic rational cubic with some $\epsilon = 10^{-\ell}$, we are able to reconstruct its singular point with the error approximately equal to $10^{-\ell-2}$.

All points ${}^i\widetilde{\mathbf{p}}$ are then used for constructing the new rational cubic curves ${}^i\widetilde{C}$ with the singular points in $\widetilde{\mathbf{p}}_i$. In Tables 2 and 3, the 25% trimmed mean of angles (23) between ${}^i\widetilde{C}$ and C and ${}^i\widetilde{C}$ mutually for 1000 different inputs (depending on ϵ) is

Table 3 The 25% trimmed mean of the angles between $\widetilde{C}_i$ in each row for 1000 different inputs

ϵ	mean $(\delta({}^1\widetilde{\mathbf{c}}, {}^2\widetilde{\mathbf{c}}), 0.25)$	mean $(\delta({}^2\widetilde{\mathbf{c}}, {}^3\widetilde{\mathbf{c}}), 0.25)$	mean $(\delta({}^1\widetilde{\mathbf{c}}, {}^3\widetilde{\mathbf{c}}), 0.25)$
10^{-1}	6.40089×10^{-3}	3.51259×10^{-3}	5.25147×10^{-3}
10^{-2}	6.43319×10^{-4}	2.83220×10^{-4}	3.41385×10^{-4}
10^{-3}	1.01985×10^{-4}	3.22014×10^{-5}	4.97543×10^{-5}
10^{-4}	4.13791×10^{-6}	1.23642×10^{-6}	4.46979×10^{-6}
10^{-5}	2.29031×10^{-7}	2.10626×10^{-7}	6.11232×10^{-7}
10^{-6}	3.23678×10^{-8}	2.68345×10^{-8}	5.23497×10^{-8}
10^{-7}	2.38730×10^{-9}	3.24679×10^{-9}	4.73451×10^{-9}

computed. Table 2 shows that the error of the guessed singularity is similar as the angle of the constructed rational curve and the original one and Table 3 implies that all three methods yield results which generically differ by $10^{-\ell-2}$.

4 Commented Examples

In this section we present three commented examples demonstrating the steps and approaches of the designed reconstruction algorithm.

Example 1 Consider a cubic curve C defined by the polynomial

$$f(x, y) = 1.00001x^3 - 4.24998x^2 + 0.00001xy + 4.68750x - 0.00001y^3 + 2.00001y^2 + 2.00000y - 1.04687. \tag{26}$$

The curve C is shown in Fig. 5 (left), and at the first sight it looks like it has a singular point and is rational. However zooming on the suspicious part of the cubic reveals that it is actually not a singularity, see again Fig. 5 (left, zoomed section), and thus condition (10) is not satisfied for any point on C. Our goal is to find a rational cubic $\widetilde{C}$ with a defining polynomial $\widetilde{f}(x, y)$ approximating $f(x, y)$.

First, we find a suitable singular point of $\widetilde{C}$ via approach (I.1). We compute a corresponding Weierstrass cubic $\mathcal{W}$ and a birational transformation ϕ mapping C to $\mathcal{W}$. We arrive at

$$y^2 = 1.59075x^3 + 0.212759x^2 + 0.00205069x + 5.127274 \times 10^{-6}. \tag{27}$$

The polynomial $p(x)$ has the following three real roots:

$$\alpha_1 \approx -0.123523, \quad \alpha_2 \approx -0.00532387, \quad \alpha_3 \approx -0.00490129. \tag{28}$$

Since α_2 and α_3 have the minimal mutual distance from all pairs we obtain a prospective singular point

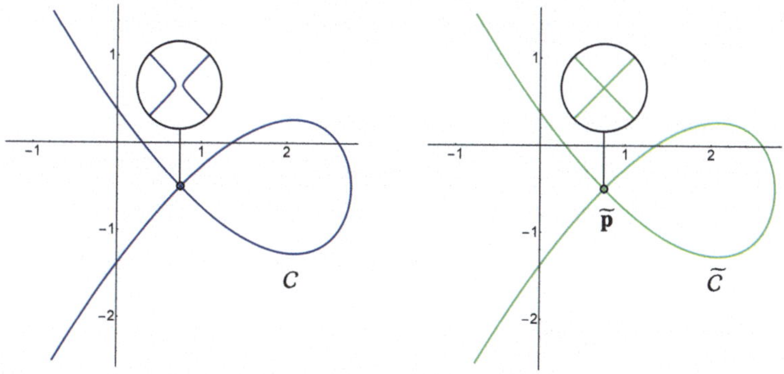

Fig. 5 A perturbed non-rational cubic C without a singular point (left), and its rational approximation $\widetilde{C}$ (right)

$$^{1}\widetilde{\mathbf{p}} = \frac{\phi^{-1}(\alpha_2, 0) + \phi^{-1}(\alpha_3, 0)}{2} \approx (0.74999, -0.49996). \tag{29}$$

Employing approach `(I.2)` yields first three different singular points of the gradient field:

$$\mathbf{p}_{1,2} \approx (1.41665, 133334), \quad \mathbf{p}_3 \approx (2.08328, -0.50000), \quad \mathbf{p}_4 \approx (0.75001, -0.49999). \tag{30}$$

Comparing the values

$$f^2(\mathbf{p}_{1,2}) \approx 1.40477 \times 10^{20}, \quad f^2(\mathbf{p}_3) \approx 1.40424, \quad f^2(\mathbf{p}_4) \approx 6.48668 \times 10^{-10}, \tag{31}$$

we choose $^{2}\widetilde{\mathbf{p}} = \mathbf{p}_4 = (0.75001, -0.49999)$.

Applying method `(I.3)`, based on computing and identifying the corresponding critical points w.r.t. 10 directions (15), we arrive at a possible singular point $^{3}\widetilde{\mathbf{p}} = (0.75000, -0.49999)$.

In the second part of the algorithm we may work either with any of the points $^{1}\widetilde{\mathbf{p}}$, $^{2}\widetilde{\mathbf{p}}$, $^{3}\widetilde{\mathbf{p}}$, or one can also compute their suitable affine combination. We round the prospective singular point w.r.t. the prescribed maximal height 1 and in all three cases we arrive at the approximate singular point $\widetilde{\mathbf{p}} = (3/4, -1/2)$. All rational cubics possessing the singular point $\widetilde{\mathbf{p}}$ are described by

$$\mathbf{L}_{\widetilde{\mathbf{p}}} = \ker \begin{pmatrix} 0 & 0 & 0 & 0 & 9 & -6 & 4 & 24 & -16 & 48 \\ 27 & -12 & 4 & 0 & 24 & -8 & 0 & 16 & 0 & 0 \\ 0 & 9 & -12 & 12 & 0 & 12 & -16 & 0 & 16 & 0 \end{pmatrix}, \tag{32}$$

see (18). Next, we project the cubic C to subspace (32). A defining polynomial of the cubic $\widetilde{C}$ can be obtained via (5), i.e., we arrive at

$$\begin{aligned}\widetilde{C}:\ &-120896908944753 + 541330479134636x \\ &- 490805891701980x^2 + 115484180839728x^3 + 230968365748992y \\ &+ 1581564218xy + 478676928x^2y + 230969124200970y^2 \\ &- 187835608xy^2 - 1117138218y^3 = 0. \end{aligned} \tag{33}$$

Finally, computing (23) yields

$$\delta(\widetilde{\mathbf{c}}, \mathbf{c}) \approx 0.000163488. \tag{34}$$

Example 2 Consider 18 affine points $\mathbf{q}_i$, $i = 1, \ldots, 18$, which were sampled from a cuspidal cubic curve C_0. The coordinates of these points were determined inexactly and moreover rounded, and thus our goal is to find a rational cubic as close to the points $\mathbf{q}_i$ as possible. First, employing the method of least squares approximation, we construct a non-rational cubic C,

$$\begin{aligned} f(x, y) = &- 1.00413x^3 + 5.28002 \times 10^{-4}x^2y + 3.00974x^2 \\ &+ 5.17398 \times 10^{-4}xy^2 + 1.67082 \times 10^{-3}xy - 3.00580x \\ &- 3.7894 \times 10^{-4}y^3 + 1.00670y^2 + 1.33963 \times 10^{-3}y + 1.00000, \end{aligned} \tag{35}$$

see Fig. 6 (left), which will become an input to our algorithm. Now we proceed as in Example 1. Using `(I.1)`, we find the Weierstrass cubic and the corresponding birational transformation ϕ. The polynomial $p(x)$ has one real root and two complex conjugate roots

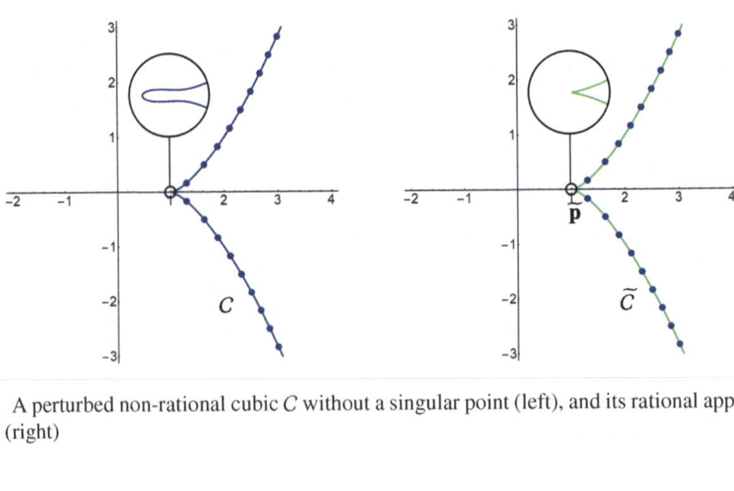

Fig. 6 A perturbed non-rational cubic C without a singular point (left), and its rational approximation $\widetilde{C}$ (right)

$$\alpha_{1,2} \approx 2.93130 \times 10^{-2} \pm 2.65934 \times 10^{-3}\mathrm{i}, \quad \alpha_3 \approx -3.52522 \times 10^{-2}. \tag{36}$$

After employing the inverse transformation ϕ^{-1} to α_1 and α_2 and putting the results together, we obtain the singular point ${}^1\widetilde{\mathbf{p}} = (1.03097, -0.00197)$. Method (I.2) yields two imaginary and two real points and the one with the minimal distance to C possesses the coordinates ${}^2\widetilde{\mathbf{p}} = (1.03016, -0.00009)$. Finally, approach (I.3) based on computing critical points yields ${}^3\widetilde{\mathbf{p}} = (1.01976, -0.00009)$. We again convert floating-point numbers to approximate rational numbers with prescribed maximal height 1 and arrive at $\widetilde{\mathbf{p}} = (1, 0)$.

Now, employing the least square approximation we construct a rational cubic $\widetilde{C}$ with the singular point $\widetilde{\mathbf{p}}$ near to the points $\mathbf{q}_i$. All cubics with the singular point $\widetilde{\mathbf{p}}$ are described by

$$\mathbf{L}_{\widetilde{\mathbf{p}}} = \ker \begin{pmatrix} 0\,0\,0\,0\,1\,0\,0\,2\,0\,3 \\ 3\,0\,0\,0\,2\,0\,0\,1\,0\,0 \\ 0\,1\,0\,0\,0\,1\,0\,0\,1\,0 \end{pmatrix}. \tag{37}$$

Among all such singular cubics we choose the resulting one employing the least squares approximation subject to the constrain $\|\boldsymbol{\lambda}\| = 1$, see Fig. 6 (right). Finally, we measure the deviance of $\widetilde{C}$ from $\mathbf{q}_i$ by (24) and obtain

$$\Delta(\widetilde{C}, \mathbf{q}_i) \approx 2.53271 \times 10^{-17}. \tag{38}$$

Example 3 Although our algorithm was designed for irreducible cubics, in this example we discuss at least shortly what can happen when a reducible cubic C_0 composed of three lines intersecting in three singular points $\mathbf{q}_1 = (0, 0)$, $\mathbf{q}_2 = (1, 0)$ and $\mathbf{q}_3 = (0, 1)$, i.e.,

$$C_0 : \ f_0(x, y) = xy(x + y - 1), \tag{39}$$

see Fig. 7 (left), serves as the input.

By perturbing the coefficients of C_0 we obtain an irreducible non-rational curve C, see Fig. 7 (right),

$$\begin{aligned} C : \ f(x, y) = 0.001x^3 + 1.001x^2y - 0.003x^2 + 0.999xy^2 - 0.999xy \\ + 0.002x - 0.001y^3 + 0.001y - 0.001. \end{aligned} \tag{40}$$

Now, we try to determine a possible singularity of C. Method (I.1) based on computing the Weierstrass curve works with the roots of polynomial $p(x)$,

$$\alpha_{1,2} = -0.503967 \pm 9.98636 \times 10^{-6}i, \quad \alpha_3 = -0.543349, \tag{41}$$

and by averaging $\alpha_{1,2}$ we arrive at ${}^1\widetilde{\mathbf{p}} = (1.00101, -0.002) \approx \mathbf{q}_2$.

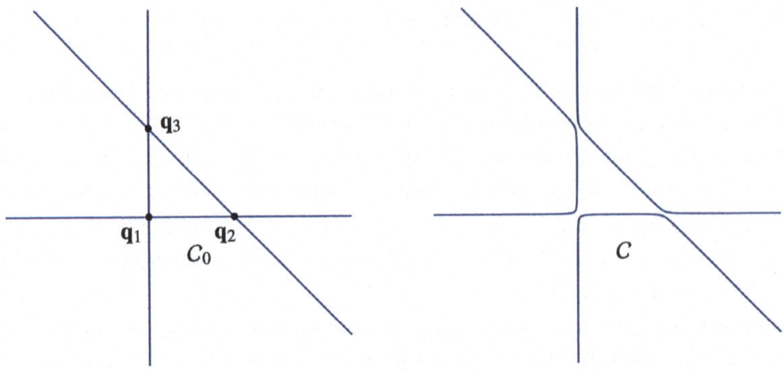

Fig. 7 A reducible cubic curve C_0 (left), and an irreducible perturbed cubic C (right)

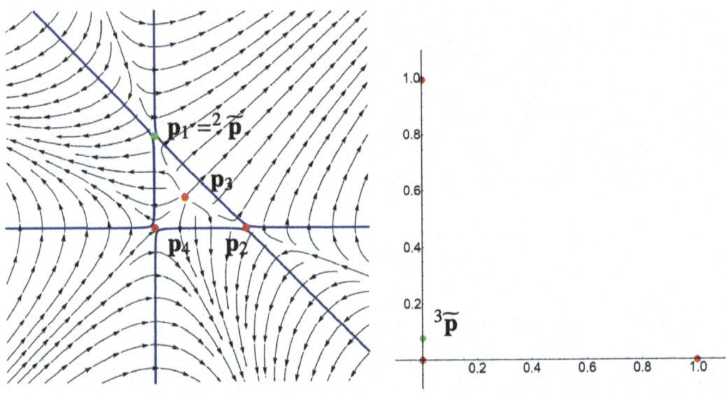

Fig. 8 Left: singular points (red and green) of the normal vector field $\nabla f(x, y)$ and the one (green) with the minimal distance from C. Right: Three loci of critical points (red) and their trimmed mean (green)

Using approach `(I.2)`, we compute first the singular points of the gradient vector field. Solving (13) yields the following real points

$$\mathbf{p}_1 = (0.00198599, 0.994018) \approx \mathbf{q}_3, \ \ \mathbf{p}_2 = (0.99499, 0.00100599) \approx \mathbf{q}_2, \\ \mathbf{p}_3 = (0.332663, 0.332329), \ \ \mathbf{p}_4 = (0.00100604, 0.00200402) \approx \mathbf{q}_1, \tag{42}$$

and their distances from C are

$$9.84175 \times 10^{-7}, \ 9.92008 \times 10^{-7}, \ 0.00138831, \ 9.95994 \times 10^{-7}, \tag{43}$$

respectively. Hence, we choose ${}^2\widetilde{\mathbf{p}} = \mathbf{p}_1 \approx \mathbf{q}_3$, however the choices ${}^2\widetilde{\mathbf{p}} = \mathbf{p}_2$ or ${}^2\widetilde{\mathbf{p}} = \mathbf{p}_3$ are also possible, see Fig. 8 (left).

Finally, method `(I.3)` based on computing the critical points of C yields three sets of points concentrating around the points $\mathbf{q}_i$. However, the trimmed mean of

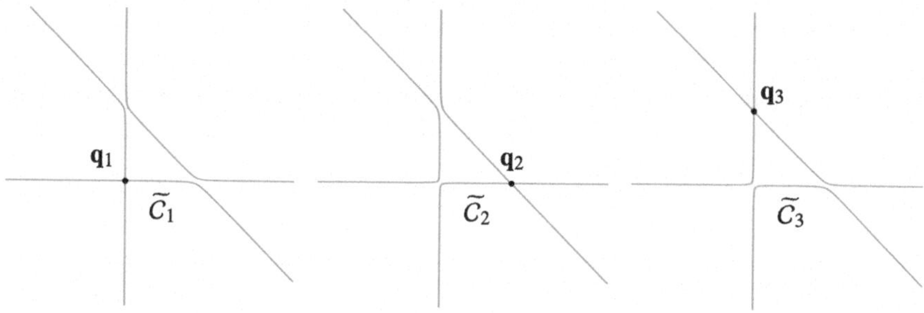

Fig. 9 Rational cubic curves $\widetilde{C}_i$ with singular points in $\mathbf{q}_i$, respectively

all points gives us ${}^3\widetilde{\mathbf{p}} = (0.00052963, 0.0775371)$, which obviously is not a clever choice, see Fig. 8 (right).

This example clearly shows that the algorithm is not very suitable for cases when a given perturbed cubic is close to a reducible cubic and these situations require a special approach. Nonetheless, the methods `(I.2)` and `(I.3)` are able to reveal all three original singular points when treated carefully. This can be also used, for instance, when additional information about degeneracy of the original cubic is known.

As a possible output, we may also construct irreducible cubic curves ${}^i\widetilde{C}$ with the singular points in $\mathbf{q}_i$, respectively which gives the following deviations (23), see also Fig. 9,

$$\delta(\widetilde{\mathbf{c}}_1, \mathbf{c}) = 0.0810552, \quad \delta(\widetilde{\mathbf{c}}_2, \mathbf{c}) = 0.0601122, \quad \delta(\widetilde{\mathbf{c}}_3, \mathbf{c}) = 0.0422905. \tag{44}$$

5 Conclusion

In this paper, we designed a simple algorithm for an approximate reconstruction of perturbed rational planar cubics. As the problem is relatively new in geometric modelling many open questions must be still answered and at least particular cases are to be solved first. So, we have focused solely on the first interesting example, namely on the case of planar cubics which are curves popular in geometric modelling.

As the input perturbed cubic is non-rational, it does not possess a singular point. Nonetheless, the original curve was by assumption singular. So the initial step of the reconstruction algorithm is to find a suitable approximate singular point by averaging suitable regular points of the perturbed curve. We presented three possible simple methods how to proceed—based on the Weierstrass form, using the gradient vector field, and applying so called critical points. Next, it was shown how to choose a suitable rational cubic sufficiently "close" to the given perturbed cubic from the subspace of all rational cubics with the given singular point.

In our future work, we would like to focus on further classes of perturbed objects important for geometric modelling, especially in connection with situations when methods of computer algebra applied on these approximate objects fail. In addition, involving the arithmetical optimality of resulting descriptions, see [19], is another challenge that is worth considering.

Acknowledgements The authors were supported by the project LO1506 of the Czech Ministry of Education, Youth and Sports. We thank all referees for their valuable comments, which helped us to improve the paper.

References

1. Hitz, M., Grabmeier, J., Kaltofen, E., Weispfenning, V. (eds.): Computer Algebra Handbook: Foundations, Applications, Systems. Springer, Berlin Heidelberg (2003)
2. Wang, D., Zhi, L.-H. (eds.): Symbolic-Numeric Computation. Trends in Mathematics. Birkhäuser, Basel (2007)
3. Langer, U., Paule, P. (eds.): Numerical and Symbolic Scientific Computing: Progress and Prospects. Texts & Monographs in Symbolic Computation. Springer, Wien (2012)
4. Robbiano, L., Abbott, J. (eds.): Approximate Commutative Algebra. Texts & Monographs in Symbolic Computation. Springer, Wien (2010)
5. Sendra, J.R., Winkler, F.: Symbolic parametrization of curves. J. Symb. Comput. **12**, 607–631 (1991)
6. Rueda, S.L., Sendra, J., Sendra, J.R.: Rational Hausdorff divisors: A new approach to the approximate parametrization of curves. J. Comput. Appl. Math. **263**, 445–465 (2014)
7. Shen, L.-Y., Pérez-Díaz, S.: Characterization of rational ruled surfaces. J. Symb. Comput. **63**, 21–45 (2014)
8. Walker, R.J.: Algebraic Curves. Springer, Berlin (1978)
9. Brieskorn, E., Knörrer, H., Stillwell, J.: Plane Algebraic Curves: Translated by John Stillwell. Modern Birkhäuser Classics. Springer, Basel (2012)
10. Silverman, J.: The Arithmetic of Elliptic Curves. Graduate Texts in Mathematics. Springer, New York (2009)
11. Gelfand, I.M., Kapranov, M., Zelevinsky, A.: Discriminants, Resultants, and Multidimensional Determinants. Mathematics: Theory and Application. Birkhäuser, Basel (1994)
12. Alcázar, J., Sendra, J.R.: Computation of the topology of real algebraic space curves. J. Symb. Comput. **39**, 719–744 (2005)
13. Bizzarri, M., Lávička, M.: A symbolic-numerical approach to approximate parameterizations of space curves using graphs of critical points. J. Comput. Appl. Math. **242**, 107–124 (2013)
14. Jüttler, B., Chalmovianský, P.: A predictor-corrector-type technique for the approximate parameterization of intersection curves. Appl. Algebr. Eng. Commun. Comput. **18**, 151–168 (2007)
15. Gonzalez-Vega, L., Necula, I., Perez-Diaz, S., Sendra, J., Sendra, J.R.: Algebraic methods in computer aided geometric design: theoretical and practical applications. In: Chen, E., Wang, D. (eds.) Geometric Computation. Lecture Notes on Computing, vol. 11, pp. 1–33. World Scientific, Singapore (2004)
16. Kotsireas, I.: Panorama of methods for exact implicitization of algebraic curves and surfaces. In: Chen, E., Wang, D. (eds.) Geometric Computation. Lecture Notes on Computing, vol. 11, pp. 126–155. World Scientific, Singapore (2004)
17. Sederberg, T.W., Chen, F.: Implicitization using moving curves and surfaces. In: Proceedings of the 22Nd Annual Conference on Computer Graphics and Interactive Techniques, SIGGRAPH '95, (New York, NY, USA), pp. 301–308. ACM (1995)

18. Sederbeg, T.W., Goldman, R., Du, H.: Implicitizing rational curves by the method of moving algebraic curves. J. Symb. Comput. **23**, 153–175 (1997)
19. Sendra, J.R., Winkler, S.M.: A heuristic and evolutionary algorithm to optimize the coefficients of curve parametrizations. J. Comput. Appl. Math. **305**, 18–35 (2016)

Quadrature Rules in the Isogeometric Galerkin Method: State of the Art and an Introduction to Weighted Quadrature

Francesco Calabrò, Gabriele Loli, Giancarlo Sangalli and Mattia Tani

Abstract In this paper we introduce the quadrature needed in the isogeometric Galerkin method. Quadrature rules affect the cost of the assembly of the discrete counterpart of the IGA method, so that the search for efficient quadrature is an active research topic. The focus of the first part is on a brief survey on the contributions available for the reduction of computational costs for such issue. We review the generalized Gaussian strategies and the reduced quadrature. Then we present the novelty of weighted quadrature, recently proposed by Calabrò, Sangalli and Tani for the efficient assembly. We detail the construction of such rules and give some examples. Finally, we end with some remarks on current work and further developments.

1 Introduction

We consider the Poisson problem

$$\begin{cases} -\nabla^2 u = f, \text{ on} & \Omega, \\ u = 0, \quad\quad \text{on} & \partial\Omega, \end{cases}$$

as a model problem. Its Galerkin approximation on a discrete space V requires the computation of the stiffness matrix (or stiffness integrals)

F. Calabrò (✉)
Dipartimento di Matematica e Applicazioni "Renato Caccioppoli", Università degli Studi di Napoli "Federico II", Naples, Italy
e-mail: calabro@unina.it

G. Loli · G. Sangalli · M. Tani
Dipartimento di Matematica, Università degli Studi di Pavia, Pavia, Italy
e-mail: gabriele.loli01@universitadipavia.it

G. Sangalli
e-mail: giancarlo.sangalli@unipv.it

M. Tani
e-mail: mattia.tani@unipv.it

C. Giannelli and H. Speleers (eds.), *Advanced Methods for Geometric Modeling and Numerical Simulation*, Springer INdAM Series 35,
https://doi.org/10.1007/978-3-030-27331-6_3

$$\int_{\Omega} \nabla R_i(\boldsymbol{x}) \nabla R_j(\boldsymbol{x}) d\boldsymbol{x}, \tag{1}$$

where R_i and R_j denote two basis functions in V. If we add a zero-order term to (1) then we also need the mass matrix (or mass integrals)

$$\int_{\Omega} R_i(\boldsymbol{x}) R_j(\boldsymbol{x}) d\boldsymbol{x}\,, \tag{2}$$

and for problems involving a transport we would have also the so-called advection matrix:

$$\int_{\Omega} R_i(\boldsymbol{x}) \nabla R_j(\boldsymbol{x}) d\boldsymbol{x}. \tag{3}$$

In IGA, Ω is given by a spline or NURBS parametrization. For the sake of simplicity, we assume Ω is given by a d-dimensional single patch spline representation:

$$\Omega = \boldsymbol{F}([0,1]^d), \text{ with } \boldsymbol{F}(\zeta) = \sum_i \boldsymbol{C}_i \hat{B}_i(\zeta),$$

where $\boldsymbol{C}_i$ are the control points and $\hat{B}_i$ are p-degree tensor-product B-spline basis functions defined on the parametric patch $[0, 1]^d$.

Being IGA based on the isoparametric paradigm, the basis functions R_i are defined as $R_i = \hat{B}_i \circ \boldsymbol{F}^{-1}$. The integrals above are then computed after change of variable, thus on the reference domain. For more details, see [22].

Here we consider the computation of the matrices. The mass matrix, that corresponds to (2) after change of variable $\zeta = \boldsymbol{F}^{-1}(\boldsymbol{x})$, is $\mathbb{M} = \{m_{i,i}\} \in \mathbb{R}^{N_{DOF} \times N_{DOF}}$ where:

$$m_{i,j} = \int_{[0,1]^d} \hat{B}_i(\zeta)\, \hat{B}_j(\zeta)\, c(\zeta)\, d\zeta\,, \tag{4}$$

with $c(\zeta) = |\hat{\boldsymbol{D}}\boldsymbol{F}(\zeta)|$.

For the stiffness matrix $\mathbb{S} = \{s_{i,j}\} \in \mathbb{R}^{N_{DOF} \times N_{DOF}}$ we have:

$$s_{i,j} = \int_{[0,1]^d} \left(\hat{\boldsymbol{D}}\boldsymbol{F}^{-T} \hat{\nabla} \hat{B}_i\right)^T \left(\hat{\boldsymbol{D}}\boldsymbol{F}^{-T} \hat{\nabla} \hat{B}_j\right) |\hat{\boldsymbol{D}}\boldsymbol{F}|\, d\zeta\,.$$

Also in this case we change notations and we write:

$$s_{i,j} = \sum_{l,m=1}^{d} \int_{[0,1]^d} \frac{\partial \hat{B}_i}{\partial \hat{x}_l}(\zeta) c_{l,m}(\zeta) \frac{\partial \hat{B}_j}{\partial \hat{x}_m}(\zeta)\, d\zeta\,, \tag{5}$$

so that $C(\zeta) := \{c_{l,m}(\zeta)\}_{l,m=1,\dots,d}$ is the matrix:

$$C(\zeta) = \left[\hat{\boldsymbol{D}}\boldsymbol{F}^{-1}(\zeta)\hat{\boldsymbol{D}}\boldsymbol{F}^{-T}(\zeta)\right]|\hat{\boldsymbol{D}}\boldsymbol{F}(\zeta)|\,.$$

For the advection matrix $\mathbb{A} = \{a_{i,j}\} \in \mathbb{R}^{N_{DOF}\times N_{DOF}}$ we have:

$$a_{i,j} = \int_{[0,1]^d} \hat{B}_i \left(\hat{\boldsymbol{D}}\boldsymbol{F}^{-T}\hat{\nabla}\hat{B}_j\right)|\hat{\boldsymbol{D}}\boldsymbol{F}|\,d\zeta\,.$$

Also in this case we change notations and we write:

$$a_{i,j} = \sum_{l=1}^{d}\int_{[0,1]^d} \hat{B}_i(\zeta)k_l(\zeta)\frac{\partial \hat{B}_j}{\partial \hat{x}_l}(\zeta)\,d\zeta\,, \tag{6}$$

so that $k(\zeta) := \{k_l(\zeta)\}_{l=1,\dots,d}$ is the vector:

$$k(\zeta) = \left[\hat{\boldsymbol{D}}\boldsymbol{F}^{-T}(\zeta)\right]|\hat{\boldsymbol{D}}\boldsymbol{F}(\zeta)|\,.$$

To clarify the use of quadrature we will consider the mass matrix $\mathbb{M} = \{m_{i,j}\}$ (4) where $\hat{B}_i$ and $\hat{B}_j$ are tensor-product B-splines, and c is a coefficient that incorporates the determinant Jacobian of the geometry mapping and other possible non-tensor product factors. Following [15, 16, 33] we know that the optimal order of convergence of the overall method is maintained if the approximate integral is computed with a quadrature rule that is exact for constant functions c.

2 Cost of Assembly

We consider in this paper a d-dimensional scalar Poisson model problem on a single-patch domain, and an isogeometric tensor-product space of degree p and total dimension N_{DOF}, with $N_{DOF} \gg p^d$. For the sake of simplicity, we focus on the case of C^{p-1} continuity, i.e., the typical setting of the so-called k-method (see e.g. [22]). The resulting stiffness matrix has $O(N_{DOF}(2p+1)^d) \approx O(N_{DOF}\,p^d)$ non-zero entries. Therefore, we assume $CN_{DOF}\,p^d$ floating point operations (FLOPs) is the (quasi)-optimal computational cost for the formation of the stiffness matrix, where C is a (reasonably small) constant[1] that does not depend on N_{DOF} and p.

The algorithms currently used in isogeometric codes are suboptimal with respect to the degree p, that is, their cost grows with respect to the degree p faster than p^d. Optimal assembly procedures are available for the ususal FEM in some cases [5, 6], while for the IGA case they are attained only with collocation [10, 47] for which also the optimal convergence rates are now available in some cases [41].

[1] Throughout the paper the constant C is in general different at each occurrence.

The majority of isogeometric codes inherit a finite element architecture, which adopts an element-wise assembly loop with element-wise standard Gaussian quadrature (SGQ). Each local stiffness matrix has dimension $(p+1)^{2d}$ and each entry is calculated by quadrature on $(p+1)^d$ Gauss points. The total cost is $O(N_{EL}p^{3d}) \approx O(N_{DOF}p^{3d})$ FLOPs, where N_{EL} is the number of elements and, for the k-method, $N_{EL} \approx N_{DOF}$.

The standard way to reduce the cost is to reduce the number of quadrature points, for example by *reduced Gaussian* (eventually corrected by variationally consistent constraints [31]) or *generalized Gaussian* quadrature (GGQ). We review the literature on these in the next section, here we clarify GGQ. Consider the mass matrix $\mathbb{M} = \{m_{i,j}\}$ (4) where $\hat{B}_i$ and $\hat{B}_j$ are tensor-product B-splines, and c is a coefficient that incorporates the determinant Jacobian of the geometry mapping and other possible non-tensor product factors. The work [32] has explored the possibility of constructing and using GGQ quadrature of the kind

$$\int_{\hat{\Omega}} c(\boldsymbol{\zeta})\,\hat{B}_i(\boldsymbol{\zeta})\hat{B}_j(\boldsymbol{\zeta})\,d\boldsymbol{\zeta} \approx \mathbb{Q}^{\mathrm{GGQ}}(c(\cdot)\,\hat{B}_i(\cdot)\hat{B}_j(\cdot)), \tag{7}$$

where the quadrature weights w_q^{GGQ} and points $\boldsymbol{x}_q^{\mathrm{GGQ}}$ of the quadrature rule $\mathbb{Q}^{\mathrm{GGQ}}(f(\cdot)) = \sum_q w_q^{\mathrm{GGQ}} f(\boldsymbol{x}_q^{\mathrm{GGQ}})$ fulfill the exactness conditions

$$\int_{\hat{\Omega}} \hat{B}_{\boldsymbol{k}}^{2p}(\boldsymbol{\zeta})d\boldsymbol{\zeta} = \mathbb{Q}^{\mathrm{GGQ}}(\hat{B}_{\boldsymbol{k}}^{2p}(\cdot)), \quad \forall \boldsymbol{k}. \tag{8}$$

Here $\{\hat{B}_{\boldsymbol{k}}^{2p}(\cdot)\}$ is the B-spline basis of degree $2p$ and continuity C^{p-1}. Exact integration of the product of a pair of p degree splines $\hat{B}_i(\cdot)\hat{B}_j(\cdot)$ is then guaranteed by (8). Since the w_q^{GGQ} and $\boldsymbol{x}_q^{\mathrm{GGQ}}$ are not known analytically, they need to be computed numerically as solution of the global non-linear problem (8), see the next section for the literature on this problem. The number of conditions in (8) is $\#\{\hat{B}_{\boldsymbol{k}}^{2p}(\cdot)\} \approx N_{DOF}(p+1)^d \approx N_{EL}(p+1)^d$, dropping the lower order terms, and therefore GGQ is expected to use about $N_{EL}\left(\frac{p+1}{2}\right)^d$ quadrature points, with a saving of a factor 2^d with respect to SGQ.

Also, one can exploit the tensor-product structure of multivariate splines by adopting the so-called sum-factorization as done in [25] and introduced in the IGA framework recently [8]. The application of sum-factorization to the usual element-wise assembly reduces costs by a factor of p^2.

With a major change of paradigm, as we will see in the final section, we can combine sum-factorization with weighted quadrature to obtain an overall cost of $O(N_{DOF}p^{d+1})$ FLOPs.

3 Bibliographic Review

First, we should mention that quadrature problems in spline spaces have an interest itself, so that there are contributions starting from the 50's [40, 43, 49, 52]. These are papers with explicit calculation of formulae with exactness conditions in the considered spline space.

The first paper dealing with the construction of quadrature rules with ad-hoc exactness for applications in IGA is [32]. In this paper the construction is based on the resolution of the nonlinear equations seen in Eq. (8) with an optimal number of points, leading to the so-called generalized Gaussian quadrature. We will refer to this procedure as the global (patch) optimization. This problem is ill-conditioned so that the computation of such optimal rules was done in limited cases. After, many attempts have been made to have solutions of the global quadrature problem. As described before, the saving when using generalized Gaussian quadrature is of a factor 2^d with respect to SGQ.

A search direction moves from the exact known results available from the papers that we have mentioned at the beginning of this section: these results are used for the construction of the exact quadrature in C^1 cubic spline spaces in [7], and quadratic and cubic splines in [14]. Then, starting from these known solutions, the same authors propose a continuation algorithm based on homothopy for the C^2 cubic case in [12] and for the odd degree spline spaces in [13]. Finally, in [11] explicit recursive algorithms are given for the C^1 quintic splines.

One way to construct the quadrature starting from the optimization problem is to search for the zero of a residual functional. Newton's algorithm and continuation for the construction of generalized Gaussian quadrature has been described in a general setting in [17, 21, 35]. This has been used in [9] for the construction of a quasi-optimal quadrature rule for IGA. In this paper, local exactness conditions are used instead of (8). For the resolution of the global problem, a difficult issue is the good choice of the starting point for the Newton iterations. This issue is considered in [29] and [34]. In the paper [34] the (global) problem (8) is effectively solved by a Newton method with an adaptive loop based on continuation, while the paper [29] is also concerned with reduced quadrature.

The strategy of reduced quadrature relies on element-wise computations that fulfills fewer conditions [3, 48] and is also introduced in order to avoid locking [1, 2]. Recently also reduced integration at superconvergent points has been proposed [26]. Reduced quadrature can save some computation for each element: usually instead of $(p+1)^d$ points on each element uses p^d or $(p-1)^d$ so that the order with respect to p is not changed. In [24, 44] reduced quadratures are introduced in order to gain optimality with respect to a dispersion measure related to spurious eigenvalues.

Very different strategies have been introduced in order to attain higher savings. In [36, 37] look-up tables are used for exact integration. Quasi-interpolation with exact computation of spline integrals is introduced in [19] and recently used also for the IGA-BEM method [18]. Variationally consistent domain integration is introduced in [31].

Finally, we mention that in [38, 39] the authors have reduced the cost of assembly via low rank approximations of the non tensor product part of the integral factors. This procedure is promising, although its good performance relies on known information on the geometry of the domain.

4 Weighted Quadrature

In [20] a new procedure has been introduced that is very general and almost optimal for the formation of the discrete counterpart of the Galerkin-IGA methods. In this paper a new algorithm which does not use the element-wise assembling loop is proposed. Instead, a loop over the matrix rows and the use of a specifically designed *weighted quadrature* (WQ) rule for each row is introduced. In particular, the quadrature rule for the $\boldsymbol{i}$-th row of $\mathbb{M}$ is as follows:

$$\int_{\hat{\Omega}} c(\boldsymbol{\zeta})\, \hat{B}_j(\boldsymbol{\zeta})\, (\hat{B}_i(\boldsymbol{\zeta}) d\boldsymbol{\zeta}) \approx \mathbb{Q}_i^{\mathrm{WQ}}(c(\cdot)\, \hat{B}_j(\cdot)), \quad \forall \boldsymbol{j}. \tag{9}$$

Unlike (7), in the right hand side of (9) the integrand function is $c(\cdot)\, \hat{B}_j(\cdot)$ since the test function is incorporated into the integral weight (measure) $(\hat{B}_i(\boldsymbol{\zeta}) d\boldsymbol{\zeta})$. The price to pay is that the quadrature weights depend on $\boldsymbol{i}$, while we select global quadrature points as suitable interpolation points that do not depend on $\boldsymbol{i}$. Again, the quadrature weights are not known analytically and need to be computed numerically as solution of the exactness conditions

$$\int_{\hat{\Omega}} \hat{B}_j(\boldsymbol{\zeta})\, (\hat{B}_i(\boldsymbol{\zeta}) d\boldsymbol{\zeta}) = \mathbb{Q}_i^{\mathrm{WQ}}(\hat{B}_j). \tag{10}$$

However, the exactness conditions (10) are linear with respect to the weights. Furthermore, (10) is a local problem as the weights outside supp$(\hat{B}_i)$ can be set to zero a priori. The knot vectors do not need to be uniform with this approach.

The number of exactness conditions of (10) is $\#\{\hat{B}_j(\cdot)\} = N_{DOF}$. This is lower than the number of conditions of (8), which is $\#\{\hat{B}_k^{2p}(\cdot)\} \approx N_{DOF}(p+1)^d$. Hence, the main advantage of the WQ with respect to GGQ is that the former requires significantly fewer quadrature points. In the case of maximum regularity only 2 points are needed in each direction sufficiently far away from the boundary, while $p+1$ points are taken on boundary knot-spans along directions that end on the boundary. Adopting sum-factorization (see [8]), the proposed algorithm has a total computational cost of $O(N_{DOF}\, p^{d+1})$ FLOPs.

In the more general case of the stiffness matrix, the exactness conditions become, [30]:

$$\int_{[0,1]^d} \partial_\alpha \hat{B}_i(\boldsymbol{\zeta})\, \partial_\beta \hat{B}_j(\boldsymbol{\zeta})\, d\boldsymbol{\zeta} = \mathbb{Q}_{\alpha,\beta,\boldsymbol{i}}^{\mathrm{WQ}}(\partial_\beta \hat{B}_j(\cdot)). \tag{11}$$

where $\alpha, \beta = 0, 1$ indicate if the derivative is done or not.

In all cases, due to the tensor product structure of the basis of B-splines, we can write the quadrature problem in the univariate case and then apply the quadrature in the multivariate case via sum-factorization, see [8]. Four different univariate WQ rules are the building blocks that we use to define the rules $\mathbb{Q}^{\mathrm{WQ}}_{\alpha,\beta,i}$ with different test and trial functions:

$$\mathfrak{Q}_i^{(0,0)}\left(\hat{B}_j(\cdot)\right) = \int_0^1 \hat{B}_i(\zeta)\hat{B}_j(\zeta)d\zeta \,;\; \mathfrak{Q}_i^{(0,1)}\left(\hat{B}'_j(\cdot)\right) = \int_0^1 \hat{B}_i(\zeta)\hat{B}'_j(\zeta)d\zeta \,; \tag{12}$$

$$\mathfrak{Q}_i^{(1,0)}\left(\hat{B}_j(\cdot)\right) = \int_0^1 \hat{B}'_i(\zeta)\hat{B}_j(\zeta)d\zeta \,;\; \mathfrak{Q}_i^{(1,1)}\left(\hat{B}'_j(\cdot)\right) = \int_0^1 \hat{B}'_i(\zeta)\hat{B}'_j(\zeta)d\zeta \,. \tag{13}$$

We prefer not to calculate four different rules for the described cases, and we introduce two different rules for the two different test functions. We start noticing that the trial functions $\hat{B}_j$, $\hat{B}'_j$ all are in the space $\mathcal{S}^p_{p-2}$. Then we can count the number of active trial functions on $p+1$ elements—that is the number of elements in the support of the test function—and this is $\leq 3p$.

Finally we search for two rules:

$$\mathfrak{Q}_i^{(0)}\left(\bar{B}_j(\cdot)\right) = \int_0^1 \hat{B}_i(\zeta)\bar{B}_j(\zeta)d\zeta, \tag{14}$$

$$\mathfrak{Q}_i^{(1)}\left(\bar{B}_j(\cdot)\right) = \int_0^1 \hat{B}'_i(\zeta)\bar{B}_j(\zeta)d\zeta, \tag{15}$$

where $\bar{B}_j(\zeta)$ are the functions of degree p and regularity $p-2$ whose support intersect the one of the test function. The quadrature rule (14) can be used also for the mass (4) and the advection (6) matrices also. Notice that the two cases (14) and (15) are different and cannot be treated with only one quadrature because the weight function changes. It is also to be noticed that in the second case the test function $\hat{B}'_i$ is sign changing.

We can choose a priori 3 nodes per direction in the interior of the elements, and on the boundary elements $p+1$ nodes and set up a linear problem for the calculation of weights that always has a solution. Notice that this leads to a construction of quadrature rules that have asymptotically $O(1)$ points per element with respect to the degree p. This is done computing the quadrature rules by means of the linear problems (14) and (15). We remark that the quadrature $\mathfrak{Q}_i^{(1)}$ could be constructed as linear combination of quadrature rules of lower degree, following the properties of the B-splines. Nevertheless, our construction is straightforward and leads to a quadrature with good approximation properties, as can be seen in the next tests. Some care has to be used when computing the solution of the linear problems especially in the case of under-determined systems.

To compute the quadrature weights for each of the univariate WQ rule we need to solve n_{DOF} linear systems of dimension $\approx(2p)^2$, where n_{DOF} indicates the number of degrees of freedom along each direction. Thus, for this systems we search for the minimum Euclidean norm solution, as proposed in [20]. This is usually done by means of QR decomposition. Since the matrices involved are often ill-conditioned,

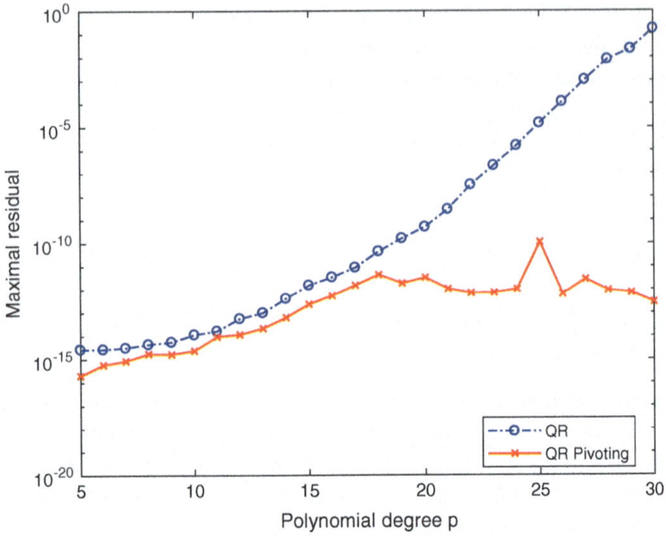

Fig. 1 Maximal residual achieved in the solution of linear systems for the computation of the quadrature weights, for maximal regularity splines on a single patch domain of 32 elements. The comparison is between the QR decomposition and the QR decomposition with column pivoting

here we test the QR decomposition performed with column pivoting so that the diagonal elements of the factor R are in non-increasing order. Numerical tests presented in Figs. 1, 2 and 3 show that the pivoting strategy greatly improves numerical accuracy.

In Fig. 1, we test the procedure with respect to the exactness attained by the quadrature rule. We compare the maximal residual achieved computing all the quadrature weights for the rule (15) by means of QR decomposition with and without column pivoting.

In Figs. 2 and 3 we have plotted the relative error in L^2-norm for the solution of the following Dirichlet problems [46]:

$$\begin{cases} -u''(x) = 4\pi^2 \sin(2\pi x) \text{ on } [0, 1], \\ u(0) = u(1) = 0, \end{cases} \tag{16}$$

and

$$\begin{cases} -\Delta u(x_1, x_2, x_3) = & (\sin(5\pi x_1)((75\pi^2 x_1^4 + x_1^2(150\pi^2 x_2^2 - 375\pi^2 - 16) + 75\pi^2 x_2^4 - \\ & x_2^2(375\pi^2 + 16) + 20(15\pi^2 + 1)) \sin(5\pi x_3) \sin(5\pi x_2) - \\ & 20\pi x_2(2x_1^2 + 2x_2^2 - 5) \sin(5\pi x_3) \cos(5\pi x_2)) - \\ & 20\pi x_1(2x_1^2 + 2x_2^2 - 5) \sin(5\pi x_3) \sin(5\pi x_2) \cos(5\pi x_1)) \text{ on } \Omega, \\ u = 0 \text{ on } \partial\Omega, & \end{cases} \tag{17}$$

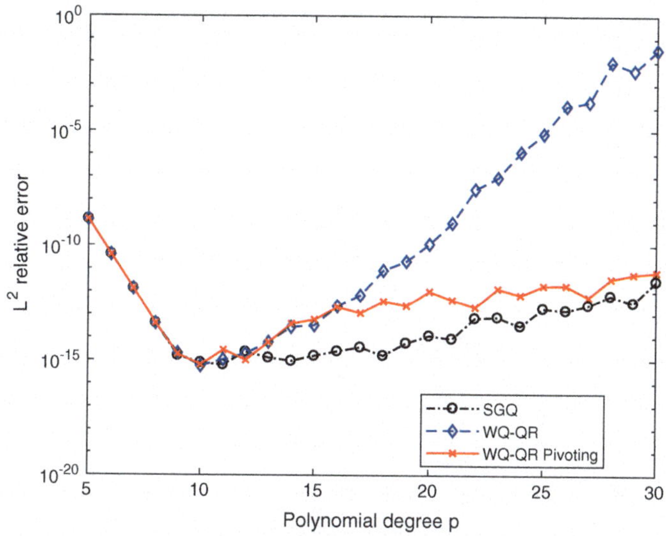

Fig. 2 Relative error in L^2-norm for the solution of problem (16) on a single patch domain of 32 elementsin the framework of isogeometric-Galerkin method with maximal regularity for various quadrature rules: SGQ, WQ with QR decomposition and WQ with QR decomposition performed with column pivoting

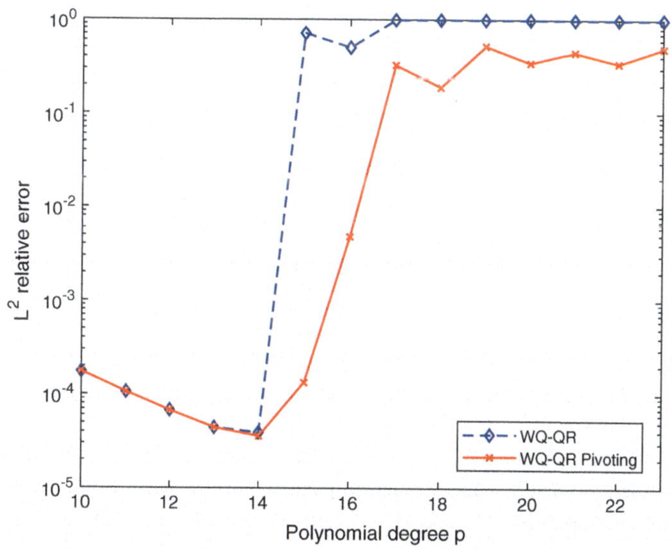

Fig. 3 Relative error in L^2-norm for the solution of problem (17) on a single patch domain of 32^3 elements in the framework of isogeometric-Galerkin method with maximal regularity for weighted quadrature rules: with QR decomposition and with QR decomposition performed with column pivoting

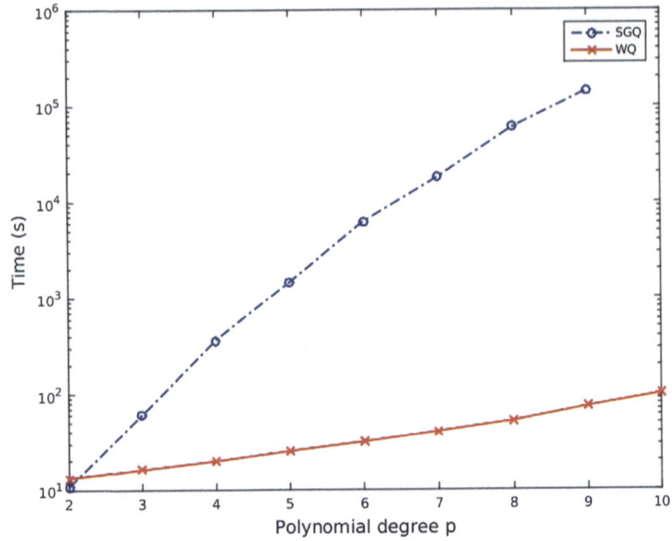

Fig. 4 Time for stiffness matrix assembly in the framework of isogeometric-Galerkin method with maximal regularity on a single patch domain of 20^3 elements. The comparison is between the WQ approach proposed and SGQ as implemented in GeoPDEs 3.0

where Ω is the thick quarter of ring described by

$$\Omega = \left\{(x_1, x_2, x_3) \in \mathbb{R}^3 | 1 \leq x_1^2 + x_2^2 \leq 4, 0 \leq x_1, x_2, \ 0 \leq x_3 \leq 1\right\}.$$

We compare standard Gaussian quadrature (only for the one-dimensional problem), Weighted quadrature through QR decomposition and Weighted quadrature through QR decomposition with column pivoting. The latter attains better accuracy with respect to the standard QR decomposition, although an increased error is noticed for very high degrees, see Fig. 3. In particular, Fig. 2 shows that, through column pivoting, Weighted quadrature reaches the same accuracy as the standard Gaussian quadrature. Indeed, at least for the one-dimensional case, we can raise the degree beyond $p = 30$ without loss of accuracy. In the three-dimensional case considered in Fig. 3 the column pivoting strategy attains better accuracy with respect to the standard QR decomposition but the case of very high degrees is more challenging. Such behavior is less related to the resolution of the linear problem, but, up to our comprehension, depends on the position of the fixed nodes: new layouts on the single element are currently under investigation.

Finally, in Fig. 4 we plot the time for assembling the stiffness matrix when using SGQ—available as standard in GeoPDEs 3.0 [23, 51]—and our new WQ. We have done computations up to degree $p = 10$ with $N_{DOF} = 20^3$. In the case $p = 9$ GeoPDEs takes more than 50 hours to form the stiffness matrix while the proposed algorithm needs only 90 sec, so that the use of high degrees is possible with WQ.

We emphasize that the WQ is well suited for matrix-free [46] and parallel implementations, but in this case, for fair comparison, the computation is serial and the whole matrices are stored. Clearly we exploit sparsity in our MATLAB implementation: we compute all the nonzero entries of matrices, the corresponding row and column indices and then call the MATLAB *sparse* function, that uses a compressed sparse column format.

5 Current Work and Further Developments

Weighted quadrature poses new issues that are currently under investigation.

First, the application of this technique has been extended to other operators. In [30] linear elasticity problems are considered. In the same paper a discussion on the localization of points is done in the case of lower—eventually non uniform—regularity.

In [4] the weighted quadrature is applied to boundary integral equations: the regular part can be treated as seen in the previous section, while the singular integrals are computed with a weighted quadrature where the singular part is incorporated in the weight. This can be done in an efficient way via modified moments and the recursive definition of B-splines, see also [18, 27, 28].

Finally, the modest computational cost attained for the construction of the discrete problem poses new issues on the final resolution of the matrix problems, that become more dense when the degree increases. Iterative solvers are well suited for such problems and recently preconditioners have been introduced for these problems, see [42, 45, 50].

References

1. Adam, C., Bouabdallah, S., Zarroug, M., Maitournam, H.: Improved numerical integration for locking treatment in isogeometric structural elements, Part I: beams. Comput. Methods Appl. Mech. Eng. **279**, 1–28 (2014)
2. Adam, C., Bouabdallah, S., Zarroug, M., Maitournam, H.: Improved numerical integration for locking treatment in isogeometric structural elements, Part II: plates and shells. Comput. Methods Appl. Mech. Eng. **284**, 106–137 (2015)
3. Adam, C., Hughes, T.J.R., Bouabdallah, S., Zarroug, M., Maitournam, H.: Selective and reduced numerical integrations for NURBS-based isogeometric analysis. Comput. Methods Appl. Mech. Eng. **284**, 732–761 (2015)
4. Aimi, A., Calabrò, F., Diligenti, M., Sampoli, M.L., Sangalli, G., Sestini, A.: Efficient assembly based on B-spline tailored quadrature rules for the IgA-SGBEM. Comput. Methods Appl. Mech. Eng. **331**, 327–342 (2018)
5. Ainsworth, M., Andriamaro, G., Davydov, O.: Bernstein-Bézier finite elements of arbitrary order and optimal assembly procedures. SIAM J. Sci. Comput. **33**, 3087–3109 (2011)

6. Ainsworth, M., Davydov, O., Schumaker, L.L.: Bernstein Bézier finite elements on tetrahedral hexahedral pyramidal partitions. Comput. Methods Appl. Mech. Eng. **304**, 140–170 (2016)
7. Ait-Haddou, R., Bartoň, M., Calo, V.M.: Explicit Gaussian quadrature rules for C^1 cubic splines with symmetrically stretched knot sequences. J. Comput. Appl. Math. **290**, 543–552 (2015)
8. Antolin, P., Buffa, A., Calabrò, F., Martinelli, M., Sangalli, G.: Efficient matrix computation for tensor-product isogeometric analysis: the use of sum factorization. Comput. Methods Appl. Mech. Eng. **285**, 817–828 (2015)
9. Auricchio, F., Calabrò, F., Hughes, T.J.R., Reali, A., Sangalli, G.: A simple algorithm for obtaining nearly optimal quadrature rules for NURBS-based isogeometric analysis. Comput. Methods Appl. Mech. Eng. **249**, 15–27 (2012)
10. Isogeometric collocation methods: Auricchio, F., Beiraõ da Veiga, L., Hughes, T.J.R., Reali, A., Sangalli, G. Math. Model. Methods Appl. Sci. **20**, 2075–2107 (2010)
11. Bartoň, M., Ait-Haddou, R., Calo, V.M.: Gaussian quadrature rules for C^1 quintic splines with uniform knot vectors. J. Comput. Appl. Math. **322**, 57–70 (2017)
12. Bartoň, M., Calo, V.M.: Gaussian quadrature for splines via homotopy continuation: Rules for C^2 cubic splines. J. Comput. Appl. Math. **296**, 709–723 (2016)
13. Bartoň, M., Calo, V.M.: Optimal quadrature rules for odd-degree spline spaces and their application to tensor-product-based isogeometric analysis. Comput. Methods Appl. Mech. Eng. **305**, 217–240 (2016)
14. Bartoň, M., Calo, V.M.: Gauss-Galerkin quadrature rules for quadratic and cubic spline spaces and their application to isogeometric analysis. Comput. Aided Des. **82**, 57–67 (2017)
15. Beiraõ da Veiga, L., Buffa, A., Rivas, J., Sangalli, G.: Some estimates for h–p–k-refinement in isogeometric analysis. Numer. Math. **118**, 271–305 (2011)
16. Mathematical analysis of variational isogeometric methods: Beiraõ da Veiga, L., Buffa, A., Sangalli, G., Vázquez, R. Acta Num. **23**, 157–287 (2014)
17. Bremer, J., Gimbutas, Z., Rokhlin, V.: A nonlinear optimization procedure for generalized Gaussian quadratures. SIAM J. Sci. Comput. **32**, 1761–1788 (2010)
18. Calabrò, F., Falini, A., Sampoli, M.L., Sestini, A.: Efficient quadrature rules based on spline quasi-interpolation for application to IGA-BEMs. J. Comput. Appl. Math. **338**, 153–167 (2018)
19. Calabrò, F., Manni, C.: The choice of quadrature in NURBS-based isogeometric analysis. In: Proceedings of the 3rd South-East European Conference on Computational Mechanics (SEECCM) (2013)
20. Calabrò, F., Sangalli, G., Tani, M.: Fast formation of isogeometric Galerkin matrices by weighted quadrature. Comput. Methods Appl. Mech. Eng. **316**, 606–622 (2017)
21. Cheng, H., Rokhlin, V., Yarvin, N.: Nonlinear optimization, quadrature, and interpolation. SIAM J. Optim. **9**, 901–923 (1999)
22. Cottrell, J.A., Hughes, T.J.R., Bazilevs, Y.: Isogeometric Analysis: Toward Integration of CAD and FEA. Wiley, Hoboken (2009)
23. De Falco, C., Reali, A., Vázquez, R.: GeoPDEs: A research tool for isogeometric analysis of PDEs. Adv. Eng. Softw. **42**, 1020–1034 (2011)
24. Deng, Q., Bartoň, M., Puzyrev, V., Calo, V.: Dispersion-minimizing quadrature rules for C^1 quadratic isogeometric analysis. Comput. Methods Appl. Mech. Eng. **328**, 554–564 (2018)
25. Eibner, T., Melenk, J.M.: Fast algorithms for setting up the stiffness matrix in hp-FEM: a comparison. HERCMA 2005 Conference Proceedings, Techn. Univ. Chemnitz, SFB 393 (2006)
26. Fahrendorf, F., De Lorenzis, L., Gomez, H.: Reduced integration at superconvergent points in isogeometric analysis. Comput. Methods Appl. Mech. Eng. **328**, 390–410 (2018)
27. Falini, A., Giannelli, C., Kanduc, T., Sampoli, M.L., Sestini, A.: An adaptive IGA-BEM with hierarchical B-splines based on quasi-interpolation quadrature schemes. Int. J. Numer. Meth. Eng. **117**(10), 1038–1058 (2019)
28. Falini, A., Kanduc, T.: Spline quasi-interpolation based quadrature rules for the isogeometric Galerkin BEM. This same collection (in press)
29. Hiemstra, R.R., Calabrò, F., Schillinger, D., Hughes, T.J.R.: Optimal and reduced quadrature rules for tensor product and hierarchically refined splines in isogeometric analysis. Comput. Methods Appl. Mech. Eng. **316**, 966–1004 (2017)

30. Hiemstra, R.R., Sangalli, G., Tani, M., Calabrò, F., Hughes, T.J.R.: Fast formation and assembly of finite element matrices with application to isogeometric linear elasticity (in preparation)
31. Hillman, M., Chen, J.S., Bazilevs, Y.: Variationally consistent domain integration for isogeometric analysis. Comput. Methods Appl. Mech. Eng. **284**, 521–540 (2015)
32. Hughes, T.J.R., Reali, A., Sangalli, G.: Efficient quadrature for NURBS-based isogeometric analysis. Comput. Methods Appl. Mech. Eng. **199**, 301–313 (2010)
33. Hughes, T.J.R., Sangalli, G.: Mathematics of isogeometric analysis: a conspectus. Encyclopedia of Computational Mechanics, 2nd edn, pp. 1–40. Wiley, Hoboken (2018)
34. Johannessen, K.A.: Optimal quadrature for univariate and tensor product splines. Comput. Methods Appl. Mech. Eng. **316**, 84–99 (2017)
35. Ma, J., Rokhlin, V., Wandzura, S.: Generalized Gaussian quadrature rules for systems of arbitrary functions. SIAM J. Numer. Anal. **33**, 971–996 (1996)
36. Mantzaflaris, A., Jüttler, B.: Exploring matrix generation strategies in isogeometric analysis. Mathematical Methods for Curves and Surfaces, pp. 364–382. Springer, Berlin (2014)
37. Mantzaflaris, A., Jüttler, B.: Integration by interpolation and look-up for Galerkin-based isogeometric analysis. Comput. Methods Appl. Mech. Eng. **284**, 373–400 (2015)
38. Mantzaflaris, A., Jüttler, B., Khoromskij, B., Langer, U.: Matrix generation in isogeometric analysis by low rank tensor approximation. Curves and Surfaces, pp. 321–340. Springer, Berlin (2014)
39. Mantzaflaris, A., Jüttler, B., Khoromskij, B.N., Langer, U.: Low rank tensor methods in Galerkin-based isogeometric analysis. Comput. Methods Appl. Mech. Eng. **316**, 1062–1085 (2017)
40. Micchelli, C.A., Pinkus, A.: Moment theory for weak Chebyshev systems with applications to monosplines, quadrature formulae and best one-sided L^1-approximation by spline functions with fixed knots. SIAM J. Math. Anal. **8**, 206–230 (1977)
41. Montardini, M., Sangalli, G., Tamellini, L.: Optimal-order isogeometric collocation at Galerkin superconvergent points. Comput. Methods Appl. Mech. Eng. **316**, 741–757 (2017)
42. Montardini, M., Sangalli, G., Tani, M.: Robust isogeometric preconditioners for the Stokes system based on the fast diagonalization method. Comput. Methods Appl. Mech. Eng. **338**, 162–185 (2018)
43. Nikolov, G.: On certain definite quadrature formulae. J. Comput. Appl. Math. **75**, 329–343 (1996)
44. Puzyrev, V., Deng, Q., Calo, V.: Dispersion-optimized quadrature rules for isogeometric analysis: modified inner products, their dispersion properties, and optimally blended schemes. Comput. Methods Appl. Mech. Eng. **320**, 421–443 (2017)
45. Sangalli, G., Tani, M.: Isogeometric preconditioners based on fast solvers for the Sylvester equation. SIAM J. Sci. Comput. **38**, A3644–A3671 (2016)
46. Sangalli, G., Tani, M.: Matrix-free weighted quadrature for a computationally efficient isogeometric k-method. Comput. Methods Appl. Mech. Eng. **338**, 117–133 (2018)
47. Schillinger, D., Evans, J.A., Reali, A., Scott, M., Hughes, T.: Isogeometric collocation: cost comparison with Galerkin methods and extension to adaptive hierarchical NURBS discretizations. Comput. Methods Appl. Mech. Eng. **267**, 170–232 (2013)
48. Schillinger, D., Hossain, S.J., Hughes, T.J.R.: Reduced Bézier element quadrature rules for quadratic and cubic splines in isogeometric analysis. Comput. Methods Appl. Mech. Eng. **277**, 1–45 (2014)
49. Schoenberg, I.J.: Spline functions, convex curves and mechanical quadrature. Bull. Am. Math. Soc. **64**, 352–357 (1958)
50. Tani, M.: A preconditioning strategy for linear systems arising from nonsymmetric schemes in isogeometric analysis. Comput. Math. Appl. **74**, 1690–1702 (2017)
51. Vázquez, R.: A new design for the implementation of isogeometric analysis in Octave and Matlab: GeoPDEs 3.0. Comput. Math. Appl. **72**, 523–554 (2016)
52. Vermeulen, A.H., Bartels, R.H., Heppler, G.R.: Integrating products of B-splines. SIAM J. Sci. Stat. Comput. **13**, 1025–1038 (1992)

Eigenvalue Isogeometric Approximations Based on B-Splines: Tools and Results

Sven-Erik Ekström and Stefano Serra-Capizzano

Abstract In this note, we focus on the spectral analysis of large matrices coming from isogeometric approximations based on B-splines of the eigenvalue problem

$$-(a(x)u'(x))' = \lambda b(x)u(x), \qquad x \in (0, 1),$$

where $u(0)$ and $u(1)$ are given. When considering the collocation case, global distribution results for the eigenvalues are available in the literature, despite the nonsymmetry of the related matrices. Here we complement such results by providing precise estimates for the extremal eigenvalues and hence for the spectral conditioning of the resulting matrices. In the Galerkin setting, the matrices are symmetric and positive definite and a more complete analysis has been conducted in the literature. In the latter case we furnish a further procedure that gives a highly accurate estimate of all the eigenvalues, starting from the knowledge of the spectral distribution symbol. The techniques involve dyadic decomposition arguments, tools from the theory of generalized locally Toeplitz sequences, and basic extrapolation methods.

1 Introduction

In this note we consider the approximation of one-dimensional elliptic eigenvalue problems by using an isogeometric either Galerkin or collocation technique with B-splines [5]. We are interested in the eigenvalues of the large matrices stemming from the considered approximation processes. In particular, we address the problem of estimating the extremal eigenvalues and of providing efficient numerical procedures for computing a good approximation of all the eigenvalues.

S.-E. Ekström (✉) · S. Serra-Capizzano
Department of Information Technology, Uppsala University, Uppsala, Sweden
e-mail: sven-erik.ekstrom@it.uu.se

S. Serra-Capizzano
Department of Humanities and Innovation, University of Insubria, Varese, Italy
e-mail: stefano.serra@it.uu.se; stefano.serrac@uninsubria.it

C. Giannelli and H. Speleers (eds.), *Advanced Methods for Geometric Modeling and Numerical Simulation*, Springer INdAM Series 35,
https://doi.org/10.1007/978-3-030-27331-6_4

In this direction, it has recently been proven that the resulting sequence of matrices, indexed with respect to the matrix size, have a canonical distribution (see [15, 17] and references therein), by using the theory of Generalized Locally Toeplitz (GLT) sequences [16].

We recall that every GLT sequence has an associated function, called the symbol, and that the uniform sampling of the symbol provides an asymptotic approximation of all the eigenvalues, if every matrix of the GLT sequence is Hermitian as it happens in our setting. However, in general the approximation is quite poor and recently some extrapolation techniques have been devised (see [9] and references therein). In the constant coefficient setting, when considering the Galerkin approach, the results presented in [10] are impressive in the sense that machine precision is obtained with very low computational cost, while for the variable coefficients further improvements are needed (see also [17]).

In the current note, when considering the Galerkin setting and variable coefficients, we propose a further numerical scheme for the computation of all the spectrum of large matrices by using the numerical computation of the eigenvalues for small matrices, underlying asymptotic expansions, and extrapolation methods as those in [9]. The numerical results are of the same quality as those produced by the best strategy in [17], especially when the problem coefficients are smooth and at least for low frequencies, which are those of highest interest for understanding the nature of the problem.

On the other hand, when dealing with the collocation approximation we obtain large nonsymmetric matrices. However, the GLT machinery can be employed and the symbol is real-valued and nonnegative [7], as in the symmetric positive definite Galerkin setting [15]. Hence we expect that most of the eigenvalues are real or with negligible imaginary part. Here we start the analysis of the collocation case, by describing a technique based on dyadic decompositions, for estimating the extreme eigenvalues and hence the asymptotic (spectral) conditioning of the involved matrix sequences. The analysis contained in Theorem 1 and Corollary 1 confirms that the conditioning grows at most as h^{-2}, h being the fineness parameter, exactly as in the case of the matrices obtained with the Galerkin approximation (see e.g [14]).

Finally, it is not clear if the delicate asymptotic expansions observed in [10, 17] holds also in the collocation setting and indeed this issue will be the subject of future investigations.

2 Preliminaries

In the following we present the notation that we use. In particular we give the definition of eigenvalue distribution, that of rearrangement, and we briefly discuss the informal meaning behind these definitions.

A matrix-sequence is any sequence of the form $\{\mathbf{X}_n\}_n$, where $\mathbf{X}_n$ is a square matrix such that $\text{size}(\mathbf{X}_n) \to \infty$ as $n \to \infty$. Let μ_d be the Lebesgue measure in $\mathbb{R}^d$ and let $C_c(\mathbb{C})$ be the space of continuous complex-valued functions with bounded support defined on $\mathbb{C}$. If $\mathbf{X}$ is an $m \times m$ matrix, the eigenvalues of $\mathbf{X}$ are denoted by $\lambda_1(\mathbf{X}), \dots, \lambda_m(\mathbf{X})$. If $\mathbf{g} : D \subset \mathbb{R}^d \to \mathbb{C}^{s\times s}$ is an $s \times s$ matrix-valued function, we say that $\mathbf{g}$ is measurable if its s^2 components $g_{ij} : D \to \mathbb{C}$, $i, j = 1, \dots, s$, are measurable.

Definition 1 Let $\{\mathbf{X}_n\}_n$ be a matrix-sequence, let $N_n := \text{size}(\mathbf{X}_n)$, and let $\mathbf{g} : D \subset \mathbb{R}^d \to \mathbb{C}^{s\times s}$ be a measurable $s \times s$ matrix-valued function defined on a set D with $0 < \mu_d(D) < \infty$. We say that $\{\mathbf{X}_n\}_n$ has an (asymptotic) eigenvalue distribution described by $\mathbf{g}$, and we write $\{\mathbf{X}_n\}_n \sim_\lambda \mathbf{g}$, if

$$\lim_{n\to\infty} \frac{1}{N_n} \sum_{i=1}^{N_n} F(\lambda_i(\mathbf{X}_n)) = \frac{1}{\mu_d(D)} \int_D \frac{\sum_{i=1}^{s} F(\lambda_i(\mathbf{g}(y_1, \dots, y_d)))}{s} \mathrm{d}y_1 \dots \mathrm{d}y_d, \quad \forall F \in C_c(\mathbb{C}),$$

where $\lambda_i(\mathbf{g}(y_1, \dots, y_d))$, $i = 1, \dots, s$, are the eigenvalues of the $s \times s$ matrix $\mathbf{g}(y_1, \dots, y_d)$.

The informal meaning behind the eigenvalue distribution $\{\mathbf{X}_n\}_n \sim_\lambda \mathbf{g}$ is the following: for all sufficiently large n, the eigenvalues of $\mathbf{X}_n$ can be subdivided into s different subsets of approximately the same cardinality; and the eigenvalues belonging to the ith subset (except possibly for $o(N_n)$ outliers) are approximated by the samples of the ith eigenvalue function $\lambda_i(\mathbf{g}(y_1, \dots, y_d))$ over a uniform grid in D (the domain of $\mathbf{g}$). For example, if $d = 1$, $N_n = ns$ and $D = [a, b]$, then the eigenvalues of $\mathbf{X}_n$ are approximately equal to

$$\lambda_i\Big(\mathbf{g}\Big(a + j\frac{b-a}{n}\Big)\Big), \qquad j = 1, \dots, n, \qquad i = 1, \dots, s,$$

for n large enough. Likewise, if $d = 2$, $N_n = n^2 s$ and $D = [a_1, b_1] \times [a_2, b_2]$, then the eigenvalues of $\mathbf{X}_n$ are approximately equal to

$$\lambda_i\Big(\mathbf{g}\Big(a_1 + j_1\frac{b_1 - a_1}{n},\ a_2 + j_2\frac{b_2 - a_2}{n}\Big)\Big), \qquad j_1, j_2 = 1, \dots, n, \qquad i = 1, \dots, s,$$

for n large enough.

Remark 1 (*rearrangement*) Let $\mathbf{g} : D \subseteq \mathbb{R}^d \to \mathbb{C}^{s\times s}$ and suppose that D is a rectangle in $\mathbb{R}^d$, say $D := [a_1, b_1] \times \dots \times [a_d, b_d]$. We also assume that the eigenvalues $\lambda_1(\mathbf{g}(y_1, \dots, y_d)), \dots, \lambda_s(\mathbf{g}(y_1, \dots, y_d))$ are real for all $(y_1, \dots, y_d) \in D$. For each positive integer r, let $\mathcal{G}_r$ be the uniform grid in D given by

$$\mathcal{G}_r := \left\{\Big(a_1 + \frac{i_1}{r}(b_1 - a_1), \dots, a_d + \frac{i_d}{r}(b_d - a_d)\Big) : i_1, \dots, i_d = 1, \dots, r\right\}.$$

Compute the samples of the eigenvalue functions

$$\lambda_1(\mathbf{g}(y_1,\ldots,y_d)),\ldots,\lambda_s(\mathbf{g}(y_1,\ldots,y_d))$$

at the points $(y_1,\ldots,y_d)\in\mathcal{G}_r$, sort them in increasing order and put them in a vector $(q_1,q_2,\ldots,q_{sr^d})$. Let $\eta_r:[0,1]\to\mathbb{R}$ be the piecewise linear non-decreasing function that interpolates the samples $(q_0:=q_1,q_1,q_2,\ldots,q_{sr^d})$ over the nodes $(0,\frac{1}{sr^d},\frac{2}{sr^d},\ldots,1)$, i.e.,

$$\eta_r\left(\frac{\ell}{sr^d}\right):=q_\ell,\qquad \ell=0,\ldots,sr^d.$$

Under certain (normally satisfied) conditions on $\mathbf{g}$, the function η_r converges (a.e.) as $r\to\infty$ to a non-decreasing function $\eta:[0,1]\to\mathbb{R}$, which is referred to as the rearranged version of $\mathbf{g}$. What is important about η is that

$$\int_D\frac{\sum_{i=1}^s F(\lambda_i(\mathbf{g}(y_1,\ldots,y_d)))}{s}\mathrm{d}y_1\ldots\mathrm{d}y_d=\int_0^1 F(\eta(t))\mathrm{d}t,\qquad \forall F\in C_c(\mathbb{C}).$$

Therefore, if we have $\{\mathbf{X}_n\}_n\sim_\lambda\mathbf{g}$, then we also have $\{\mathbf{X}_n\}_n\sim_\lambda\eta$.

3 Isogeometric Galerkin Discretization of Variable-Coefficient Eigenvalue Problems

Consider the following one-dimensional variable-coefficient eigenvalue problem:

$$\begin{cases}-(a(x)u_j'(x))'=\lambda_j b(x)u_j(x), & x\in\Omega,\\ u_j(x)=0, & x\in\partial\Omega,\end{cases}\tag{1}$$

where Ω is an open interval in $\mathbb{R}$ and $a,b\in L^1(\Omega)$ are such that $a,b>0$ a.e. on Ω. The corresponding weak formulation reads as follows: find eigenvalues $\lambda_j\in\mathbb{R}^+$ and eigenfunctions $u_j\in H_0^1(\Omega)$, for $j=1,2,\ldots,\infty$, such that, for all $v\in H_0^1(\Omega)$,

$$\mathrm{a}(u_j,v)=\lambda_j(b\,u_j,v),$$

where

$$\mathrm{a}(u_j,v):=\int_\Omega a(x)u_j'(x)v'(x)\mathrm{d}x,\qquad (b\,u_j,v):=\int_\Omega b(x)u_j(x)v(x)\mathrm{d}x.$$

In the isogeometric Galerkin method, we assume that the physical domain Ω is described by a global geometry map $G:[0,1]\to\overline{\Omega}$, which is invertible and satisfies $G(\partial([0,1]))=\partial\overline{\Omega}$. We fix a set of basis functions $\{\varphi_1,\ldots,\varphi_{N_n}\}$ defined on the reference (parametric) domain [0, 1] and vanishing on the boundary $\partial([0, 1])$. We consider the basis functions

$$\psi_i(x) := \varphi_i(G^{-1}(x)) = \varphi_i(t), \qquad x = G(t), \qquad i = 1, \dots, N_n,$$

which are defined on the physical domain Ω, and we define the approximation space $\mathscr{W}_n := \mathrm{span}(\psi_1, \dots, \psi_{N_n}) \subset H_0^1(\Omega)$. Finally, we find approximations to the exact eigenpairs (λ_j, u_j), $j = 1, 2, \dots, \infty$, by solving the following Galerkin problem: find $\lambda_{j,n} \in \mathbb{R}^+$ and $u_{j,n} \in \mathscr{W}_n$, for $j = 1, \dots, N_n$, such that, for all $v_n \in \mathscr{W}_n$,

$$\mathrm{a}(u_{j,n}, v_n) = \lambda_{j,n}(b\, u_{j,n}, v_n). \tag{2}$$

Assuming that both the exact and numerical eigenvalues are arranged in non-decreasing order, the pair $(\lambda_{j,n}, u_{j,n})$ is taken as an approximation of the pair (λ_j, u_j) for all $j = 1, \dots, N_n$. The numbers $\lambda_{j,n}/\lambda_j - 1$, $j = 1, \dots, N_n$, are referred to as the (relative) eigenvalue errors.

In view of the canonical identification of each function $v_n \in \mathscr{W}_n$ with its coefficient vector with respect to the basis $\{\varphi_1, \dots, \varphi_{N_n}\}$, solving the Galerkin problem (2) is equivalent to solving the generalized eigenvalue problem

$$\mathbf{K}_n(a, G)\mathbf{u}_{j,n} = \lambda_{j,n}\mathbf{M}_n(b, G)\mathbf{u}_{j,n}, \tag{3}$$

where $\mathbf{u}_{j,n}$ is the coefficient vector of $u_{j,n}$ with respect to $\{\varphi_1, \dots, \varphi_{N_n}\}$ and

$$\begin{aligned} \mathbf{K}_n(a, G) := [\mathrm{a}(\psi_j, \psi_i)]_{i,j=1}^{N_n} &= \left[\int_\Omega a(x)\psi_j'(x)\psi_i'(x)\mathrm{d}x\right]_{i,j=1}^{N_n} \\ &= \left[\int_0^1 \frac{a(G(t))}{|G'(t)|}\varphi_j'(t)\varphi_i'(t)\mathrm{d}t\right]_{i,j=1}^{N_n}, \end{aligned} \tag{4}$$

$$\begin{aligned} \mathbf{M}_n(b, G) := [(b\,\psi_j, \psi_i)]_{i,j=1}^{N_n} &= \left[\int_\Omega b(x)\psi_j(x)\psi_i(x)\mathrm{d}x\right]_{i,j=1}^{N_n} \\ &= \left[\int_0^1 b(G(t))|G'(t)|\varphi_j(t)\varphi_i(t)\mathrm{d}t\right]_{i,j=1}^{N_n}. \end{aligned} \tag{5}$$

The matrices $\mathbf{K}_n(a, G)$ and $\mathbf{M}_n(b, G)$ are, respectively, the stiffness and mass matrices. Due to our assumption that $a, b > 0$ a.e. on Ω, both $\mathbf{K}_n(a, G)$ and $\mathbf{M}_n(b, G)$ are always symmetric positive definite, regardless of the chosen basis functions $\varphi_1, \dots, \varphi_{N_n}$ and the map G. Moreover, it is clear from (3) that the numerical eigenvalues $\lambda_{j,n}$, $j = 1, \dots, N_n$, are just the eigenvalues of the matrix

$$\mathbf{L}_n(a, b, G) := (\mathbf{M}_n(b, G))^{-1}\mathbf{K}_n(a, G). \tag{6}$$

Therefore, if $a, b \in L^1(\Omega)$ and the basis functions $\varphi_1, \ldots, \varphi_{N_n}$ are chosen as the B-splines of degree p and global smoothness C^k defined on the interval $[0, 1]$ partitioned into n equal subintervals (see [17, Sect. 2.1]) and denoted by

$$B_{2,[p,k]}, \ldots, B_{n(p-k)+k,[p,k]}, \tag{7}$$

then

$$\mathbf{K}_n(a, G) = \left[\int_0^1 \frac{a(G(t))}{|G'(t)|} B'_{j+1,[p,k]}(t) B'_{i+1,[p,k]}(t)\mathrm{d}t\right]_{i,j=1}^{n(p-k)+k-1}, \tag{8}$$

$$\mathbf{M}_n(b, G) = \left[\int_0^1 b(G(t))|G'(t)| B_{j+1,[p,k]}(t) B_{i+1,[p,k]}(t)\mathrm{d}t\right]_{i,j=1}^{n(p-k)+k-1}. \tag{9}$$

From the *-algebra structure of GLT sequences, as proven in [16, 17], we know that

$$\left\{\frac{1}{n}\mathbf{K}_n(a, G)\right\}_n \sim_\lambda \kappa_{[p,k]}, \tag{10}$$

$$\left\{n\mathbf{M}_n(b, G)\right\}_n \sim_\lambda \xi_{[p,k]}, \tag{11}$$

$$\left\{\frac{1}{n^2}\mathbf{L}_n(a, b, G)\right\}_n \sim_\lambda \zeta_{[p,k]}, \tag{12}$$

where, according to Remark 1, $\kappa_{[p,k]}, \xi_{[p,k]}, \zeta_{[p,k]}$ are the rearranged versions of

$$\frac{a(G(t))}{|G'(t)|}\mathbf{f}_{[p,k]}(\theta), \quad b(G(t))|G'(t)|\mathbf{h}_{[p,k]}(\theta), \quad \frac{a(G(t))}{b(G(t))|G'(t)|^2}\mathbf{e}_{[p,k]}(\theta),$$

respectively, with

$$\mathbf{f}_{[p,k]}(\theta) := \mathbf{K}^{[0]}_{[p,k]} + \sum_{\ell=1}^{\eta-1}\Big(\mathbf{K}^{[\ell]}_{[p,k]}\mathrm{e}^{\mathrm{i}\ell\theta} + (\mathbf{K}^{[\ell]}_{[p,k]})^{\mathrm{T}}\mathrm{e}^{-\mathrm{i}\ell\theta}\Big), \tag{13}$$

$$\mathbf{h}_{[p,k]}(\theta) := \mathbf{M}^{[0]}_{[p,k]} + \sum_{\ell=1}^{\eta-1}\Big(\mathbf{M}^{[\ell]}_{[p,k]}\mathrm{e}^{\mathrm{i}\ell\theta} + (\mathbf{M}^{[\ell]}_{[p,k]})^{\mathrm{T}}\mathrm{e}^{-\mathrm{i}\ell\theta}\Big), \tag{14}$$

$$\mathbf{e}_{[p,k]}(\theta) := (\mathbf{h}_{[p,k]}(\theta))^{-1}\mathbf{f}_{[p,k]}(\theta), \tag{15}$$

where the blocks $\mathbf{K}^{[\ell]}_{[p,k]}$ and $\mathbf{M}^{[\ell]}_{[p,k]}$, of size $p-k$, are defined in [17]. The analytical predictions of the eigenvalue errors are obtained through the following sampling procedure:

$$\frac{\lambda_{j,n}}{\lambda_j} - 1 \approx \frac{n^2\zeta_{[p,k]}\left(\frac{j}{n(p-k)}\right)}{\lambda_j} - 1, \qquad j = 1, \ldots, m, \tag{16}$$

where $m := \min(n(p-k), n(p-k)+k-1)$.

However, such a procedure is not completely satisfactory especially in low frequencies (see [17]), when variable coefficients are considered. More precisely, we recall that the smallest eigenvalues are those related to slowly oscillating modes, that is the related eigenvectors belong to the subspace generated by low frequencies. Since the eigenvalues of the continuous operator appear in the denominator in formula (16), the relative error could be higher even if the absolute approximation is of high precision.

We address this issue in the next section.

For notational completeness, we stress that in the following the geometrical map G is always the identity and hence the matrices we consider are those denoted as $\mathbf{L}_n(a,b) := \mathbf{L}_n(a,b,I)$ with $\mathbf{L}_n(a,b,G)$ as in (6); furthermore, the regularity will be maximal that is $k = p-1$ and hence $e_p(\theta)$, $f_p(\theta)$, $h_p(\theta)$ will denote $\mathbf{e}_{[p,p-1]}(\theta)$, $\mathbf{f}_{[p,p-1]}(\theta)$, $\mathbf{h}_{[p,p-1]}(\theta)$, respectively.

4 Global Distribution Results and Extrapolation

As discussed in [17], the mismatch between the analytical predictions and the eigenvalue errors essentially occurs only for small eigenvalues and a way to significantly reduce it has already been illustrated in that paper. Drawing inspiration from [1, 9, 10, 12, 13], we here describe—in the case of an isogeometric Galerkin discretization based on B-splines of the eigenvalue problem (1)—an alternative interpolation–extrapolation procedure to considerably improve the analytical predictions for small eigenvalues. We also illustrate the performance of this procedure by two examples.

1. We assume that there exists a function $c_p : [0,\pi] \to \mathbb{R}$, depending only on p and the coefficients a, b of the considered eigenvalue problem (1), such that, when using an isogeometric p-degree C^{p-1} B-spline discretization, the following property holds: independently of the parameter n, all the eigenvalues of the matrix $\frac{1}{n^2}\mathbf{L}_n(a,b)$, sorted in increasing order, satisfy

$$\frac{\lambda_{j,n}}{n^2} \approx c_p(\theta_{j,n}) e_p(\theta_{j,n}), \qquad j = 1, \ldots, \min(n+p-2, n), \tag{17}$$

where

$$\begin{aligned} e_p = \frac{f_p(\theta)}{h_p(\theta)} &= \frac{(2-2\cos\theta)h_{p-1}(\theta)}{h_p(\theta)} \\ &= \frac{(2-2\cos\theta)\left(\phi_{[2p-1]}(p) + 2\sum_{\ell=1}^{p-1}\phi_{[2p-1]}(p-\ell)\cos(\ell\theta)\right)}{\phi_{[2p+1]}(p+1) + 2\sum_{\ell=1}^{p}\phi_{[2p+1]}(p+1-\ell)\cos(\ell\theta)} \end{aligned}$$

is defined in [17], and

$$\theta_{j,n} = \frac{j\pi}{n}, \qquad j = 1, \ldots, n.$$

2. Compute the eigenvalues of the matrix $\frac{1}{n_1^2}\mathbf{L}_{n_1}(a, b)$ corresponding to a small value of n_1, i.e.,

$$\frac{\lambda_{j_1,n_1}}{n_1^2}, \qquad j_1 = 1, \ldots, n_1 + p - 2.$$

   Since n_1 is small, this computation can be efficiently performed by any standard eigensolver (e.g., the MATLAB `eig` function).
3. Under the assumption in item 1,

$$c_p(\theta_{j_1,n_1}) \approx \frac{\lambda_{j_1,n_1}}{n_1^2 e_p(\theta_{j_1,n_1})}, \qquad j_1 = 1, \ldots, \min(n_1 + p - 2, n_1).$$

   This means that we have an approximation of the unknown function c_p over the coarse uniform grid consisting of the points θ_{j_1,n_1}.
4. Interpolate the data

$$\left(\theta_{j_1,n_1}, \frac{\lambda_{j_1,n_1}}{n_1^2 e_p(\theta_{j_1,n_1})}\right), \qquad j_1 = 1, \ldots, \min(n_1 + p - 2, n_1),$$

   by using, e.g., the MATLAB `interp1` function with the `'spline'` option, so as to obtain an approximation of c_p over the whole interval $[0, \pi]$. We call this approximation $\tilde{c}_p$. Note that, by construction,

$$\tilde{c}_p(\theta_{j_1,n_1}) = \frac{\lambda_{j_1,n_1}}{n_1^2 e_p(\theta_{j_1,n_1})}, \qquad j_1 = 1, \ldots, \min(n_1 + p - 2, n_1).$$

5. Given a large n, compute approximations of the numerical eigenvalues $\lambda_{j,n}$ by replacing c_p with $\tilde{c}_p$ in (17):

$$\lambda_{j,n} \approx n^2 \tilde{c}_p(\theta_{j,n}) e_p(\theta_{j,n}), \qquad j = 1, \ldots, \min(n + p - 2, n).$$

6. Compute analytical predictions for the eigenvalue errors as follows:

$$\frac{\lambda_{j,n}}{\lambda_j} - 1 \approx \frac{n^2 \tilde{c}_p(\theta_{j,n}) e_p(\theta_{j,n})}{\lambda_j} - 1, \qquad j = 1, \ldots, \min(n + p - 2, n).$$

The next two examples show that the interpolation–extrapolation algorithm defined by items 1–6 may lead to analytical predictions of the eigenvalue errors which perform better (for small eigenvalues) than the analytical predictions proposed in (16).

Table 1 Example 1 [Linear C^0 B-spline discretization with $a(x) = 2 + 0.5\,x$ and $b(x) = 1$]: The five smallest eigenvalues and eigenvalue approximations, scaled by 10^2, for $n = 200$, $n' = 1500$, $p' = 5$, and $r = 10000$

p	$\lambda_{j,n'}$	$\lambda_{j,n}$	$n^2\zeta_r\left(\frac{j}{n}\right)$	$n^2\tilde{c}_1(\theta_{j,n})e_1(\theta_{j,n})$
1	0.221308917999822	0.221313437064828	0.225759339802366	0.221202202766985
	0.885445782386009	0.885517680038136	0.894411370606914	0.884899608283264
	1.992343949797817	1.992707562710680	2.006135775082463	1.991307942612523
	3.542008222229894	3.543157060383101	3.560895188355945	3.540749097918976
	5.534445409962895	5.537249919330117	5.559588944750918	5.533651310530473

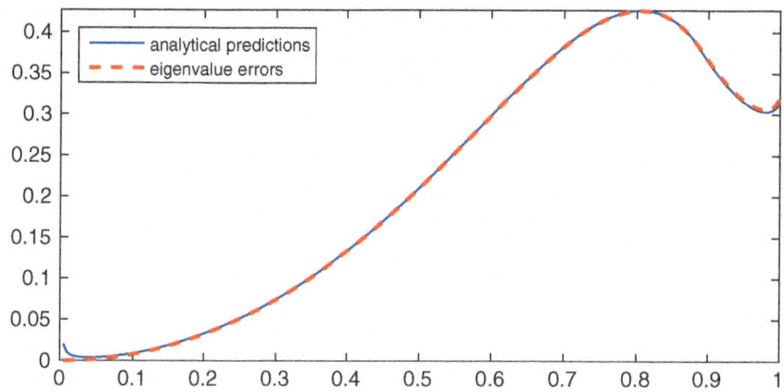

Fig. 1 Example 1 [Linear C^0 B-spline discretization with $a(x) = 2 + 0.5\,x$ and $b(x) = 1$]: analytical predictions $n^2\zeta_r(\frac{j}{n})/\lambda_{j,n'} - 1$ and eigenvalue errors $\lambda_{j,n}/\lambda_{j,n'} - 1$ versus j/N_n, $j = 1, \ldots, N_n$ ($N_n = n - 1$, $n = 200$, $n' = 1500$, $p' = 5$, $r = 10000$)

Example 1 Let $p = 1$, $n = 200$, $a(x) = 2 + 0.5\,x$ and $b(x) = 1$. Let $n' = 1500 \gg n$, $p' = 5$, and take the first $n - 1$ eigenvalues of $\mathbf{L}_{n'}(a, b)$, namely $\lambda_{1,n'}, \ldots, \lambda_{n-1,n'}$, as approximations of the unknown exact eigenvalues $\lambda_1, \ldots, \lambda_{n-1}$. In Table 1 we report the five smallest eigenvalues and approximated eigenvalues for $p = 1$. As is clear the new approximation, $n^2\tilde{c}_1(\theta_{j,n})e_1(\theta_{j,n})$, performs better than $n^2\zeta_r\left(\frac{j}{n}\right)$ to approximate $\lambda_{j,n}$, where the function ζ_r introduced in [17, Sect. 3.1] is defined in the following way. Sample $\frac{a(x)}{b(x)}e_1(\theta)$ at the grid points $(x, \theta) \in \mathcal{G}_r$, for a chosen r, where

$$\mathcal{G}_r = \left\{ \left(\frac{i}{r}, \frac{j\pi}{r}\right) : \quad i, j = 1, \ldots, r \right\}.$$

The samples are ordered in increasing order in a vector $(z_1, z_2, \ldots, z_{r^2})$. Let $\zeta_r : [0, 1] \to \mathbb{R}$ be a piecewise linear non-decreasing function that interpolate the samples $(z_0 := z_1, z_2, \ldots, z_{r^2})$ over the nodes $(0, \frac{1}{r^2}, \frac{2}{r^2}, \ldots, 1)$.

In Fig. 1 we present the (approximate) analytical predictions $n^2\zeta_r(\frac{j}{n})/\lambda_{j,n'} - 1$, with $r = 10000$, together with the (approximate) eigenvalue errors $\lambda_{j,n}/\lambda_{j,n'} - 1$.

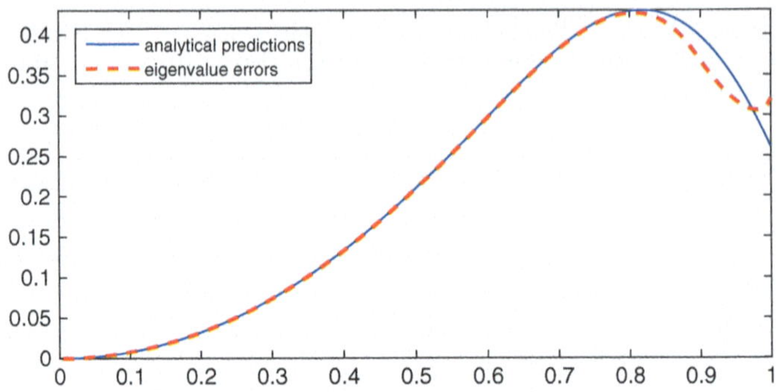

Fig. 2 Example 1 [Linear C^0 B-spline discretization with $a(x) = 2 + 0.5\,x$ and $b(x) = 1$]: analytical predictions $n^2\tilde{c}_1(\theta_{j,n})e_1(\theta_{j,n})/\lambda_{j,n'} - 1$ and eigenvalue errors $\lambda_{j,n}/\lambda_{j,n'} - 1$ versus j/N_n, $j = 1, \ldots, N_n$ ($N_n = n - 1$, $n = 200$, $n' = 1500$, $p' = 5$, $n_1 = 10$)

Table 2 Example 2 [p-degree C^{p-1} B-spline discretization for $p = 2, 3, 4, 5$ with $a(x) = 2.1 \cdot 10^9 + 1.05 \cdot 10^9\,x$ and $b(x) = 8000$]: The five smallest eigenvalues and eigenvalue approximations, scaled by 10^7, for $n = 200$, $n' = 1500$, $p' = 5$, and $r = 10000$

p	$\lambda_{j,n'}$	$\lambda_{j,n}$	$n^2\zeta_{r,[p,p-1]}\left(\frac{j}{n}\right)$	$n^2\tilde{c}_p(\theta_{j,n})e_p(\theta_{j,n})$
2	0.320242073856735	0.320242073906214	0.326994616633570	0.319596822473787
	1.281965164064426	1.281965165832994	1.295209495587940	1.278576943997320
	2.884839278092422	2.884839298406320	2.904497248558355	2.877209798346387
	5.128863345308352	5.128863459872621	5.154973083201467	5.115745284850879
	8.014037230782240	8.014037668616439	8.046649299586466	7.994414445665448
3	0.320242073856735	0.320242073878847	0.326994616613888	0.319613348810547
	1.281965164064426	1.281965164062972	1.295209493290653	1.278636625771722
	2.884839278092422	2.884839278083705	2.904497230696836	2.877330529920427
	5.128863345308352	5.128863345316139	5.154972936807892	5.115937346593204
	8.014037230782240	8.014037230847546	8.046648648420712	7.994681508249463
4	0.320242073856735	0.320242073879743	0.326994616613781	0.319603273140503
	1.281965164064426	1.281965164063456	1.295209493290564	1.278600044346683
	2.884839278092422	2.884839278081923	2.904497230695864	2.877256125504316
	5.128863345308352	5.128863345304529	5.154972936792049	5.115818311034016
	8.014037230782240	8.014037230780470	8.046648648302227	7.994514965707993
5	0.320242073856735	0.320242073879344	0.326994616614055	0.319620210926112
	1.281965164064426	1.281965164062812	1.295209493290633	1.278661525969442
	2.884839278092422	2.884839278081491	2.904497230695859	2.877381146957867
	5.128863345308352	5.128863345303335	5.154972936792168	5.116018280099667
	8.014037230782240	8.014037230779344	8.046648648302266	7.994794677407651

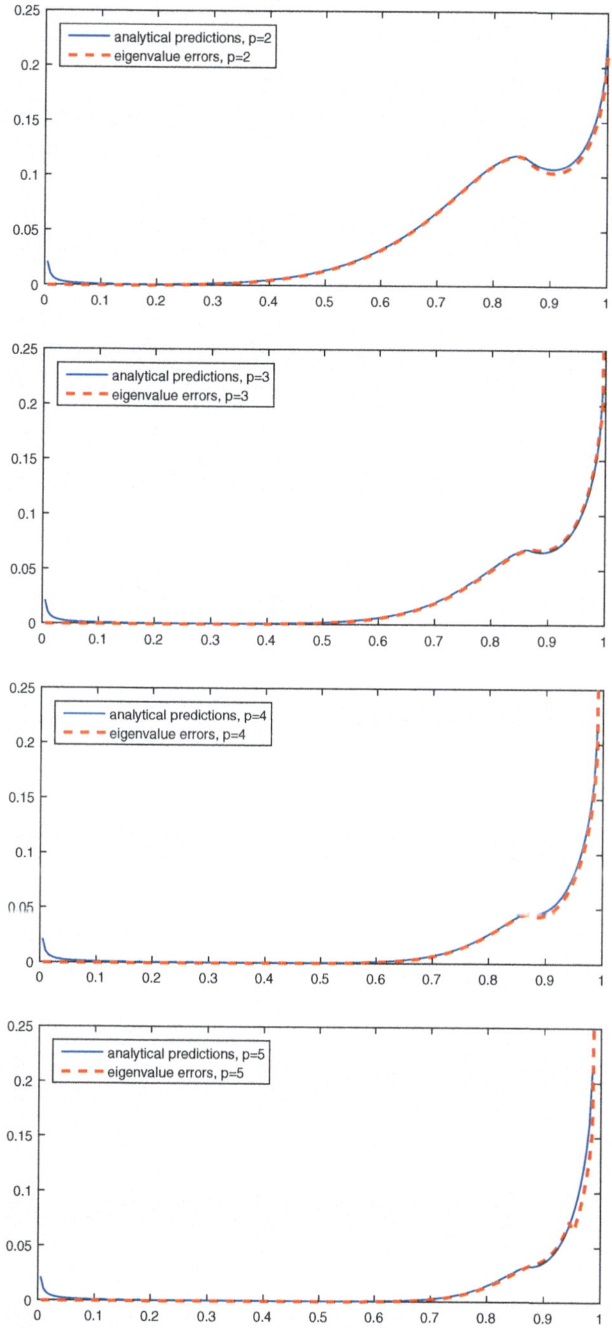

Fig. 3 Example 2 [p-degree C^{p-1} B-spline discretization for $p = 2, 3, 4, 5$ with $a(x) = 2.1 \cdot 10^9 + 1.05 \cdot 10^9 x$ and $b(x) = 8000$]: analytical predictions $n^2 \zeta_{r,[p,p-1]}(\frac{j}{n})/\lambda_{j,n'} - 1$ and eigenvalue errors $\lambda_{j,n}/\lambda_{j,n'} - 1$ versus j/N_n, $j = 1, \ldots, N_n$ ($N_n = n + p - 2$, $n = 200$, $n' = 1500$, $p' = 5$, $r = 10000$)

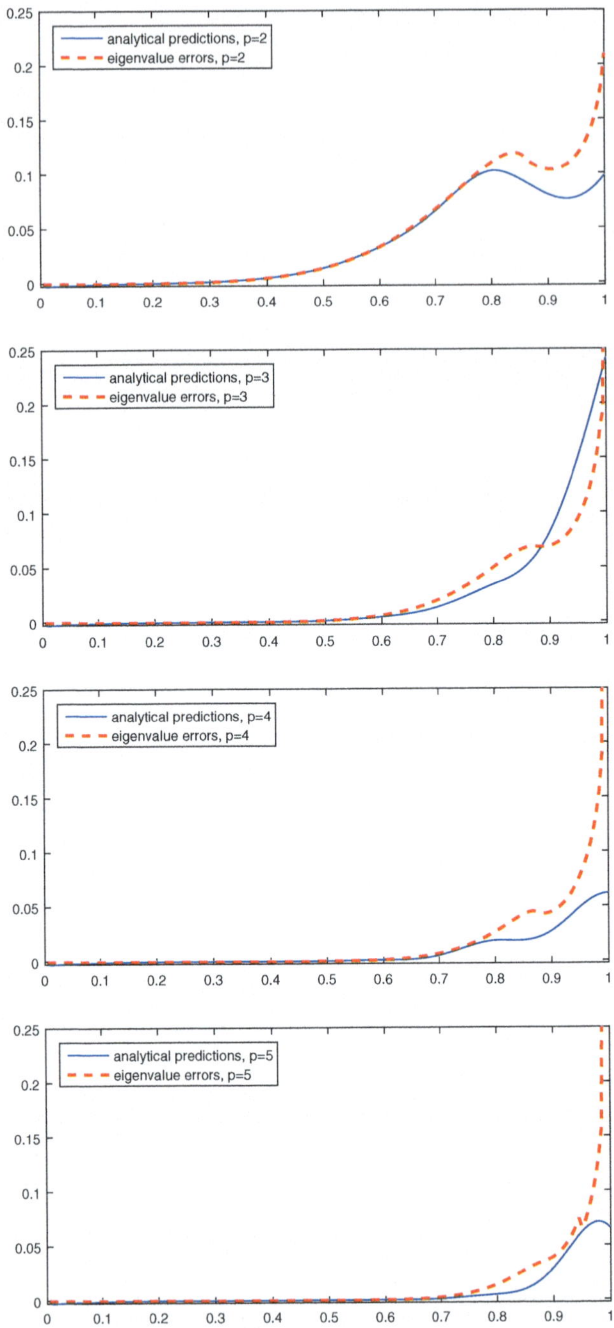

Fig. 4 Example 2 [p-degree C^{p-1} B-spline discretization for $p = 2, 3, 4, 5$ with $a(x) = 2.1 \cdot 10^9 + 1.05 \cdot 10^9 x$ and $b(x) = 8000$]: analytical predictions $n^2\tilde{c}_p(\theta_{j,n})e_p(\theta_{j,n})/\lambda_{j,n'} - 1$ and eigenvalue errors $\lambda_{j,n}/\lambda_{j,n'} - 1$ versus j/N_n, $j = 1, \ldots, N_n$ ($N_n = n + p - 2$, $n = 200$, $n' = 1500$, $p' = 5$, $n_1 = 10$)

We note the mismatch for the smallest eigenvalues. In Fig. 2 we plot the (approximate) analytical predictions $n^2\tilde{c}_1(\theta_{j,n})e_1(\theta_{j,n})/\lambda_{j,n'} - 1$, obtained from the above interpolation–extrapolation algorithm for $n_1 = 10$, and the (approximate) eigenvalue errors $\lambda_{j,n}/\lambda_{j,n'} - 1$ versus $j/(n-1)$, for $j = 1, \ldots, n-1$. We clearly see that in Fig. 2 the slight mismatch for small eigenvalues observed Fig. 1 has been reduced.

Example 2 Let $p = 2, 3, 4, 5, n = 200, a(x) = 2.1 \cdot 10^9 + 1.05 \cdot 10^9 x$ and $b(x) = 8000$. The approximation parameters n' and n_1 have been chosen as $n' = 1500$ and $n_1 = 10$, respectively, and the eigenvalues $\lambda_{j,n'}$ have been chosen correspondingly, in the sense that $\lambda_{j,n'}$ is computed as follows: we compute the eigenvalues $\lambda_{j,n'}$ for a matrix $\mathbf{L}_{n'}(a, b)$ of order $n' = 1500$ and B-spline degree $p' = 5$. Then we assume that the first $n + p - 2$ (for $n = 200$) of these eigenvalues, that is $j = 1, \ldots, n + p - 2$, are accurate representations of the true eigenvalues λ_j; see [17] for a motivation and a discussion on this matter.

In Table 2 we present the five smallest eigenvalues and approximated eigenvalues for $p = 2, 3, 4, 5$. As is clear the new approximation, $n^2\tilde{c}_p(\theta_{j,n})e_p(\theta_{j,n})$, performs better than $n^2\zeta_{r,[p,p-1]}\left(\frac{j}{n}\right)$ to approximate $\lambda_{j,n}$. Figure 3 shows the comparison between the (approximate) analytical predictions

$$\frac{n^2\zeta_{r,[p,p-1]}\left(\frac{j}{n}\right)}{\lambda_{j,n'}} - 1, \qquad j = 1, \ldots, n,$$

and the (approximate) eigenvalue errors

$$\frac{\lambda_{j,n}}{\lambda_{j,n'}} - 1, \qquad j = 1, \ldots, n + p - 2,$$

whereas Fig. 4 shows the comparison between the (approximate) analytical predictions

$$\frac{n^2\tilde{c}_p(\theta_{j,n})e_p(\theta_{j,n})}{\lambda_{j,n'}} - 1, \qquad j = 1, \ldots, n,$$

with the (approximate) eigenvalue errors. We see from Fig. 4 that the mismatch for small eigenvalues observed in Fig. 3 has been lowered.

5 Dyadic Decomposition Argument and Extreme Eigenvalues

While the distribution results are available both in the Galerkin [15] and in the collocation setting [7], the use of extrapolation methods, as those described in the previous section and in [10, 11], has been developed only in the Galerkin setting. The reason is the inherent nonsymmetry of the collocation matrices. However, this issue has to be investigated in the future, even if few preliminary experiments seem quite promising. In this section, we start by analyzing the extreme eigenvalues of the (nonsymmetric) stiffness matrices in the collocation setting. We use a dyadic decomposition argument already employed for symmetric structures in several contexts (see [3, 20] and references therein).

Consider the one-dimensional Poisson problem

$$\begin{cases} -u''(x) = \mathrm{f}(x), & x \in (0,1), \\ u(x) = 0, & x \in \{0,1\}, \end{cases} \tag{18}$$

where $\mathrm{f} \in C([0,1])$. Suppose we approximate (18) by using the isogeometric collocation method based on uniform B-splines (see [7, Sect. 4] for the details on this method). Then, the resulting discretization matrix is

$$n^2 \tilde{\mathbf{K}}_n^{[p]} := \left[-N''_{j,[p]}(\xi_{i,[p]})\right]_{i,j=2}^{n+p-1}, \tag{19}$$

where $\xi_{i,[p]}$, $i = 2, \ldots, n+p-1$, are the Greville abscissae defined in [7, Eq. (4.6)], while $N_{j,[p]}$, $j = 2, \ldots, n+p-1$, are the usual B-spline basis functions of maximal regularity $k = p-1$. In other words, with reference to (7), we have

$$N_{j,[p]} = B_{j,[p,p-1]}, \quad j = 2, \ldots, n+p-1.$$

We denote by $\Re(\tilde{\mathbf{K}}_n^{[p]})$ the real part of $\tilde{\mathbf{K}}_n^{[p]}$ (the real part of a square complex matrix X [4] is by definition $(X + X^{\mathrm{H}})/2$, X^{H} being the transpose conjugate). In Theorem 1 we prove for $p = 3$ the following result:

$$\exists\, c_p > 0 : \ \Re(\tilde{\mathbf{K}}_n^{[p]}) \geq c_p\, \tau_{n+p-2}(2 - 2\cos\theta), \quad \forall n \geq 2, \tag{20}$$

where

$$\Re(\tilde{\mathbf{K}}_n^{[3]}) = \begin{bmatrix} \frac{11}{2} & -\frac{4}{3} & -\frac{1}{6} & & & & & & \\ -\frac{4}{3} & \frac{5}{2} & -1 & & & & & & \\ -\frac{1}{6} & -1 & 2 & -1 & & & & & \\ & & -1 & 2 & -1 & & & & \\ & & & \ddots & \ddots & \ddots & & & \\ & & & & -1 & 2 & -1 & & \\ & & & & & -1 & 2 & -1 & -\frac{1}{6} \\ & & & & & & -1 & \frac{5}{2} & -\frac{4}{3} \\ & & & & & & -\frac{1}{6} & -\frac{4}{3} & \frac{11}{2} \end{bmatrix}$$

and

$$\tau_{n+1}(2-2\cos\theta)=\begin{bmatrix} 2 & -1 & & & \\ -1 & \ddots & \ddots & & \\ & \ddots & & \ddots & \\ & & \ddots & \ddots & -1 \\ & & & -1 & 2 \end{bmatrix}.$$

Theorem 1 *There exists a constant $c_3 > 0$ such that*

$$\Re(\tilde{\mathbf{K}}_n^{[3]}) \geq c_3\, \tau_{n+1}(2-2\cos\theta), \quad \forall n \geq 2.$$

Proof For $n = 2, \ldots, 5$, a direct verification shows that $\Re(\tilde{\mathbf{K}}_n^{[3]})$ is positive definite. Therefore, the theorem is proved if there exists a constant $c > 0$ such that

$$\Re(\tilde{\mathbf{K}}_n^{[3]}) \geq c\, \tau_{n+1}(2-2\cos\theta), \qquad \forall n \geq 6. \tag{21}$$

For $n \geq 6$ and $c > 0$, the matrix $\Re(\tilde{\mathbf{K}}_n^{[3]}) - c\,\tau_{n+1}(2-2\cos\theta)$ is explicitly given by

$$\begin{bmatrix} \frac{11}{2}-2c & -\frac{4}{3}+c & -\frac{1}{6} & & & & & & \\ -\frac{4}{3}+c & \frac{5}{2}-2c & -1+c & & & & & & \\ -\frac{1}{6} & -1+c & 2-2c & -1+c & & & & & \\ & & -1+c & 2-2c & -1+c & & & & \\ & & & \ddots & \ddots & \ddots & & & \\ & & & & -1+c & 2-2c & -1+c & & \\ & & & & & -1+c & 2-2c & -1+c & -\frac{1}{6} \\ & & & & & & -1+c & \frac{5}{2}-2c & -\frac{4}{3}+c \\ & & & & & & -\frac{1}{6} & -\frac{4}{3}+c & \frac{11}{2}-2c \end{bmatrix}$$

and can be decomposed as follows:

$$\boxed{\begin{matrix} r & -\frac{4}{3}+c \\ -\frac{4}{3}+c & s \end{matrix}}_{\,2}^{\,1} + \boxed{\begin{matrix} \frac{11}{2}-2c-r & 0 & -\frac{1}{6} \\ 0 & 0 & 0 \\ -\frac{1}{6} & 0 & t \end{matrix}}_{\,3}^{\,1} + \boxed{\begin{matrix} \frac{5}{2}-2c-s & -1+c \\ -1+c & 1-c-t \end{matrix}}_{\,3}^{\,2}$$

$$+\ \boxed{\begin{matrix} 1-c & -1+c & & & \\ -1+c & 2-2c & -1+c & & \\ & \ddots & \ddots & \ddots & \\ & & -1+c & 2-2c & -1+c \\ & & & -1+c & 1-c \end{matrix}}_{\,n-1}^{\,3}$$

$$+\boxed{\begin{matrix} 1-c-t & -1+c \\ -1+c & \frac{5}{2}-2c-s \end{matrix}}_{n}^{n-1} + \boxed{\begin{matrix} t & 0 & -\frac{1}{6} \\ 0 & 0 & 0 \\ -\frac{1}{6} & 0 & \frac{11}{2}-2c-r \end{matrix}}_{n+1}^{n-1} + \boxed{\begin{matrix} s & -\frac{4}{3}+c \\ -\frac{4}{3}+c & r \end{matrix}}_{n+1}^{n}.$$

In the above decomposition, r, s, and t are arbitrary real numbers, whereas each term of the summation is a $(n+1) \times (n+1)$ matrix whose only nonzero entries are shown in the associated box and are contained in the principal submatrix corresponding to the rows from the superscript to the subscript. For instance,

$$\boxed{\begin{matrix} \frac{11}{2}-2c-r & 0 & -\frac{1}{6} \\ 0 & 0 & 0 \\ -\frac{1}{6} & 0 & t \end{matrix}}_{3}^{1} = \left[\begin{array}{ccc|ccc} \frac{11}{2}-2c-r & 0 & -\frac{1}{6} & 0 & \cdots & 0 \\ 0 & 0 & 0 & \vdots & & \vdots \\ -\frac{1}{6} & 0 & t & \vdots & & \vdots \\ \hline 0 & \cdots & \cdots & 0 & & \vdots \\ \vdots & & & & & \vdots \\ 0 & \cdots & \cdots & \cdots & \cdots & 0 \end{array}\right].$$

If each term of the above summation is a nonnegative definite matrix, then $\Re(\tilde{\mathbf{K}}_n^{[3]}) - c\,\tau_{n+1}(2-2\cos\theta)$ is nonnegative definite as well. The following conditions ensure that each term of the summation is nonnegative definite:

$$\begin{cases} r > 0,\ s > 0,\ rs \geq \left(\frac{4}{3}-c\right)^2, \\ \frac{11}{2}-2c-r > 0,\ t > 0,\ \left(\frac{11}{2}-2c-r\right)t \geq \frac{1}{36}, \\ \frac{5}{2}-2c-s > 0,\ 1-c-t > 0, \\ (\frac{5}{2}-2c-s)(1-c-t) \geq (1-c)^2, \\ c \leq 1. \end{cases}$$

These conditions are satisfied, for instance, with

$$c = \frac{58}{59},\quad s = \frac{1}{2},\quad r = \frac{\left(\frac{4}{3}-c\right)^2}{s},\quad t = \frac{\frac{1}{36}}{\frac{11}{2}-2c-r}.$$

Hence, (21) holds with $c = 58/59$. □

We verified that the dyadic decomposition argument used in the proof of Theorem 1 can also be used to prove (20) for $p = 2, 4$. Although this argument becomes quite difficult to apply for $p \geq 5$, we have reason to believe that a careful application of it could prove (20) for any given $p \geq 2$.

The above theorem and the relations in (20) have interesting consequences on the conditioning measured with respect to the induced Euclidean norm as reported in Corollary 1, whose proof requires a technical but interesting in itself lemma.

Lemma 1 *Given X square matrix, assuming $\Re(X)$ positive definite, we have*

$$\lambda_{\min}(\Re(X)) \leq \sigma_{\min}(X).$$

Proof As a first step define the following matrix $\hat{X}$ of order $2n$

$$\hat{X} = \begin{bmatrix} 0 & X \\ X^{\mathrm{H}} & 0 \end{bmatrix}.$$

By taking into account the standard singular value decomposition, it is easy to see (and it is well-known) that the singular values $\sigma_j(X)$ of X, $j = 1, \ldots, n$, coincide with the n largest eigenvalues of $\hat{X}$, since the eigenvalues of the Hermitian matrix $\hat{X}$ are $\pm\sigma_j(X)$, $j = 1, \ldots, n$. Therefore, we infer that

$$\sigma_{\min}(X) = \lambda_{n+1}(\hat{X}),$$

with $\lambda_1(\hat{X}) \leq \lambda_2(\hat{X}) \leq \cdots \leq \lambda_{2n}(\hat{X})$. From the minimax characterization (see [4]) of the $(n+1)$-th eigenvalue of $\hat{X}$, we obtain

$$\sigma_{\min}(X) = \max_{\dim(U)=n} \min_{u \in U} \frac{u^{\mathrm{H}} \hat{X} u}{u^{\mathrm{H}} u}, \tag{22}$$

where the maximum is taken over all subspaces U with codimension n in $\mathbb{C}^{2n}$ or, equivalently, over all subspaces U of dimension n. Let U_* be the n-dimensional subspace of $\mathbb{C}^{2n}$ made of all vectors y of the form

$$y = \begin{pmatrix} u \\ u \end{pmatrix}, \quad u \in \mathbb{C}^n, \quad u = \begin{pmatrix} u_1 \\ u_2 \\ \vdots \\ u_n. \end{pmatrix}.$$

If $y \in U_*$ is partitioned into blocks as above, it holds

$$\frac{y^{\mathrm{H}} \hat{X} y}{y^{\mathrm{H}} y} = \frac{u^{\mathrm{H}} \Re(X) u}{u^{\mathrm{H}} u}, \tag{23}$$

and since U_* is a particular subspace of dimension n, from (22) the thesis follows. □

Corollary 1 *Assume that* (20) *is satisfied for a given p. Then*

$$\mu_2(\Re(\tilde{\mathbf{K}}_n^{[p]})) \sim n^2, \tag{24}$$

and

$$\mu_2(\tilde{\mathbf{K}}_n^{[p]}) \leq C_p n^2, \tag{25}$$

with C_p a positive constant depending only on p and $\mu(X) = \|X\|\|X^{-1}\|$ being the conditioning of an invertible matrix X measured with respect to the induced Euclidean norm $\|\cdot\|$.

Proof First of all we observe that both $\Re(\tilde{\mathbf{K}}_n^{[p]})$ and $\tilde{\mathbf{K}}_n^{[p]}$ are banded matrices with coefficients not depending on the parameter n. Hence we have that their induced Euclidean norm is asymptotic to a a constant not depending on n that is

$$\|\Re(\tilde{\mathbf{K}}_n^{[p]})\| \sim 1, \tag{26}$$

$$\|\tilde{\mathbf{K}}_n^{[p]}\| \sim 1. \tag{27}$$

We first prove (24). By (20) we have

$$\Re(\tilde{\mathbf{K}}_n^{[p]}) \geq c_p\, \tau_{n+p-2}(2 - 2\cos\theta), \quad \forall n \geq 2,$$

with $c_p > 0$ and hence $\Re(\tilde{\mathbf{K}}_n^{[p]})$ is positive definite and its minimal eigenvalue is bounded from below by

$$c_p 4\sin^2\left(\frac{\pi}{2(n+p-1)}\right),$$

$4\sin^2\left(\frac{\pi}{2(n+p-1)}\right)$ being the minimal eigenvalue of $\tau_{n+p-2}(2 - 2\cos\theta)$ (see [19] and references therein). On the other hand, $\Re(\tilde{\mathbf{K}}_n^{[p]})$ contains as a principal submatrix a Toeplitz matrix of size proportional to n and generated by f_p being nonnegative and having a unique zero of order two (see [7, 14]). Hence such a Toeplitz matrix has the minimal eigenvalue l_n asymptotic to n^{-2} (see again [19] and references therein) and by the Cauchy interlacing theorem the minimal eigenvalue of $\Re(\tilde{\mathbf{K}}_n^{[p]})$ is bounded from above by $l_n \sim n^{-2}$. Therefore

$$\lambda_{\min}(\Re(\tilde{\mathbf{K}}_n^{[p]})) \sim n^{-2}.$$

But $\Re(\tilde{\mathbf{K}}_n^{[p]})$ is positive definite and hence

$$\left\|[\Re(\tilde{\mathbf{K}}_n^{[p]})]^{-1}\right\| = [\lambda_{\min}(\Re(\tilde{\mathbf{K}}_n^{[p]}))]^{-1} \sim n^2,$$

from which, using (26), we deduce

$$\mu_2(\Re(\tilde{\mathbf{K}}_n^{[p]})) \sim n^2.$$

We now prove (25). Given the known fact $\|X^{-1}\| = [\sigma_{\min}(X)]^{-1}$, statement (25) is a consequence of (27), of (20), of Lemma 1, and of the fact that the minimal eigenvalue of $\tau_{n+p-2}(2 - 2\cos\theta)$ is

$$\sin^2\left(\frac{\pi}{2(n+p-1)}\right) \sim n^{-2}.$$

□

Furthermore, we have reasons to conjecture that the spectral conditioning of the collocation stiffness matrices grows asymptotically as n^2, exactly as in the Galerkin setting [14], since $\tilde{\mathbf{K}}_n^{[p]}$ contains as a principal submatrix a Toeplitz matrix of size proportional to n and generated by f_p having a unique zero of order two (as $\Re(\tilde{\mathbf{K}}_n^{[p]})$, see the proof of Corollary 1).

We finally stress that (20) is also important in a multigrid context for proving the optimality of the related two-grid and multigrid techniques (see [2, 6, 8, 19]), by following the theory reported by Ruge and Stüben in [18].

6 Conclusions

In this note, we have considered the spectral analysis of large matrices coming from the isogeometric approximations based on B-splines of the eigenvalue problem

$$-(a(x)u'(x))' = \lambda b(x)u(x), \qquad x \in (0, 1),$$

where $u(0)$ and $u(1)$ are given. In the collocation setting, we complemented global eigenvalue distribution results, available in the literature [7], with precise estimates for the extremal eigenvalues and hence for the spectral conditioning of the resulting matrices. In the Galerkin setting, we have designed an efficient matrix-less procedure (see [9]) that gives a highly accurate estimate of the all the eigenvalues, starting from the knowledge of the spectral GLT distribution symbol. Possible extensions include a more systematic treatment of the collocation case, both via the use of dyadic decomposition arguments and via the use of proper matrix-less extrapolation techniques. This last item is completely new and represents a real challenge, thus it will be the subject of a future investigation.

References

1. Ahmad, F., Al-Aidarous, E.S., Abdullah Alrehaili, D., Ekström, S.-E., Furci, I., Serra-Capizzano, S.: Are the eigenvalues of preconditioned banded symmetric Toeplitz matrices known in almost closed form? Numer. Algorithms **78**, 867–893 (2018)
2. Aricò, A., Donatelli, M., Serra-Capizzano, S.: V-cycle optimal convergence for certain (multilevel) structured linear systems. SIAM J. Matrix Anal. Appl. **26**, 186–214 (2004)
3. Beckermann, B., Serra-Capizzano, S.: On the asymptotic spectrum of finite element matrix sequences. SIAM J. Numer. Anal. **45**, 746–769 (2007)
4. Bhatia, R.: Matrix Analysis. Springer, New York (1997)

5. Cottrell, J.A., Hughes, T.J.R., Bazilevs, Y.: Isogeometric Analysis: Toward Integration of CAD and FEA. Wiley, Chichester (2009)
6. Donatelli, M., Garoni, C., Manni, C., Serra-Capizzano, S., Speleers, H.: Two-grid optimality for Galerkin linear systems based on B-splines. Comput. Vis. Sci. **17**, 119–133 (2015)
7. Donatelli, M., Garoni, C., Manni, C., Serra-Capizzano, S., Speleers, H.: Spectral analysis and spectral symbol of matrices in isogeometric collocation methods. Math. Comput. **85**, 1639–1680 (2016)
8. Donatelli, M., Garoni, C., Manni, C., Serra-Capizzano, S., Speleers, H.: Symbol-based multigrid methods for Galerkin B-spline isogeometric analysis. SIAM J. Numer. Anal. **55**, 31–62 (2017)
9. Ekström, S.-E.: Matrix-less methods for computing eigenvalues of large structured matrices. Ph.D. Thesis, Uppsala University (2018)
10. Ekström, S.-E., Furci, I., Garoni, C., Manni, C., Serra-Capizzano, S., Speleers, H.: Are the eigenvalues of the B-spline isogeometric analysis approximation of $-\Delta u = \lambda u$ known in almost closed form? Numer. Linear Algebr. Appl. **25**, e2151 (2018)
11. Ekström, S.-E., Furci, I., Serra-Capizzano, S.: Exact formulae and matrix-less eigensolvers for block banded symmetric Toeplitz matrices. BIT Numer. Math. **58**, 937–968 (2018)
12. Ekström, S.-E., Garoni, C.: A matrix-less and parallel interpolation-extrapolation algorithm for computing the eigenvalues of preconditioned banded symmetric Toeplitz matrices. Numer. Algorithms **80**(3), 819–848 (2019)
13. Ekström, S.-E., Garoni, C., Serra-Capizzano, S.: Are the eigenvalues of banded symmetric Toeplitz matrices known in almost closed form? Exp. Math. **27**, 478–487 (2018)
14. Garoni, C., Manni, C., Pelosi, F., Serra-Capizzano, S., Speleers, H.: On the spectrum of stiffness matrices arising from isogeometric analysis. Numer. Math. **127**, 751–799 (2014)
15. Garoni, C., Manni, C., Serra-Capizzano, S., Sesana, D., Speleers, H.: Spectral analysis and spectral symbol of matrices in isogeometric Galerkin methods. Math. Comput. **86**, 1343–1373 (2017)
16. Garoni, C., Serra-Capizzano, S.: Generalized Locally Toeplitz Sequences: Theory and Applications. Springer Monographs in Mathematics, vol. 1. Springer, Cham (2017)
17. Garoni, C., Speleers, H., Ekström, S.-E., Hughes, T.J.R., Reali, A., Serra-Capizzano, S.: Symbol-based analysis of finite element and isogeometric B-spline discretizations of eigenvalue problems: Exposition and review. Arch. Comput. Methods Eng. (In press) https://doi.org/10.1007/s11831-018-9295-y
18. Ruge, J.W., Stüben, K.: Algebraic multigrid. In: McCormick, S. (ed.) Multigrid Methods, pp. 73–130. SIAM Publications, Philadelphia (1987)
19. Serra-Capizzano, S.: Convergence analysis of two-grid methods for elliptic Toeplitz and PDEs matrix-sequences. Numer. Math. **92**, 433–465 (2002)
20. Serra-Capizzano, S., Tablino-Possio, C.: Spectral and structural analysis of high precision finite difference matrices for elliptic operators. Linear Algebra Appl. **293**, 85–131 (1999)

Lofting with Patchwork B-Splines

Nora Engleitner and Bert Jüttler

Abstract Lofting—also denoted as surface skinning—is one of the fundamental operations for creating free-form surfaces in Computer Aided Design. This process generates a surface from a given sequence of section curves. It is particularly useful for airfoils and turbine blades, since these shapes are often defined by cross sections with a family of auxiliary surfaces. The use of tensor-product B-splines, which is currently the standard technology, leads to large data volumes if section curves with incompatible knot vectors are used. We adopt the framework of Patchwork B-splines, which supports very flexible refinement strategies, and apply it to the construction of lofting surfaces. This approach not only reduces the resulting data volume but also limits the propagation of derivative discontinuities.

1 Introduction

The process of lofting, which is also called surface skinning, is one of the fundamental operations for the construction of free-form surfaces in geometric design, and its origins can be traced back to the early days for Computer-Aided Design [14]. It creates surfaces from given sequences of section curves, and it is particularly useful for airfoils and turbine blades, since these shapes are often defined by cross sections with a family of auxiliary surfaces.

More precisely, we consider a sequence of space curves $c_k(u)$, $k = 0, \ldots, N$, which are all defined over the parameter interval $[0, 1]$. Each curve is represented as B-spline curve of degree p_k,

$$c_k(u) = \sum_{i=0}^{m_k} d_{i,k} N_{i,p_k,T_k}(u), \quad \text{for } k = 0, \ldots, N,$$

N. Engleitner (✉) · B. Jüttler
Institute of Applied Geometry, Johannes Kepler University, Linz, Austria
e-mail: nora.engleitner@jku.at

B. Jüttler
e-mail: bert.juettler@jku.at

C. Giannelli and H. Speleers (eds.), *Advanced Methods for Geometric Modeling and Numerical Simulation*, Springer INdAM Series 35,
https://doi.org/10.1007/978-3-030-27331-6_5

with control points $d_{i,k} \in \mathbb{R}^3$ and B-splines N_{i,p_k,T_k} of degree p_k defined on the knot vector T_k. These curves will be denoted as *section curves*, since they represent the intersection of an unknown surface (the lofting surface) with a family of auxiliary surfaces. Note that the degrees and the knot vectors are potentially different for each of the section curves.

A lofting surface $s(u, v)$ is a surface with parameter domain $[0, 1]^2$, that interpolates the section curves at certain parameters $\bar{v}_k$, i.e.,

$$s(u, \bar{v}_k) = c_k(u), \quad \text{for all } k = 0, \ldots, N.$$

This paper is devoted to procedures that generate a lofting surface from the given section curves.

The construction of lofting surfaces has been analyzed in a substantial number of publications. In order to keep the paper focused, we list only a few representative references.

The basic definition and construction of tensor-product B-spline lofting surfaces via interpolation of spline curves is described by Piegl and Tiller [11]. The same authors also introduced an approximate approach [10], which helps to reduce the required data volume. In fact, the number of control points of tensor-product spline lofting surfaces will be quite large if the section curves possess incompatible knot vectors, and this motivates the use of approximate methods. In another paper, they presented several improvements and extensions of the skinning algorithm [12].

In recent years, research has focused on particular geometric problems related to skinning and lofting. Bizzarri et al. [1] describe techniques for skinning and blending with rational envelope surfaces. Kunkli and Hoffmann [8] discuss the skinning of sequences of circles and spheres.

Besides tensor-product spline spaces, it appears to be promising to use spline spaces that support adaptive refinement. As one of the first contributions in this direction, Yang and Zhen [15] applied the T-spline technology to perform approximate surface skinning. Li et al. [9] provide an explicit method for surface skinning using periodic T-spline surfaces.

Hierarchical splines [7] are another well-established approach to adaptive refinement of tensor-product splines. Recently, the original construction of a basis was enhanced by introducing the truncation mechanism [4, 5]. This has prepared the ground for the definition of Patchwork B-splines (PB-splines), which form a generalization of hierarchical splines that provides further possibilities for adaptive refinement [3].

The present paper applies this new construction of adaptive splines to the construction of lofting surfaces. We are able to perform exact interpolation of the section curves and obtain a smaller data volume than tensor-product spline lofting surfaces. Moreover, the use of PB-splines also helps to limit the propagation of derivative discontinuities. In addition, the construction is very flexible, supporting not only knot vectors containing knots with varying multiplicities, but even section curves with different polynomial degrees.

The remainder of the paper is structured into five sections. The next section recalls the standard approach, which employs tensor-product splines, based on the technology described in [11]. Section 3 introduces a very simple blending-based solution, which is not useful for applications but helps to prove the existence of PB-spline lofting surfaces. The fourth section adapts PB-splines to the lofting problem and describes the new method for lofting. Several computational results are presented in Sect. 5, in order to compare the tensor-product B-spline and PB-spline lofting surfaces. Finally, we conclude the paper.

2 Lofting with Tensor-Product B-Splines

We construct a tensor-product B-spline surface of degree (p, q) with knot vectors U and V,

$$s_{\text{tp}}(u, v) = \sum_{i=0}^{m}\sum_{j=0}^{n} c_{i,j}\, N_{i,p,U}(u) N_{j,q,V}(v),$$

that interpolates the section curves $c_k(u)$ of degrees p_k with knot vectors T_k, i.e.,

$$\sum_{i=0}^{m}\sum_{j=0}^{N} c_{i,j}\, N_{i,p,U}(u) N_{j,q,V}(\bar{v}_k) = \sum_{i=0}^{m_k} d_{i,k} N_{i,p_k,T_k}(u), \quad \text{for all } k = 0, \ldots, N.$$

In order to generate such a tensor-product spline surface, we create a knot vector U defining a spline space of degree

$$p = \max_{k=0,\ldots,N} p_k,$$

which contains the spline spaces associated with the section curves. Consequently, each section curve possesses a representation

$$c_k(u) = \sum_{i=0}^{m} d^{\star}_{i,k} N_{i,p,U}(u), \quad \text{for } k = 0, \ldots, N. \tag{1}$$

In addition, we choose a suitable degree q, the parameters $\bar{v}_k$ and a corresponding knot vector $V = \{v_i\}_{i=0,\ldots,N+q+1}$. Finally, we apply a curve interpolation algorithm to the rows of control points $d^{\star}_{i,0}, \ldots, d^{\star}_{i,N}$, to compute the control points $c_{i,j}$ of the lofting surface. More precisely, $c_{i,j}$ is the jth control point of the B-spline curve that interpolates the points $d^{\star}_{i,0}, \ldots, d^{\star}_{i,N}$.

Now we discuss the procedure in more detail, summarizing the approach presented in [11]. We start by constructing the common knot vector U. First, we have to perform degree elevation on the curves until all curves have the same degree p. Second, we apply the knot insertion algorithm to the different, degree elevated knot vectors until we arrive at the common knot vector U with basis functions $N_{i,p,U}$, $i = 0, \ldots, m$, for all curves. Hence, we construct U as the union of all knots, with associated multiplicities, that appear in one of the knot vectors T_k after the degree elevation. The knot insertion algorithm also provides the new control points $d^\star_{i,k}$ of the curves in (1), such that the shape of the section curves is preserved.

Next we have to choose a degree q, compute the parameters $\bar{v}_k$ and define a knot vector V in v-direction. The degree can be chosen freely as long as it satisfies $q \leq N$. For computing the parameters $\bar{v}_k$ and the knots $v_i \in V$ we use an averaging approach based on chord-length parameterization. The parameters $\bar{v}_k$ are chosen as

$$\bar{v}_0 = 0, \quad \bar{v}_N = 1,$$

and

$$\bar{v}_k = \bar{v}_{k-1} + \frac{1}{m+1} \sum_{i=0}^{m} \frac{|d^\star_{i,k} - d^\star_{i,k-1}|}{L_i}, \qquad \text{for } k = 1, \ldots, N-1,$$

with $L_i = \sum_{k=1}^{N} |d^\star_{i,k} - d^\star_{i,k-1}|$. The knots v_i are obtained by averaging the parameter values,

$$v_0 = \cdots = v_q = 0, \quad v_{N+1} = \cdots = v_{N+q+1} = 1,$$

and

$$v_i = \frac{1}{q} \sum_{k=i-q}^{i-1} \bar{v}_k, \quad \text{for } i = q+1, \ldots, N.$$

Finally we compute the control points $c_{i,j}$ for $i = 0, \ldots, m$ and $j = 0, \ldots, n$. Interpolating over the rows of control points $d^\star_{i,0}, \ldots, d^\star_{i,N}$ results in $m+1$ systems of equations with $N+1$ unknowns each. More precisely, for all $i = 0, \ldots, m$ we solve the $N+1$ equations

$$\sum_{j=0}^{n} c_{i,j} N_{j,q,V}(\bar{v}_k) = d^\star_{i,k}, \quad \text{for } k = 0, \ldots, N,$$

for the $N+1$ control points $c_{i,j}$ with $j = 0, \ldots, N$. As stated in [11], the choice of parameters and knots v_i guarantees that the systems are regular.

The method can be generalized by considering larger knot vectors V, which provide additional degrees of freedom. These degrees of freedom can be used to satisfy additional constraints or to optimize the shape of the resulting lofting surfaces. Our experimental results are based on the implementation in the Parasolid™ CAD kernel.

Generating a lofting surface with tensor-product B-splines is a relatively simple but highly effective tool and is therefore used in virtually all CAD systems. However, as a consequence of the degree elevation and knot insertion, the resulting surface may possess a considerable number of control points, especially when using a large number of section curves or curves with different knots and varying degrees.

3 Blending-Based Lofting

A particularly simple solution to the lofting problem can be obtained by the blending approach. This approach is not based on a globally defined tensor-product spline space, hence it provides a coarse representation of the lofting surface.

We define blending B-spline functions,

$$B_i(v) = \sum_{r=0}^{q} N_{i(q+1)+r,q,V}(v),$$

for $i = 0, \ldots, N$. The corresponding lofting surface is then defined as

$$s_{\text{blend}}(u, v) = \sum_{k=0}^{N} c_k(u) B_k(v).$$

Again we have to determine the degree q, the parameters $\bar{v}_k$ and the knot vector V. As in the previous section, the degree q can be choosen arbitrarily. The parameters are computed with an averaging procedure similar to the tensor-product B-spline case. However, we do not compute the distances between the control points here, since we do not have a one-to-one correspondence between the control points of adjacent section curves. Instead we use sample points on the section curves to obtain the parameter values $\bar{v}_k$. Hence, we evaluate all curves at the parameter values

$$r_i = \frac{i}{\hat{m}}, \quad \text{for } i = 0, \ldots, \hat{m}, \quad \hat{m} = \max_{k=0,\ldots,N} m_k,$$

and compute the parameters as

$$\bar{v}_0 = 0, \quad \bar{v}_N = 1, \tag{2}$$

and

$$\bar{v}_k = \bar{v}_{k-1} + \frac{1}{\hat{m}+1} \sum_{i=0}^{\hat{m}} \frac{|c_k(r_i) - c_{k-1}(r_i)|}{\hat{L}_i}, \tag{3}$$

with

$$\hat{L}_i = \sum_{k=1}^{N} |c_k(r_i) - c_{k-1}(r_i)|.$$

We define the knots $v_i \in V$ such that each parameter $\bar{v}_k$ is contained in the knot span $[v_{i_k}, v_{i_k+1}]$ and that v_{i_k+1} and $v_{i_{k+1}}$ are separated by q knot spans. To be more precise, we choose the the first and last $q + 1$ knots as

$$v_0 = \cdots = v_q = 0, \quad \text{and} \quad v_{n-q} = \cdots = v_n = 1, \tag{4}$$

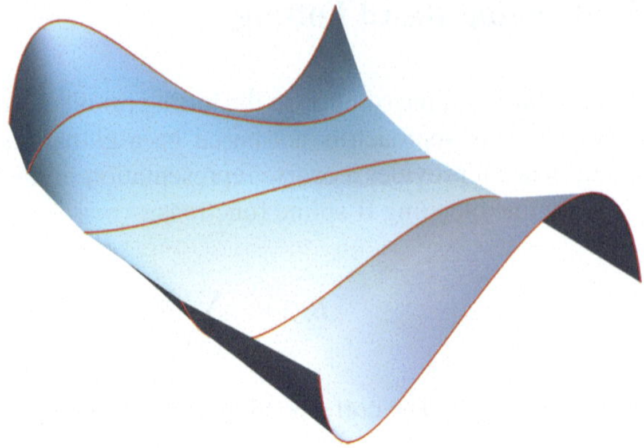

Fig. 1 Example for a blending-based lofting surface over 5 section curves (colored in red)

with $n = (N+2)(q+1)$. The remaining knots are then defined as

$$v_{k(q+1)+i} = \begin{cases} \bar{v}_{k-1} + (i+1)h_k, & \text{if } k = 1, \\ \bar{v}_{k-1} + (i+\frac{1}{2})h_k, & \text{else,} \end{cases} \tag{5}$$

with

$$h_k = \begin{cases} \frac{\bar{v}_k - \bar{v}_{k-1}}{q+\frac{3}{2}}, & \text{if } k = 1 \text{ or } k = N, \\ \frac{\bar{v}_k - \bar{v}_{k-1}}{q+1}, & \text{else,} \end{cases}$$

for $k = 1, \ldots, N$ and $i = 0, \ldots, q$. Figure 1 shows an example of a lofting surface obtained by this simple blending approach. As an advantage, the lofting surface is obtained without the need for solving systems of linear equations. However, the section curves correspond to singular curves on the lofting surface. Clearly, the quality of this result is not sufficient for applications. Nevertheless, we will use the blending approach to prove the existence of solutions for the lofting surface defined by PB-splines.

4 Lofting with PB-Splines

After discussing tensor-product B-spline and the blending-based lofting, we introduce a novel lofting method that provides high-quality surfaces while keeping the number of degrees of freedom small. The construction is based on Patchwork B-splines (PB-splines) that are defined on sequences of partially nested tensor-product spline spaces with associated patches. This enables us to employ independently chosen spline spaces in the vicinity of the section curves, thereby eliminating redundant control points.

4.1 Patchwork B-Splines

We recall the definition of Patchwork B-splines via a patchwork hierarchy that defines a patchwork spline space. PB-splines are defined by suitably generalizing the selection mechanism for hierarchical B-splines. We present certain assumptions on the patchwork hierarchy which guarantee linear independence and that the PB-splines form a basis for the patchwork spline space.

The Patchwork Hierarchy

The patchwork hierarchy combines a sequence of tensor-product spline spaces with a corresponding sequence of patches.

First, we consider a finite sequence of spline spaces S^ℓ that are spanned by tensor-product B-splines,

$$S^\ell = \text{span}\{N_{i,p_\ell,U_\ell}\, N_{j,q_\ell,V_\ell}\}_{(i,j)\in\mathcal{J}^\ell},$$

for the *levels* $\ell = 0, \ldots, 2N$ with the index set

$$\mathcal{J}^\ell = \{(i, j) \ : \ i = 0, \ldots, m_\ell, \ j = 0, \ldots, n_\ell\}.$$

The basis functions are defined over knot vectors U_ℓ and V_ℓ of degrees p_ℓ and q_ℓ in u- and v-direction, respectively. At a point $\mathbf{x} = (u, v)$, the basis functions possess the smoothness

$$\mathbf{s}^\ell(\mathbf{x}) = (s_1^\ell(u), s_2^\ell(v)) = (p_\ell - m_1^\ell(u), q_\ell - m_2^\ell(v)),$$

where $m_1^\ell(u)$ is the multiplicity of u in U_ℓ and $m_2^\ell(v)$ the multiplicity of v in V_ℓ. Note that the spline spaces S^ℓ are not required to be nested, i.e., S^ℓ is not necessarily a subspace of $S^{\ell+1}$.

Second, we assign a corresponding *patch* π^ℓ to every spline space S^ℓ. The patches are mutually disjoint open rectangles in the $\mathbb{R}^2$ that cover the *domain* Ω,

$$\Omega = \bigcup_{\ell=0}^{2N} \overline{\pi^\ell} = [0, 1]^2.$$

Furthermore, we assume that the boundaries of a patch π^ℓ are aligned with the knot lines of the corresponding spline space.[1]

Finally, we introduce the *patchwork spline space* P, which contains functions $f \in C([0, 1]^2)$ with the following two properties.

[1]Note that the PB-spline construction defined in [3] allows arbitrary domains and admits more general open subsets of the $\mathbb{R}^d$ as patches. However, for the lofting problem this simpler setting is sufficient.

(i) The restriction of f to one of the patches is contained in the corresponding spline space on that patch, i.e.,

$$f|_{\pi^\ell} \in S^\ell|_{\pi^\ell}, \quad \text{for } \ell = 0, \ldots, 2N.$$

(ii) The restriction of f to $\overline{(\pi^\ell \cap \pi^k)}$ possesses the smoothness $\mathbf{s}^{\max(\ell,k)}(\mathbf{x})$ for all $\mathbf{x} \in \overline{\pi^\ell} \cap \overline{\pi^k}$.

Clearly, the properties of the functions in the patchwork spline space are determined by the chosen patchwork hierarchy, i.e., by the sequence of tensor-product spline spaces and associated patches.

The Selection Mechanism

In order to construct a basis for the patchwork spline space we adapt Kraft's selection mechanism [7], which is based on the relation between the support of a B-spline and certain subdomains. Instead of this classical approach, we work with the constraining boundary Γ^ℓ of a patch π^ℓ, which is defined as the part of the boundary $\partial\pi^\ell$ that is shared with patches of a lower level,

$$\Gamma^\ell = \bigcup_{k=0}^{\ell-1} \overline{\pi^k} \cap \overline{\pi^\ell}.$$

We use a constraining boundary-based selection mechanism on the patchwork hierarchy to obtain a basis. For each level ℓ we select the bivariate basis functions $N_{i,p_\ell,U_\ell}\, N_{j,q_\ell,V_\ell}$ that are active on the corresponding patch π^ℓ while vanishing at the constraining boundary Γ^ℓ,

$$\mathcal{K}^\ell = \{(i,j) \in \mathcal{J}^\ell \;:\; (N_{i,p_\ell,U_\ell}\, N_{j,q_\ell,V_\ell})|_{\pi^\ell} \neq 0 \text{ and } (N_{i,p_\ell,U_\ell}\, N_{j,q_\ell,V_\ell})|_{\Gamma^\ell} = 0\}. \tag{6}$$

The supports of the selected basis functions of level ℓ define the *shadow* $\hat{\pi}^\ell$ of a patch,

$$\hat{\pi}^\ell = \bigcup_{(i,j)\in\mathcal{K}^\ell} \operatorname{supp}(N_{i,p_\ell,U_\ell}\, N_{j,q_\ell,V_\ell}), \quad \ell = 0, \ldots, 2N.$$

Finally, collecting the selected basis functions from all levels gives us the *Patchwork B-splines* (PB-splines),

$$K = \bigcup_{\ell=0}^{2N} \{(N_{i,p_\ell,U_\ell}\, N_{j,q_\ell,V_\ell}) \;:\; (i,j) \in \mathcal{K}^\ell\}.$$

Linear Independence and Space Characterization

Two assumptions allow us to characterize the space that is spanned by the PB-splines:

Assumption S2C. The patches and spline spaces possess the property of *Simple Shadow Compatibility*:

(i) The shadow $\hat{\pi}^\ell$ of a patch π^ℓ does not intersect patches of a lower level $k < \ell$, i.e.,

$$\hat{\pi}^\ell \cap \pi^k = \emptyset, \quad \text{for all } k < \ell.$$

(ii) If the shadow $\hat{\pi}^k$ of a level $k < \ell$ intersects the patch π^ℓ then the corresponding spaces are nested, i.e.,

$$\hat{\pi}^k \cap \pi^\ell \neq \emptyset \implies S^k \subseteq S^\ell, \quad \text{for all } k < \ell.$$

Assumption SMC. All patches and associated spline spaces fulfill the *Smoothness Monotonicity Condition*: The smoothness across the constraining boundary Γ^ℓ in transversal direction does not increase when moving from a lower to a higher level, i.e., for $k < \ell$ it holds that

$$s_i^\ell(\mathbf{x}) \leq s_i^k(\mathbf{x}) \qquad \text{for all} \quad \mathbf{x} \in \Gamma^\ell \cap \overline{\pi^k},$$

where the ith coordinate direction is transversal with respect to Γ^ℓ at $\mathbf{x}$.

As shown in [3], these two assumptions imply two fundamental results:

Theorem 1 *The PB-splines are linearly independent on Ω if the patchwork hierarchy satisfies Assumption S2C.*

Consequently, the selected B-splines from all levels form a basis.

Theorem 2 *The PB-splines span the patchwork spline space P if Assumptions S2C and SMC are both satisfied.*

Therefore, we have two different characterizations of the patchwork spline space P. On the one hand, there is the implicit definition that characterizes the space by the properties of the functions it contains. On the other hand, we have a constructive definition that describes P as the linear hull of its basis, namely the PB-splines.

Note that the PB-splines do not form a partition of unity. In order to restore this property we have to introduce a truncation mechanism [3]. A detailed discussion is beyond the scope of the present paper.

4.2 The Patchwork Hierarchy for Lofting

Recall that we want to construct a PB-spline surface $s_{\text{pb}}(u, v)$, which satisfies the interpolation conditions $s_{\text{pb}}(u, \bar{v}_k) = c_k(u)$ for certain parameter values $\bar{v}_0, \ldots, \bar{v}_N$, i.e.,

$$s_{\text{pb}}(u, \bar{v}_k) = \sum_{\ell=0}^{2N} \sum_{(i,j)\in\mathcal{K}^\ell} c_{i,j}^\ell \, N_{i,p_\ell,U_\ell}(u) N_{j,q_\ell,V_\ell}(\bar{v}_k) = \sum_{i=0}^{m_k} d_{i,k} N_{i,p_k,T_k}(u),$$

for $k = 0, \ldots, N$. We choose the same spline space in v-direction for all levels, i.e., $V_\ell = V$ and $q_\ell = q$. The degree, the parameter values $\bar{v}_k$ and the elements of the knot vector $V = \{v_i\}$ are chosen in the same way as for the blending-based loft. Consequently, the locations of the parameters $\bar{v}_0, \ldots, \bar{v}_N$ are determined by (2) and (3), and the knots v_i take the values (4) and (5). How to construct the vectors U_ℓ will be discussed later.

Curve Patches and Intermediate Patches

All patches are chosen as axis-aligned boxes. We denote the southwest (lower left) and northeast (upper right) vertex of the rectangular patch π^ℓ by

$$\mathbf{r}_{\text{sw}}^\ell = (r_{\text{sw},1}^\ell, r_{\text{sw},2}^\ell) \quad \text{and} \quad \mathbf{r}_{\text{ne}}^\ell = (r_{\text{ne},1}^\ell, r_{\text{ne},2}^\ell),$$

respectively. The first $N+1$ patches possess the vertices

$$\mathbf{r}_{\text{sw}}^\ell = (0, v_{(\ell+1)(q+1)-1}) \quad \text{and} \quad \mathbf{r}_{\text{ne}}^\ell = (1, v_{(\ell+1)(q+1)}), \tag{7}$$

for $\ell = 0, \ldots, N$. We denote these patches as *curve patches* since each of them contains the parameter value which is associated with one of the section curves,

$$(0, 1) \times \{\bar{v}_\ell\} \subset \pi^\ell .$$

Consecutive curve patches π^ℓ and $\pi^{\ell+1}$ are connected by the *intermediate patch* $\pi^{\ell+N+1}$ with the vertices

$$\mathbf{r}_{\text{sw}}^{\ell+N+1} = (0, r_{\text{ne},2}^\ell) \quad \text{and} \quad \mathbf{r}_{\text{ne}}^{\ell+N+1} = (1, r_{\text{sw},2}^{\ell+1}),$$

for $\ell = 0, \ldots, N-1$.

Summing up, all the patches are horizontal stripes with width 1. The height of the curve patches is determined by a single knot span of the knot vector V, while the height of the intermediate patches is equal to q knot spans.

Now we discuss the choice of the knot vector U_ℓ and the associated degree p_ℓ for each patch. Again we distinguish between curve and intermediate patches.

The curve patch π^ℓ, $\ell = 0, \ldots, N$, simply inherits the degree p_ℓ and the knot vector T_ℓ from the corresponding section curve $c_\ell(u)$. The intermediate patch π^ℓ, $\ell = N+1, \ldots, 2N$, has the degree

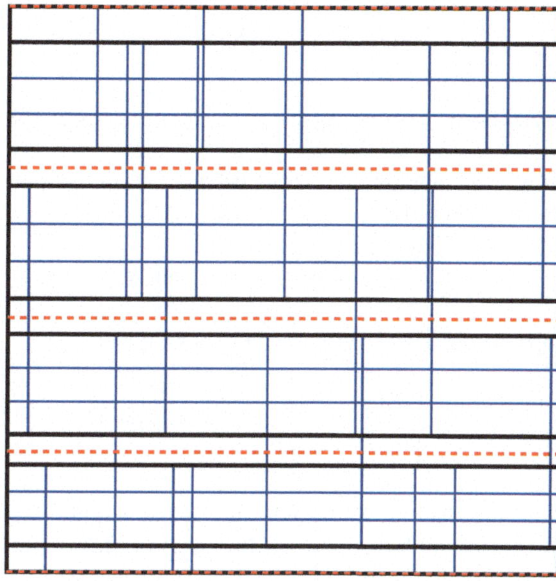

Fig. 2 Example for a simple patchwork hierarchy H. Blue lines represent the knot lines, black lines are patch boundaries and dashed red lines illustrate the parameter lines $v = \bar{v}_k$

$$p_\ell = \max\{p_{\ell-N-1}, p_{\ell-N}\}.$$

The corresponding knot vector is obtained by first applying degree elevation to the knot vectors $U_{\ell-N-1}$ and $U_{\ell-N}$ and then performing knot insertion until one arrives at a knot vector U_ℓ that is the union of the degree elevated knot vectors, where each knot is considered with multiplicity.

The resulting patchwork hierarchy, which is fully determined by the given section curves, will be denoted as H. Figure 2 shows an instance of such a hierarchy for 5 section curves of maximum smoothness and degrees $p_k = q = 3$.

Basis Construction

Applying the selection mechanism (6) to the patchwork hierarchy H generates sets of basis functions with associated index sets

$$\mathcal{K}^\ell = \{(i, j) \; : \; i = 0, \ldots, m_\ell, \; j = \ell(q+1), \ldots, (\ell+1)(q+1) - 1\},$$

for $\ell = 0, \ldots, N$ and

$$\mathcal{K}^\ell = \emptyset,$$

for $\ell > N$. In fact, we select $m_\ell(q+1)$ basis functions from each curve patch and none from the intermediate patches. The total number of selected functions is equal to

$$|K| = (q+1)\sum_{\ell=0}^{N} m_\ell.$$

Corollary 1 *The patchwork hierarchy H satisfies S2C and SMC.*

Proof. The shadow of a curve patch π^ℓ is an axis-aligned box with lower-left and upper-right corners

$$\hat{\mathbf{r}}_{\mathrm{sw}}^\ell = (0, v_{\ell(q+1)}) \quad \text{and} \quad \hat{\mathbf{r}}_{\mathrm{ne}}^\ell = (1, v_{(\ell+2)(q+1)-1}),$$

respectively. These vertices can be rewritten as

$$\hat{\mathbf{r}}_{\mathrm{sw}}^\ell = \begin{cases} (0,0), & \text{if } \ell = 0, \\ (0, r_{\mathrm{ne},2}^{\ell-1}), & \text{otherwise,} \end{cases} \quad \text{and} \quad \hat{\mathbf{r}}_{\mathrm{ne}}^\ell = \begin{cases} (1,1), & \text{if } \ell = N, \\ (1, r_{\mathrm{sw},2}^{\ell+1}), & \text{otherwise.} \end{cases}.$$

Thus, the shadow $\hat{\pi}^\ell$ extends only to the neighboring intermediate patches, which possess a level larger than ℓ. Moreover, the shadows of the intermediate patches are empty. The first part of S2C is therefore satisfied.

For the second part of S2C we note that the construction of the vectors U_ℓ implies that $S^{\ell-N-1} \subseteq S^\ell$ and $S^{\ell-N} \subseteq S^\ell$ for $\ell = N+1, \ldots, 2N$. Thus, S2C is fulfilled by the patchwork hierarchy H.

Now we consider SMC. Note that S2C implies SMC for all pairs of neighboring patches π^k and π^ℓ with $k < \ell$ and $\hat{\pi}^k \cap \pi^\ell \neq \emptyset$. Since no other pairs of neighboring patches exist in H, we conclude that SMC is also satisfied. □

Theorem 3 *The Patchwork B-splines defined on H form a partition of unity,*

$$\sum_{k=0}^{2N} \sum_{(i,j)\in\mathcal{K}^k} N_{i,p_k,U_k}(u) N_{j,q,V}(v) = 1, \quad \textit{for } (u,v) \in \Omega. \tag{8}$$

Proof. Recall that the tensor-product B-splines $N_{i,p_\ell,U_\ell}\, N_{j,q,V}$ with $(i,j) \in \mathcal{J}^\ell$ form a partition of unity on $[0,1]^2$. We consider the restriction to π^ℓ and obtain

$$\sum_{\substack{(i,j)\in\mathcal{J}^\ell \\ (N_{i,p_\ell,U_\ell} N_{j,q,V})|_{\pi^\ell} \neq 0}} N_{i,p_\ell,U_\ell}(u) N_{j,q,V}(v) = 1, \quad \text{for } (u,v) \in \pi^\ell. \tag{9}$$

This summation considers the indices

$$\{(i,j) \,:\, i = 0,\ldots,m_\ell,\ j = \ell(q+1),\ldots,(\ell+1)(q+1)-1\} \tag{10}$$

for $\ell = 0, \ldots, N$, and

$$\{(i, j) : i = 0, \ldots, m_\ell,\ j = (\ell - N - 1)(q + 1) + 1, \ldots, (\ell - N + 1)(q + 1) - 2\} \tag{11}$$

for $\ell = N + 1, \ldots, 2N$.

First, we show the partition of unity on a curve patch π^ℓ, with $\ell = 0, \ldots, N$. The left-hand side of (8) evaluates to

$$\sum_{k=0}^{2N} \sum_{(i,j)\in\mathcal{K}^k} N_{i,p_k,U_k}(u) N_{j,q,V}(v) = \sum_{(i,j)\in\mathcal{K}^\ell} N_{i,p_\ell,U_\ell}(u) N_{j,q,V}(v), \quad \text{for } (u, v) \in \pi^\ell.$$

Since the index set $\mathcal{K}^\ell$ coincides with the set in (10), the partition of unity is implied by (9).

Second, when considering the restriction to an intermediate patch π^ℓ, with $\ell = N + 1, \ldots, 2N$, we observe that the only functions that contribute to the sum on the left-hand side in (8) possess the indices

$$\bigcup_{k=\ell-N-1}^{\ell-N} \{(i, j) \in \mathcal{J}^k : (N_{i,p_k,U_k} N_{j,q,V})|_{\overline{\pi^k}\cap\overline{\pi^\ell}} \neq 0\}.$$

Therefore, this left-hand side can be rewritten as

$$\left(\sum_{i=0}^{m_{\ell-N-1}} N_{i,p_{\ell-N-1},U_{\ell-N-1}}(u)\right)\left(\sum_{j=(\ell-N-1)(q+1)+1}^{(\ell-N)(q+1)-1} N_{j,q,V}(v)\right) +$$
$$\left(\sum_{i=0}^{m_{\ell-N}} N_{i,p_{\ell-N},U_{\ell-N}}(u)\right)\left(\sum_{j=(\ell-N)(q+1)}^{(\ell-N+1)(q+1)-2} N_{j,q,V}(v)\right), \quad \text{for } (u, v) \in \pi^\ell.$$

The two sums with index i are equal to one, according to the partition of unity property of univariate B-splines. Merging the remaining two sums confirms (8) on π^ℓ, since the summation with respect to j is in agreement with the index set defined in (11). □

Since the selected B-splines defined by H satisfy S2C and SMC, they form a basis of the associated patchwork spline space according to Theorem 2. This fact enables us to obtain the following result.

Theorem 4 *The patchwork spline space P defined by the hierarchy H contains the coordinate functions of the lofting surface $s_{\text{blend}}(u, v)$, which is generated by the blending-based approach.*

Proof. We show that the restrictions of $s_{\text{blend}}(u, v)$ to the patches π^ℓ belong to the associated spline spaces S^ℓ. Recall that

$$s_{\text{blend}}(u, v) = \sum_{k=0}^{N}\sum_{i=0}^{m_k}\sum_{r=0}^{q} d_{i,k} N_{i,p_k,T_k}(u) N_{k(q+1)+r,q,V}(v).$$

First we consider a curve patch π^ℓ, $\ell = 0, \ldots, N$. We obtain

$$s_{\text{blend}}(u, v) = \sum_{i=0}^{m_\ell}\sum_{r=0}^{q} d_{i,\ell} N_{i,p_\ell,T_\ell}(u) N_{\ell(q+1)+r,q,V}(v), \quad \text{for } (u, v) \in \pi^\ell.$$

By comparing the knot vectors we find

$$s_{\text{blend}}(u, v) = \sum_{(i,j)\in\mathcal{K}^\ell} d_{i,\ell} N_{i,p_\ell,U_\ell}(u) N_{j,q,V}(v), \quad \text{for } (u, v) \in \pi^\ell,$$

which implies that the restriction of $s_{\text{blend}}(u, v)$ to the curve patch π^ℓ is contained in S^ℓ for $\ell = 0, \ldots, N$.

Second, for the intermediate patches π^ℓ, $\ell = N + 1, \ldots, 2N$, we obtain

$$s_{\text{blend}}(u, v) = \sum_{k=\ell-N-1}^{\ell-N}\ \sum_{(i,j)\in\mathcal{K}^k} d_{i,k} N_{i,p_k,U_k}(u) N_{j,q,V}(v), \quad \text{for } (u, v) \in \pi^\ell.$$

Assumption S2C guarantees that

$$S^{\ell-N-1} \subseteq S^\ell \quad \text{and} \quad S^{\ell-N} \subseteq S^\ell, \tag{12}$$

which implies that the restriction of $s_{\text{blend}}(u, v)$ to the intermediate patch π^ℓ is contained in S^ℓ for $\ell = N + 1, \ldots, 2N$.

In addition, we need to verify the smoothness property (ii). This follows immediately from (12) as well and completes the proof. □

4.3 Comparison with Other Adaptive Splines

For section curves of uniform degrees $p = p_\ell$, where $\ell = 0, \ldots, N$, the basis constructed on the patchwork hierarchy H can equivalently be generated by other adaptive spline constructions. We focus on (analysis-suitable) T-splines and LR B-splines, e.g., see [2, 13].

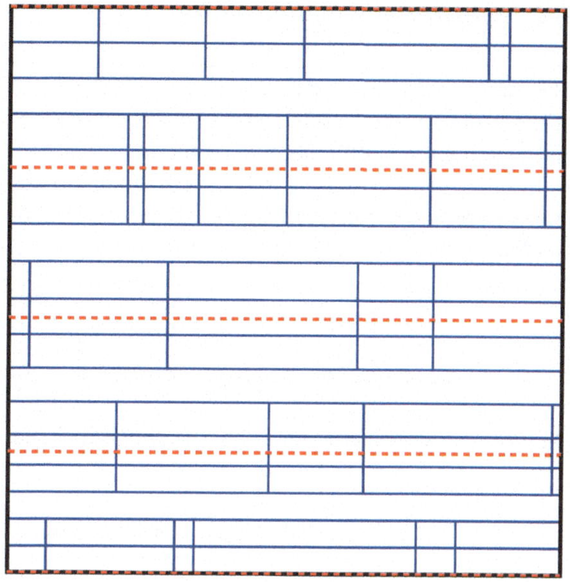

Fig. 3 The T-mesh that defines the same basis as the patchwork hierarchy in Fig. 2

First, we consider T-splines on the T-mesh defined by horizontal lines with start points $(0, v_i)$ and end points $(1, v_i)$ for all knot values $v_i \in V$, where V is computed as for the PB-splines, and vertical lines with start and end points

$$\left(\omega, r^{\ell}_{\mathrm{sw},2} - \left\lfloor \frac{q}{2} \right\rfloor\right) \quad \text{and} \quad \left(\omega, r^{\ell}_{\mathrm{ne},2} + \left\lfloor \frac{q}{2} \right\rfloor\right),$$

for all inner knot values $\omega \in U_\ell \setminus \{0, 1\}$, $\ell = 0, \ldots, N$. The values $r^{\ell}_{\mathrm{sw},2}$ and $r^{\ell}_{\mathrm{ne},2}$ are defined in (7). Thus, the knot lines of the section curves extend over q knot spans with respect to V when considering odd degrees and $q + 1$ knot spans for even degrees. The T-splines defined on this T-mesh are analysis-suitable since no horizontal T-junction extensions exist. Figure 3 shows the T-mesh that defines the same basis as the patchwork hierarchy in Fig. 2.

Second, we construct a collection of LR B-splines, starting from a tensor-product mesh that possesses no inner knots in u-direction and a knot vector V as for the PB-splines. We perform meshline insertions with start point $(\omega, r^{\ell-1}_{\mathrm{ne},2})$ and end point $(\omega, r^{\ell+1}_{\mathrm{ne},2})$ for each inner knot $\omega \in U_\ell \setminus \{0, 1\}$, for $\ell = 0, \ldots, N$. The resulting mesh is the same that defines the patchwork hierarchy H. Note that the insertion of the meshlines can be formulated as a series of *primitive meshline extensions*, where we insert a new meshline spanning q elements and extend it $(q + 1)$ times by a single element. Therefore, linear independence is guaranteed by the theory of LR B-splines [6].

Note that this equivalence with T-splines and LR B-splines is limited to the case of uniform degrees for all section curves, since these constructions do not support non-uniform degrees.

4.4 Constrained Optimization

Theorem 4 guarantees that the patchwork spline space P contains at least one solution of the lofting problem. Uniqueness, however, is not guaranteed. We use numerical optimization to identify the 'best' solution.

In order to construct the PB-spline lofting surface we solve a constrained optimization problem where we minimize an energy term subject to the interpolation conditions,

$$\begin{aligned} &\text{minimize} \quad J(s) = \int_{\Omega} (\partial_{vv} s(u,v))^2, \\ &\text{subject to} \quad s(u, \bar{v}_k) = c_k(u), \quad \text{for all } k = 0, \dots, N. \end{aligned}$$

On the patchwork hierarchy H, the interpolation conditions can be rewritten as

$$\sum_{j=k(q+1)}^{(k+1)(q+1)-1} c_{i,j}^k N_{j,q,V}(\bar{v}_k) = d_{i,k}, \quad \text{for all } i = 0, \dots, m_k, \quad \text{and} \quad k = 0, \dots, N.$$

By using the method of Lagrange multipliers we obtain the system

$$\begin{pmatrix} E & B^T \\ B & 0 \end{pmatrix} \begin{pmatrix} c \\ \lambda \end{pmatrix} = \begin{pmatrix} 0 \\ d \end{pmatrix},$$

of linear equations, which is solved for the unknowns

$$c = (c_{i,j}^{\ell})_{\ell=0,\dots,N;\ (i,j)\in\mathcal{K}^{\ell}},$$

and the Lagrange multipliers

$$\lambda = (\lambda_{i,k})_{k=0,\dots,N;\ i=0,\dots,m_k}.$$

The vector

$$d = (d_{i,k})_{k=0,\dots,N;\ i=0,\dots,m_k}$$

contains the control points of the given section curves. The $|K| \times |K|$ energy matrix E possesses the elements

$$e_{(\ell_1,i_1,j_1),(\ell_2,i_2,j_2)} = 2 \iint_{\Omega} N_{i_1,p_{\ell_1},U_{\ell_1}}(u) N_{i_2,p_{\ell_2},U_{\ell_2}}(u) \partial_{vv} N_{j_1,q,V}(v) \partial_{vv} N_{j_2,q,V}(v) du\, dv,$$

for $\ell_k = 0, \dots, 2N$ and $(i_k, j_k) \in \mathcal{K}^{\ell_k}$. The integrals in the above expression can be solved exactly by a Gaussian quadrature rule with $\lceil \frac{p+q-1}{2} \rceil$ points in each coordinate direction where $p = \max_{k=0,\dots,N} p_k$. For the $(\sum_{k=0}^{N} m_k + 1) \times |K|$ constraints matrix B we obtain

$$b_{(\ell,i),(\ell,i,j)} = N_{j,q,\mathcal{V}}(\bar{v}_\ell),$$

for $\ell = 0, \ldots, N, i = 0, \ldots, m_\ell$ and $(i, j) \in \mathcal{K}^\ell$. Note that the elements of B depend only on ℓ and j, therefore, $b_{(\ell,i_1),(\ell,i_1,j)} = b_{(\ell,i_2),(\ell,i_2,j)}$ for any $i_1, i_2 = 0, \ldots, m_\ell$.

5 Results

This section presents four lofting examples where we compare the results of using tensor-product B-splines with PB-splines. The blending-based lofting is not included, since the surface quality of this method is not satisfactory.

We start with two academic examples. In the first one, we consider curves of uniform degrees with highest order smoothness. For the second example, we use curves of varying degrees and knot multiplicities larger than 1. Furthermore, we present two real world examples, provided by our industrial partner MTU Aero Engines, where we construct a surface from a wireframe model consisting of 41 periodic curves with 153 control points each. In this case we use periodic PB-splines, which are straightforwardly obtained in the same way as periodic tensor-product B-splines.

We choose degree $q = 3$ for all examples.

First Example

Consider the 21 section curves of uniform degrees $p_k = 3$ and maximum smoothness, which are shown in Fig. 4, left. The two lofting surfaces by tensor-product splines and PB-splines are virtually identical, and therefore we visualize only one of them, see Fig. 4, right. The maximum deviation between the surfaces $s_{\text{tp}}(u, v)$ and $s_{\text{pb}}(u, v)$ does not exceed $2.5e-4\%$ of the diameter of the bounding box. The tensor-product surface has 2507 (100%) control points, whereas the PB-spline surface needs only 756 (30.2%). Figure 5 depicts the meshes in the parameteric domain for the tensor-product B-splines (left) and PB-splines (right), respectively. The knot lines of the tensor-product mesh extend over the entire unit square, while the patchwork hierarchy contains the knots of the curves only in a certain region around the associated parameter lines $v = \bar{v}_k$.

Second Example

Consider the 11 section curves of varying degrees p_k between 2 and 4, see Fig. 6, left. Moreover, the knot vectors T_k contain knots with higher multiplicities, which even causes the loss of differentiability. In this case, the maximum deviation between the two results is smaller than 0.35% of the diameter of the bounding box. The tensor-product surface has 1599 (100%) control points, whereas the PB-spline surface needs

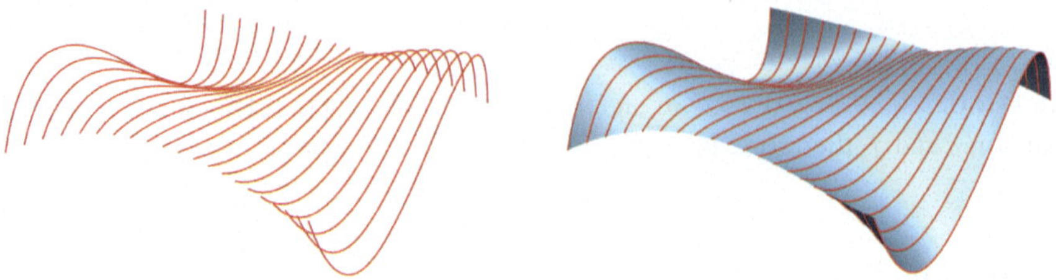

Fig. 4 Section curves and corresponding lofting surface $s_{pb}(u, v)$ for the first example

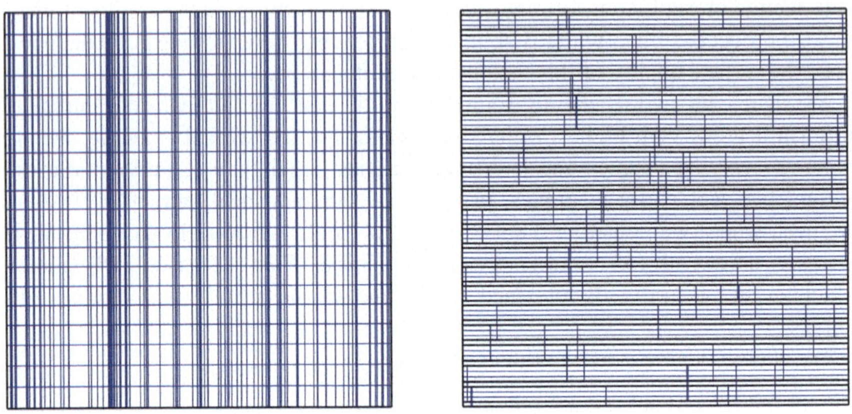

Fig. 5 2D meshes for tensor-product B-splines (*left*) and PB-splines (*right*). The black lines in the right picture represent the patch boundaries

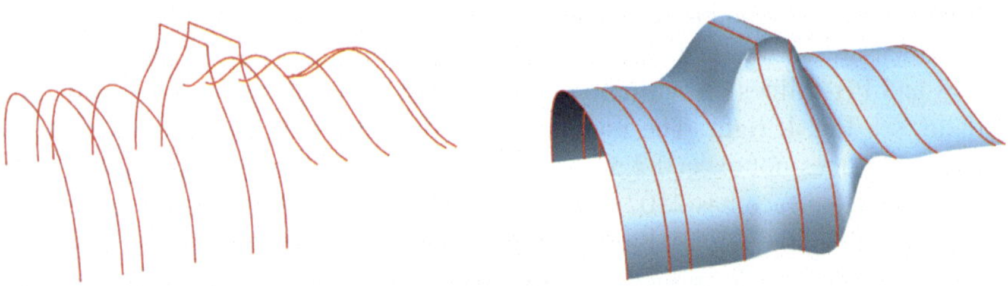

Fig. 6 Section curves and corresponding lofting surface $s_{pb}(u, v)$ for the second example

only 508 (32%). Furthermore, the tangent discontinuities of the section curves extend to the entire tensor-product spline surface, while affecting only some patches of the PB-spline surface, see Fig. 7. The 2D meshes for both methods are illustrated in Fig. 8. Note that the different colors of the knot lines correspond to different orders of smoothness.

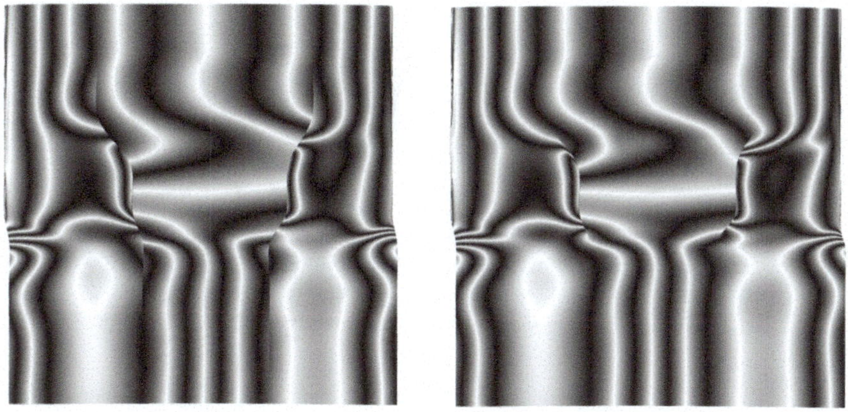

Fig. 7 Reflection lines on the tensor-product spline surface (*left*) and the PB-spline surface (*right*) illustrate the propagation of the tangential discontinuities

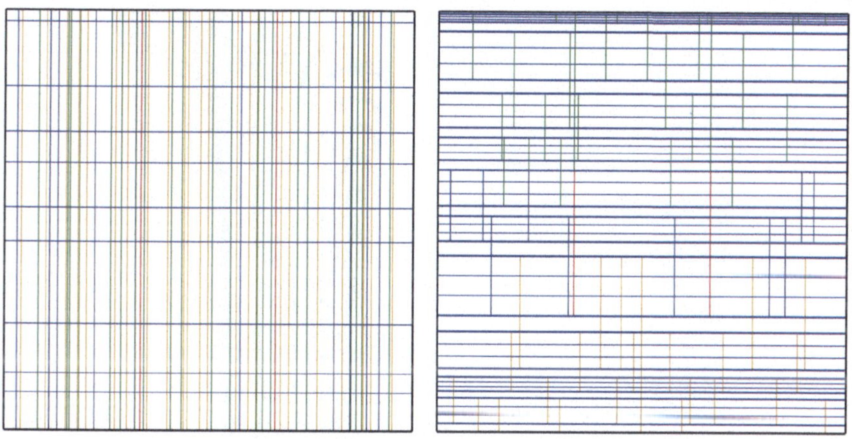

Fig. 8 2D meshes for tensor-product B-splines (*left*) and PB-splines (*right*) for the second example. The colors encode the degree of smoothness across knot lines, C^0 (*red*), C^1 (*orange*), C^2 (*blue*) and C^3 (*green*). Bold lines represent patch boundaries

Third Example

In the first industrial example we consider the wireframe model of a bulky airfoil with 41 periodic curves of degree 3 with maximum smoothness. Lofting with tensor-product B-splines requires 262,816 (100%) control points. When using PB-splines we obtain a surface with only 24,600 (9.4%) control points. The maximum deviation between the airfoils with a total height of 0.15m is 0.04μm. Figure 9 shows the section curves on the left, followed by the lofting surface with reflection line analysis.

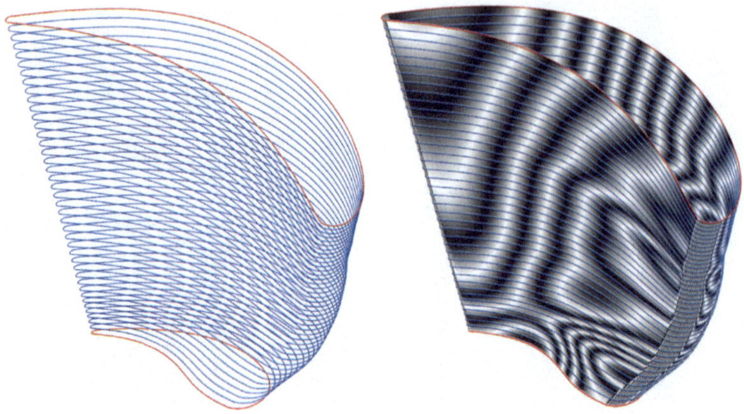

Fig. 9 Section curves and the resulting lofting surface with reflection lines for the third example

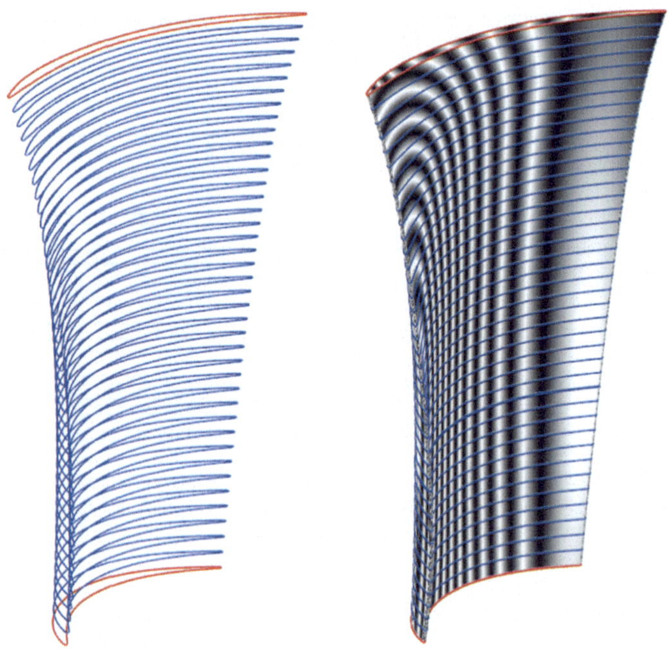

Fig. 10 Section curves and the resulting lofting surface with reflection lines for the fourth example

Fourth Example

Example 4 is similar to the previous example. The only difference lies in the shape of the airfoil, which is rather thin and elongated. The tensor-product surface possesses 262,773 (100%) control points, while the PB-spline surface is defined by 24,600

(9.4%) control points. As before, the differences in the surfaces are marginal with a maximum deviation of 0.016μm for an airfoil height of 0.117m. Figure 10 depicts the section curves and the lofting surface with reflection line analysis.

6 Conclusion

We applied the recently introduced framework of Patchwork B-splines (PB-splines) [3] to the construction of lofting surfaces. Especially, when considering a large number of section curves with many control points defined on disjoint knot vectors, the PB-splines offer a substantial advantage: The adaptivity of the PB-splines enables us to generate high-quality lofting surfaces with fewer control points compared to the standard tensor-product B-spline approach, where repeated global knot insertion is required. Moreover, the patchwork structure helps to limit the propagation of derivative discontinuities.

In contrast to the algorithms proposed in [10, 15] the PB-spline lofting provides an exact interpolation of the section curves. Also, while [9] presents an algorithm for periodic cubic splines of C^0- and C^2-smoothness, we provide a very flexible method that works for arbitrary section curves. More precisely, these can be periodic, defined on knot vectors with varying multiplicities and possess different polynomial degrees. The use of an energy term, which was mentioned but not yet implemented in [9], leads to visually pleasing surfaces without bumps between the section curves.

Furthermore, a PB-spline surface can be exported as collections of standard tensor-product spline surfaces, thereby providing a simple way to use them in an existing CAD environment. The resulting model consists of $2N + 1$ faces. Clearly, the built-in smoothness will not be preserved when performing additional control point modifications within CAD.

Finally, in the case of uniform degrees, we observed that the PB-splines used for constructing the lofting surface can equivalently be seen as analysis-suitable T-splines [13] or LR B-splines [2], defined by suitable meshes. The extension to non-uniform degrees, however, is not covered by these two existing constructions.

Acknowledgements Supported by the Austrian Science Fund (FWF) though NFN S117 "Geometry + Simulation". The authors thank MTU Aero Engines AG for kindly providing the airfoil data sets.

References

1. Bizzarri, M., Lávička, M., Kosinka, J.: Skinning and blending with rational envelope surfaces. Comput. Aided Des. **87**, 41–51 (2017)
2. Dokken, T., Lyche, T., Pettersen, K.F.: Polynomial splines over locally refined box-partitions. Comput. Aided Geom. Des. **30**, 331–356 (2013)

3. Engleitner, N., Jüttler, B.: Patchwork B-spline refinement. Comput. Aided Des. **90**, 168–179 (2017)
4. Giannelli, C., Jüttler, B., Kleiss, S.K., Mantzaflaris, A., Simeon, B., Špeh, J.: THB-splines: an effective mathematical technology for adaptive refinement in geometric design and isogeometric analysis. Comput. Methods Appl. Mech. Eng. **299**, 337–365 (2016)
5. Giannelli, C., Jüttler, B., Speleers, H.: THB-splines: The truncated basis for hierarchical splines. Comput. Aided Geom. Des. **29**, 485–498 (2012)
6. Johannessen, K.A., Kvamsdal, T., Dokken, T.: Isogeometric analysis using LR B-splines. Comput. Methods Appl. Mech. Eng. **269**, 471–514 (2014)
7. Kraft, R.: Adaptive und linear unabhängige Multilevel B-splines und ihre Anwendungen. Ph.D. thesis, University of Stuttgart (1998)
8. Kunkli, R., Hoffmann, M.: Skinning of circles and spheres. Comput. Aided Geom. Des. **27**, 611–621 (2010)
9. Li, Y., Chen, W., Cai, Y., Nasri, A., Zheng, J.: Surface skinning using periodic T-spline in semi-NURBS form. J. Comput. Appl. Math. **273**, 116–131 (2015)
10. Piegl, L., Tiller, W.: Algorithm for approximate NURBS skinning. Comput. Aided Des. **28**, 699–706 (1996)
11. Piegl, L., Tiller, W.: The NURBS Book, 2nd edn. Springer, New York (1997)
12. Piegl, L., Tiller, W.: Surface skinning revisited. Vis. Comput. **18**, 273–283 (2002)
13. Scott, M., Li, X., Sederberg, T., Hughes, T.: Local refinement of analysis-suitable T-splines. Comput. Methods Appl. Mech. Eng. **213–216**, 206–222 (2012)
14. Woodward, C.D.: Cross-sectional design of B-spline surfaces. Comput. Graph. **11**, 193–201 (1987)
15. Yang, Y., Zheng, J.: Approximate T-spline surface skinning. Comput. Aided Des. **44**, 1269–1276 (2012)

A Study on Spline Quasi-interpolation Based Quadrature Rules for the Isogeometric Galerkin BEM

Antonella Falini and Tadej Kanduč

Abstract The two recently introduced quadrature schemes in [7] are investigated for regular and singular integrals, in the context of boundary integral equations arising in the isogeometric formulation of the Galerkin Boundary Element Method (BEM). In the first scheme, the regular part of the integrand, consisting of a B-spline and of an auxiliary function, is approximated by a suitable quasi-interpolant spline. In the second scheme, the auxiliary function is approximated by applying the quasi-interpolation operator and then the product of the two resulting splines is expressed as a linear combination of particular B-splines. The two schemes are tested and compared against other standard and novel methods available in the literature to evaluate different types of integrals arising in the Galerkin formulation. When h-refinement is performed, numerical tests reveal that under reasonable assumptions, the second scheme provides the optimal order of convergence, even with a small amount of quadrature nodes. The quadrature schemes are validated also in numerical examples to solve 2D Laplace problems with Dirichlet boundary conditions.

1 Introduction

The Boundary Element Method (BEM) is a numerical technique to transform the differential problem into an integral one, where the unknowns are defined only on the boundary of the computational domain (see, e.g., [9, 25] and references therein). The main two advantages of the method are the dimension reduction of the problem and the simplicity to treat external problems. As a major drawback, the integral formulation involves Boundary Integral Equations (BIEs), which contain singular kernel functions. Therefore, robust and precise quadrature formulae are necessary

A. Falini · T. Kanduč (✉)
Istituto Nazionale di Alta Matematica, Unità di Ricerca di Firenze c/o DiMaI "U. Dini", Università degli Studi di Firenze, Florence, Italy
e-mail: tadej.kanduc@unifi.it; tadej.kanduc@fmf.uni-lj.si

A. Falini
e-mail: antonella.falini@unisi.it

C. Giannelli and H. Speleers (eds.), *Advanced Methods for Geometric Modeling and Numerical Simulation*, Springer INdAM Series 35,
https://doi.org/10.1007/978-3-030-27331-6_6

to provide an accurate numerical evaluation. The solution of the considered BIE is then obtained by collocation or Galerkin procedures.

The isogeometric formulation of boundary element method (IgA-BEM) has been successfully applied to 2D and 3D problems, such as linear elasticity [27], fracture mechanics [22], acoustic [28] and Stokes flows [16]. Recently, the IgA paradigm has been combined with the Symmetric Galerkin Boundary Element Method (IgA-SGBEM), which has revealed to be very effective among BEM schemes (see [3, 4, 13, 14, 21] and references therein). Moreover, the full potential of B-splines over the more common Lagrangian basis has been recently exploited in the assembly stage of the system matrix in [1].

In this work we frame the two quadrature procedures in [7] in a Galerkin IgA-BEM for the 2D Laplace problem with Dirichlet boundary conditions. In particular, the derived quadrature formulae are obtained using a *quasi-interpolation* (*QI*) operator, firstly introduced in [18] and then applied to construct quadrature rules for regular integrals in [19]. The second procedure has been successfully applied in a Galerkin adaptive BEM using hierarchical B-splines in [11]. The authors also provided some theoretical results on the convergence order of the quadrature rule, when h-refinement is performed.

In this paper we experimentally test both procedures in [7] for the regular and weakly singular integrals occurring in the Galerkin formulation. We compare the achieved accuracy with other quadratures available in the literature and suitable for the evaluation of the assayed boundary integrals; namely the methods in [1, 5, 8, 15, 29]. Moreover, we recall some results on perturbed Galerkin BEM to provide an estimate for the asymptotic accuracy of the quadratures required to obtain the optimal order of convergence. The accuracy of the methods is compared also by computing Galerkin solutions for different 2D Laplace problems with Dirichlet boundary conditions.

The structure of the paper is as follows. Section 2 explains the isogeometric Galerkin BEM. Section 3 is devoted to the adopted quadrature rules based on quasi-interpolation. Section 4 deals with the accuracy of the considered quadratures for different types of integrals. Three numerical examples to model 2D Laplace problems are presented in Sect. 5. Finally, some conclusions follow in Sect. 6.

2 BEM Formulation for Interior and Exterior Laplace Problems

In Sect. 2.1 we summarise the main features of the BEM formulation for the 2D Laplace problem with Dirichlet boundary conditions and we derive the considered boundary integral equations. Then, in Sect. 2.2, following the isogeometric paradigm, both the boundary representation and the approximate solution are expressed in a B-spline basis. In Sect. 2.3, the governing boundary integrals for the Galerkin discretization are recalled.

2.1 BEM Formulation for Laplace Problems

In the present work two different domains are considered: open, bounded, simply connected domains $\Omega \subset \mathbb{R}^2$ and unbounded ones, external to an open bounded arc Γ, $\Omega = \mathbb{R}^2 \setminus \Gamma$. The solution of the Laplace problem is to find $u \in H^1(\Omega)$ that satisfies

$$\begin{cases} \Delta u = 0 \text{ in } \Omega\,, \\ u = u_D \text{ on } \Gamma\,. \end{cases} \tag{1}$$

Given the differential problem (1), the boundary element method provides an integral formulation, where the unknown is defined only on the boundary of the considered computational domain [9, 25]; in particular, in the rest of the subsection we follow the derivation presented in [9]. In potential theory we can express the solution u in terms of double layer and single layer potentials using the representation formula,

$$u(\mathbf{x}) = -\frac{1}{2\pi}\int_\Gamma \partial_{\mathbf{n}_\mathbf{y}} U(\mathbf{x},\mathbf{y})\,[u(\mathbf{y})]\,d\Gamma_\mathbf{y} + \frac{1}{2\pi}\int_\Gamma U(\mathbf{x},\mathbf{y})\,[\partial_{\mathbf{n}_\mathbf{y}} u(\mathbf{y})]\,d\Gamma_\mathbf{y}, \quad \mathbf{x} \in \mathbb{R}^2 \setminus \Gamma. \tag{2}$$

The symbol $\partial_\mathbf{n}$ denotes the normal derivative with respect to the exterior unit normal vector $\mathbf{n}$ on Γ. The harmonic function u is supposed to be regular both in Ω and in $\mathbb{R}^2 \setminus \overline{\Omega}$ with different boundary values on both sides of Γ. The jump of a function v across Γ is defined as

$$[v(\mathbf{x})] := v|_{\mathbb{R}^2\setminus\Omega}(\mathbf{x}) - v|_{\overline{\Omega}}(\mathbf{x}), \quad \mathbf{x} \in \Gamma.$$

The function $-\frac{1}{2\pi}U$ is the fundamental solution of the 2D Laplace operator,

$$U(\mathbf{x},\mathbf{y}) := \log \|\mathbf{x}-\mathbf{y}\|_2.$$

By applying the trace operator to Eq. (2) we can derive a specific *Boundary Integral Equation* (*BIE*) according to certain assumptions on $\mathbb{R}^2 \setminus \overline{\Omega}$. The derived BIE allows us to compute the missing boundary datum.

When Γ is a bounded open curve, which does not separate $\mathbb{R}^2$ into two disjoint domains, the external problem is modeled by setting $[u]_\Gamma \equiv 0$ and the resulting BIE can be written as

$$u_D(\mathbf{x}) = -\frac{1}{2\pi}\int_\Gamma U(\mathbf{x},\mathbf{y})\,\phi(\mathbf{y})\,d\Gamma_\mathbf{y}, \quad \mathbf{x} \in \Gamma. \tag{3}$$

In (3) the unknown density function $\phi(\mathbf{y}) := [\partial_{\mathbf{n}_\mathbf{y}} u(\mathbf{y})] \in \tilde{H}^{-1/2}(\Gamma)$ represents the jump of the flux of the solution u. The space $\tilde{H}^{-1/2}(\Gamma)$ is the dual space of the fractional Sobolev space $H^{1/2}(\Gamma)$, where the duality is defined with respect to the usual $L^2(\Gamma)$-scalar product.

In case of an interior problem we assume $u|_{\mathbb{R}^2\setminus\Omega} \equiv 0$. The resulting BIE reads

$$\frac{1}{2}u_D(\mathbf{x}) = \frac{1}{2\pi}\int_\Gamma \partial_{\mathbf{n}_\mathbf{y}} U(\mathbf{x},\mathbf{y})\, u_D(\mathbf{y})\, d\Gamma_\mathbf{y} - \frac{1}{2\pi}\int_\Gamma U(\mathbf{x},\mathbf{y})\, \phi(\mathbf{y})\, d\Gamma_\mathbf{y}, \quad \mathbf{x} \in \Gamma, \tag{4}$$

and the unknown function $\phi(\mathbf{y}) := \partial_{\mathbf{n}_\mathbf{y}} u(\mathbf{y}) \in \tilde{H}^{-1/2}(\Gamma)$ is the flux of u.

Thus, we have reformulated the original Laplace problem (1) in terms of BIE (3) and (4). The latter formulation has a clear physical meaning in terms of variables and it is called the *direct approach*.

Both integral Eqs. (3) and (4) are referred to as the Symm's integral equation

$$V\phi(\mathbf{x}) = f(\mathbf{x}), \quad \mathbf{x} \in \Gamma, \tag{5}$$

where the operator $V : \tilde{H}^{-1/2}(\Gamma) \to H^{1/2}(\Gamma)$ is defined by

$$V\phi(\mathbf{x}) := -\frac{1}{2\pi}\int_\Gamma U(\mathbf{x},\mathbf{y})\, \phi(\mathbf{y})\, d\Gamma_\mathbf{y}.$$

The operator is elliptic if the domain is properly scaled, e.g., its diameter is less than 1. The right-hand side f in (5) is given by u_D in the case of the exterior problem (3), or as

$$\frac{1}{2}u_D(\mathbf{x}) - \frac{1}{2\pi}\int_\Gamma \partial_{\mathbf{n}_\mathbf{y}} U(\mathbf{x},\mathbf{y})\, u_D(\mathbf{y})\, d\Gamma_\mathbf{y}$$

in the case of the interior problem (4).

2.2 B-Splines

A *knot vector* $\mathbf{T} = \{t_1, \ldots, t_{N+d+1}\}$ is defined as a non-decreasing sequence of *knots*:

$$t_1 \le \cdots \le t_{d+1} =: a < t_{d+2} \le \cdots \le t_N < t_{N+1} =: b \le \cdots \le t_{N+d+1}.$$

By removing repeated knots from $\mathbf{T}$ we obtain the *breakpoints* of $\mathbf{T}$. The vector $\mathbf{T}$ defines a univariate B-spline basis on $[a, b]$ of cardinality N and polynomial degree d; the basis is defined by the well-known recursion formula [6]:

$$B_{i,0}(t) := B_{i,0}^{(\mathbf{T})}(t) := \begin{cases} 1, & \text{if } t_i \le t < t_{i+1}, \\ 0, & \text{otherwise,} \end{cases}$$
$$B_{i,r}(t) := \omega_{i,r}(t)\, B_{i,r-1}(t) + \left(1 - \omega_{i+1,r}(t)\right) B_{i+1,r-1}(t), \quad r = 1, \ldots, d,$$

where

$$\omega_{i,r}(t) := \begin{cases} \frac{t-t_i}{t_{i+r}-t_i}, & \text{if } t_i < t_{i+r}, \\ 0, & \text{otherwise.} \end{cases}$$

B-splines span a space of splines $\hat{S}$, whose smoothness depends on the multiplicities of the breakpoints in $\mathbf{T}$. An interior breakpoint t_i has multiplicity $m \geq 1$ if its value appears m times in $\mathbf{T}$; the space $\hat{S}$ has a regularity C^{d-m} at t_i. If $m > 1$, then t_i is called multiple, otherwise it is simple.

It is common to assume that the boundary Γ can be parametrized by a parametric B-spline curve $\mathbf{F} : [a, b] \to \Gamma \subset \mathbb{R}^2$, written in the *B-form*,

$$\mathbf{F} := \sum_{i=1}^{N} \mathbf{d}_i \, B_{i,d}.$$

An ordered set of *control points* in $\mathbb{R}^2$ is denoted by $\{\mathbf{d}_i\}_{i=1,\ldots,N}$.

To recover the interpolation of the first and the last control point, an *open knot vector* is constructed by setting $t_1 = \cdots = t_d = a$ and $t_{N+2} = \cdots = t_{N+d+1} = b$. This is a standard choice to define open curves.

For a closed curve, i.e., $\mathbf{F}(a) = \mathbf{F}(b)$, it is more convenient to introduce a periodic definition of the auxiliary knots. To address also the case of a knot vector with multiple knots, we introduce $\rho := d - m + 1$, where m is the multiplicity of the knot a. In particular, when the knots are simple, $\rho = d$. The splines are defined on a periodic knot vector, thus we identify each pair $\{B_{i,d}, B_{N-\rho+i,d}\}$, $i = 1, 2, \ldots, \rho$, as one periodic basis function, denoted by $B_{i,d}$ and defined on two disjoint intervals on $[a, b]$. For the periodic compatibility it is sufficient that the 2ρ knot differences on the left are identical to the 2ρ ones on the right, $t_{i+1} - t_i = t_{N-\rho+i+1} - t_{N-\rho+i}$ for $i = 1, \ldots, 2\rho$. Furthermore, $\mathbf{d}_i = \mathbf{d}_{N-\rho+i}$ for $i = 1, \ldots, \rho$. See [12, Sect. 10.7] for more details.

2.3 Galerkin Formulation

For both exterior and interior problems, the solution ϕ of the considered BIE (3) and (4) belongs to the Sobolev space $\tilde{H}^{-1/2}(\Gamma)$. The variational formulation of (5) is (see [30]):

$$\text{for } u_D \in H^{1/2}(\Gamma), \text{ find } \phi \in \tilde{H}^{-1/2}(\Gamma) \text{ such that } \mathcal{A}(\phi, \psi) = \mathcal{F}(\psi), \quad \forall \psi \in H^{1/2}(\Gamma), \tag{6}$$

where the bilinear form $\mathcal{A}(\phi, \psi)$ and right-hand side $\mathcal{F}(\psi)$ are defined as

$$\mathcal{A}(\phi, \psi) := \int_\Gamma \psi(\mathbf{x}) \, V\phi(\mathbf{x}) \, d\Gamma_x, \quad \mathcal{F}(\psi) := \int_\Gamma \psi(\mathbf{x}) \, f(\mathbf{x}) \, d\Gamma_x. \tag{7}$$

When applying the Galerkin method on (6), the infinite dimensional solution space $\tilde{H}^{-1/2}(\Gamma)$ in (6) is approximated by a finite dimensional subspace S_h. The parameter h is related to the discretization step size of the subspace S_h, and the subspace is generated by the lifted B-splines defined on $\mathbf{T}_\mathrm{h}$,

$$S_h := \left\langle B_{1,d}^{(\mathbf{T}_\mathrm{h})} \circ \mathbf{F}^{-1}, B_{2,d}^{(\mathbf{T}_\mathrm{h})} \circ \mathbf{F}^{-1}, \dots, B_{N,d}^{(\mathbf{T}_\mathrm{h})} \circ \mathbf{F}^{-1} \right\rangle. \tag{8}$$

Let us introduce coordinates $s,\ t \in [a,b] \subset \mathbb{R}$ in the parametric domain,

$$s := \mathbf{F}^{-1}(\mathbf{x}), \quad t := \mathbf{F}^{-1}(\mathbf{y}).$$

Then the weak form of the exterior problem (3) reads

$$\begin{aligned} &\int_{D_i} B_{i,d}^{(\mathbf{T})}(s)\, J(s)\, u_D(\mathbf{F}(s))\, ds \\ &= -\frac{1}{2\pi} \int_{D_i} B_{i,d}^{(\mathbf{T})}(s)\, J(s) \int_a^b U(\mathbf{F}(s), \mathbf{F}(t))\, \phi_h(\mathbf{F}(t))\, J(t)\, dt\, ds, \quad i = 1, \dots, N, \end{aligned} \tag{9}$$

where $B_{i,d}^{(\mathbf{T})}$ are the test functions of the problem defined on a knot vector $\mathbf{T}$ for some fixed h, $D_i := \operatorname{supp} B_{i,d}^{(\mathbf{T})}$ and $\phi_h \in S_h$ is the approximate solution of ϕ. The function J denotes the parametric speed of the curve,

$$J(\cdot) := \|\mathbf{F}'(\cdot)\|_2.$$

For the interior problem, the corresponding BIE follows from (4),

$$\begin{aligned} &\frac{1}{2} \int_{D_i} B_{i,d}^{(\mathbf{T})}(s)\, J(s)\, u_D(\mathbf{F}(s))\, ds \\ &= \frac{1}{2\pi} \int_{D_i} B_{i,d}^{(\mathbf{T})}(s)\, J(s) \int_a^b \partial_{\mathbf{n}_t} U(\mathbf{F}(s), \mathbf{F}(t))\, u_D(\mathbf{F}(s))\, J(t)\, dt\, ds \\ &- \frac{1}{2\pi} \int_{D_i} B_{i,d}^{(\mathbf{T})}(s)\, J(s) \int_a^b U(\mathbf{F}(s), \mathbf{F}(t))\, \phi_h(\mathbf{F}(t))\, J(t)\, dt\, ds, \quad i = 1, \dots, N. \end{aligned} \tag{10}$$

To separate the geometrical influence from the singular contribution of the kernel $U(\mathbf{F}(s), \mathbf{F}(t)) = \log \|\mathbf{F}(s) - \mathbf{F}(t)\|_2$, we split U into two functions: K_1 regular and K_2 weakly singular, defined as

$$K_1(s,t) = \frac{1}{2} \log \frac{\|\mathbf{F}(s) - \mathbf{F}(t)\|_2^2}{\delta^2(s,t)}, \quad K_2(s,t) = \log \delta(s,t).$$

Thus $U(\mathbf{F}(s), \mathbf{F}(t)) =: K_1(s,t) + K_2(s,t)$. To obtain a regular K_1, the function δ needs to be chosen according to the type of domain. Following the idea from [11], it is defined as:

$$\delta(s,t) := \begin{cases} |s-t|, & \text{if } \Gamma \text{ is an open curve,} \\ |s-t|\dfrac{|(s-t)^2-\gamma^2|}{\gamma^2}, & \text{if } \Gamma \text{ is a closed curve,} \end{cases} \tag{11}$$

with $\gamma = b - a$ being the length of the parametric interval.

By writing the approximate solution $\phi_h \in S_h$ in terms of B-splines as

$$\phi_h = \sum_{j=1}^{N} \alpha_j \, B_{j,d}^{(\mathbf{T})} \circ \mathbf{F}^{-1},$$

and by substituting K_1 and K_2 in place of U, we can rearrange Eqs. (9) and (10) in a linear system $A\boldsymbol{\alpha} = \boldsymbol{\beta}$ with unknowns $\boldsymbol{\alpha} := (\alpha_j)_{j=1}^N$. In particular, the matrix A consists of entries $A_{i,j} = A_{i,j}^{(1)} + A_{i,j}^{(2)}$ given as

$$A_{i,j}^{(\ell)} := -\frac{1}{2\pi} \int_{D_i} B_{i,d}^{(\mathbf{T})}(s)\, J(s)\, ds \int_{D_j} K_\ell(s,t)\, B_{j,d}^{(\mathbf{T})}(t)\, J(t)\, dt, \tag{12}$$

for $i, j = 1, \dots, N$ and $\ell = 1, 2$. The entries of the right-hand side $\boldsymbol{\beta} := (\beta_i)_{i=1}^N$ are computed according to the type of problem:

- For the exterior problems,

$$\beta_i := \int_{D_i} B_{i,d}^{(\mathbf{T})}(s)\, J(s)\, u_D(\mathbf{F}(s))\, ds, \quad i = 1, \dots, N. \tag{13}$$

- For the interior problems, $\boldsymbol{\beta}$ consists of two terms, $\boldsymbol{\beta} = \frac{1}{2}\boldsymbol{\beta}^{(1)} - \frac{1}{2\pi}\boldsymbol{\beta}^{(2)}$. We compute the entries of $\boldsymbol{\beta}^{(1)}$ by (13), while the elements of $\boldsymbol{\beta}^{(2)}$ as

$$\beta_i^{(2)} := \int_{D_i} B_{i,d}^{(\mathbf{T})}(s)\, J(s)\, ds \int_a^b \bar{K}(s,t)\, u_D(\mathbf{F}(t))\, dt, \quad i = 1, \dots, N. \tag{14}$$

Note that the kernel function $\bar{K}(s,t) := \partial_{\mathbf{n}_t} U(\mathbf{F}(s), \mathbf{F}(t))\, J(t)$ is regular everywhere assuming $\mathbf{F} = (F_1, F_2)$ to be C^2 smooth. More precisely,

$$\bar{K}(s,t) = \frac{(F_2(s) - F_2(t))\, F_1'(t) - (F_1(s) - F_1(t))\, F_2'(t)}{\|\boldsymbol{F}(s) - \boldsymbol{F}(t)\|_2^2}, \quad s \neq t,$$

$$\lim_{s \to t} \bar{K}(s,t) = \frac{F_1'(t)\, F_2''(t) - F_2'(t)\, F_1''(t)}{2\, J^2(t)},$$

see [1] for further details.

3 Quadrature Rules

In this section we summarize the two spline QI-based quadrature rules introduced in [7]. The quadratures are adopted to evaluate regular and singular integrals that appear in the system matrix (12) and in the right-hand side vector (13) and (14). Some implementation aspects are explained afterwards.

3.1 The QI-Based Schemes

The core idea of the quadrature procedures is to express the integrand functions in terms of simpler functions that can be efficiently integrated. Double integrals in (12) are split into two single ones. Regular non-piecewise polynomial parts are approximated by particular quasi-interpolation splines. To simplify and speed up the implementation of the quadratures we can consider the quadrature nodes to be uniformly spaced on the integration domain.

To evaluate entries in (12)–(14), the computation of the following two types of integrals needs to be addressed,

$$I_{B_i}[g] := \int_{D_i} B_{i,d}^{(\mathbf{T})}(t)\, g(t)\, dt, \tag{15}$$

$$I_{w_i^s}[g] := \int_{D_i} K_2(s,t)\, B_{i,d}^{(\mathbf{T})}(t)\, g(t)\, dt, \tag{16}$$

where we denote $w_i^s := K_2(s,\cdot)\, B_{i,d}^{(\mathbf{T})}(\cdot)$ and we assume $g \in C(\overline{D}_i)$. Integrals (16) are considered weakly singular if $s \in D_i$ and nearly singular if $s \notin D_i$ but the distance between s and D_i is sufficiently small. When the distance is sufficiently large, the integral (16) is regular and it can be considered as a type of (15), where $K_2(s,\cdot)$ is hidden inside g.

We recall the basic ideas of the spline QI quadrature procedures in [7]. In the so-called `procedure 1` the whole product $\tilde{g} := B_{i,d}^{(\mathbf{T})}\, g$ is approximated by a quasi-interpolant spline $\sigma_{\tilde{g}}$ of a chosen degree p. The QI space is defined on an open knot vector $\boldsymbol{\tau} := \{\tau_{-p}, \ldots, \tau_{n+p}\}$, with uniform knots $\tau_{-p} = \cdots = \tau_0 < \cdots < \tau_n = \cdots = \tau_{n+p}$ and it is constructed locally on the support $D_i =: [\tau_0, \tau_n]$ of every basis function $B_{i,d}^{(\mathbf{T})}$:

$$\hat{S}_{\boldsymbol{\tau}} := \langle B_{-p,p}^{(\boldsymbol{\tau})}, \ldots, B_{n-1,p}^{(\boldsymbol{\tau})} \rangle.$$

The breakpoints of $\boldsymbol{\tau}$ define $n+1$ quadrature nodes localised at D_i. By replacing $\tilde{g}$ with the quasi-interpolant $\sigma_{\tilde{g}}$,

$$\sigma_{\tilde{g}} = \sum_{k=-p}^{n-1} \lambda_k(\tilde{g})\, B_{k,p}^{(\boldsymbol{\tau})} \approx \tilde{g},$$

where $\lambda_k(\tilde{g})$ are suitable coefficients, the integrals (15) and (16) are approximated by

$$I_{B_i}[g] \approx \sum_{k=-p}^{n-1} \lambda_k(\tilde{g}) \int_{D_i} B_{k,p}^{(\tau)}(t)\, dt = \sum_{k=-p}^{n-1} \lambda_k(\tilde{g}) \frac{\left|\text{supp}\, B_{k,p}^{(\tau)}\right|}{p+1}, \quad (17)$$

$$I_{w_i^s}[g] \approx \sum_{k=-p}^{n-1} \lambda_k(\tilde{g}) \int_{D_i} K_2(s,t)\, B_{k,p}^{(\tau)}(t)\, dt, \quad (18)$$

where $|\cdot|$ stands for the size of the region. In case of integrals (15) the evaluation gets reduced to the computation of integrals of B-splines using the expression (17) (for the closed form expressions for integrals of B-splines see, e.g., [26, Sect. 5.4]). For integrals (16) a preliminary computation of the so-called *modified moments* $\mu_{k,p}^{(i)}(s)$ is needed:

$$\mu_{k,p}^{(i)}(s) := \int_{D_i} K_2(s,t)\, B_{k,p}^{(\tau)}(t)\, dt. \quad (19)$$

Explicit formulae to compute the moments are derived in [1, 7] for $\delta = |s-t|$. We refer to [11], when δ takes the second form in (11), suitable for a closed boundary curve.

The `procedure 2` differs from the `procedure 1` in the initial step, where only the function g is approximated by a QI spline σ_g in the local space $\hat{S}_{\tau}$ of degree p:

$$\sigma_g = \sum_{k=-p}^{n-1} \lambda_k(g)\, B_{k,p}^{(\tau)} \approx g.$$

Thereafter, the product $B_{i,d}^{(\mathbf{T})}(\cdot)\, \sigma_g(\cdot)$ of splines is expressed as a linear combination of B-splines basis functions spanning the product space Π of degree $d+p$,

$$B_{i,d}^{(\mathbf{T})}\, \sigma_g = \sum_{k=-p}^{n-1} \lambda_k(g)\, B_{i,d}^{(\mathbf{T})}\, B_{k,p}^{(\tau)} = \sum_m \eta_m\, B_{m,d+p}^{(\tau_\Pi)}.$$

The spline space Π is defined on a knot vector $\boldsymbol{\tau}_\Pi$ constructed locally on D_i and η_m are the appropriate coefficients in the new basis. For the details on the construction of the B-spline product space and the representation of the product in the new basis we refer to [20]. By applying `procedure 2`, the integrals (15) and (16) are approximated by the following expressions,

$$I_{B_i}[g] \approx \int_{D_i} \sum_{k=-p}^{n-1} \lambda_k(g)\, B_{i,d}^{(\mathbf{T})}(t)\, B_{k,p}^{(\tau)}(t)\, dt = \sum_m \eta_m \frac{\left|\text{supp}\, B_{m,d+p}^{(\tau_\Pi)}\right|}{d+p+1}, \quad (20)$$

$$I_{w_i^s}[g] \approx \int_{D_i} K_2(s,t) \sum_{k=-p}^{n-1} \lambda_k(g)\, B_{i,d}^{(\mathbf{T})}(t)\, B_{k,p}^{(\boldsymbol{\tau})}(t)\, dt = \sum_m \eta_m \int_{D_i} K_2(s,t)\, B_{m,d+p}^{(\boldsymbol{\tau}\Pi)}(t)\, dt$$
$$= \sum_m \eta_m\, \mu_{m,d+p}^{(i)}(s). \tag{21}$$

Clearly, choosing a good QI operator is of fundamental importance to obtain accurate quadrature rules. In our study we adopt the Hermite type QI introduced in [18] and its derivative free variant [19], already framed in a singular integrals context in [7].

The quasi-interpolant spline σ_g, obtained by the adopted QI operator for a given function g, is defined on a knot vector $\boldsymbol{\tau}$ with $n+1$ breakpoints and it can be written in B-form as

$$\sigma_g = \sum_{j=-p}^{n-1} \lambda_j(g)\, B_{j,p}^{(\boldsymbol{\tau})}. \tag{22}$$

For the scheme in [18] the coefficients $\lambda_j(g)$ in (22) are defined as a suitable linear combination of a local subset of values of g and g' at the spline breakpoints. For instance, for $p = 2$ they can be computed as

$$\begin{aligned} \lambda_j(g) &= \tfrac{1}{2}\left(g(\tau_{j+2}) + g(\tau_{j+1})\right) - \tfrac{\tau_{j+2}-\tau_{j+1}}{4}\left(g'(\tau_{j+2}) - g'(\tau_{j+1})\right), \quad j = -1, \ldots, n-2, \\ \lambda_{-2}(g) &= g(\tau_0), \quad \lambda_{n-1}(g) = g(\tau_n). \end{aligned}$$

This p-degree quasi-interpolation scheme is a projector on the considered spline space and has the optimal approximation order $p+1$ in the infinity norm for $g \in C^{p+1}([\tau_0, \tau_n])$.

If the variant scheme in [19] is chosen, then the g' values necessary to compute λ_j in (22), are approximated using a suitable finite difference formula. Also this scheme has optimal approximation order, but it is not a projector.

The asymptotic approximation order does not directly depend on n, however n should satisfy $n \geq p+1$ if p is odd and $n \geq p+2$ if p even. This condition gives us the lowest number of quadrature nodes that can be utilized in the two considered QI-based quadrature schemes. When $B_{i,d}^{(\mathbf{T})}$ has the maximum regularity, then if d is odd the quadrature nodes can be chosen as the knots of $B_{i,d}^{(\mathbf{T})}$. For the derivative free variant and even d an additional node is needed to correctly utilize the finite difference formula; it can be chosen as the midpoint between the two central knots.

As reported in [19], the introduced QI-based quadrature formulae for regular integrals are competitive with respect to other QI-based schemes (see for instance [23, 24]), thanks to the usage of the additional derivative information. The quadrature rules for singular integrals based on the variant QI scheme [19] also exhibit a competitive or superior behaviour when compared to others QI-based schemes, see [7] for more details.

3.2 Implementation Aspects

When evaluating the modified moments, numerical instability issues might occur. In this subsection we provide an approach to circumvent this problem. A simple observation that can speed up the computation of the modified moments is given afterwards. Some insights regarding additional restrictions related to `procedure 1` in the BEM context are stressed at the end.

The derived quadrature techniques (18) and (21) can be applied to evaluate integrals $I_{w_i^s}[g]$ also in case of regular integrals. Nevertheless, we experimentally observed that the computation of the modified moments in (19) exhibits numerical instability as the distance between s and D_i increases; similarly as it was observed for Legendre polynomial-based modified moments [2]. Furthermore, the instability increases at a fixed distance, when D_i gets smaller and when the spline degree p in $\mu_{k,p}^{(i)}(s)$ increases. Therefore, when $I_{w_i^s}[g]$ is regular, it is advisable to adopt regular-based quadrature rules (17) and (20).

When computing the analytical expression of a regular modified moment in [1, 7, 11] in finite arithmetic, a loss of significance in the computation might occur, since the operations contain addends of similar sizes but different signs. Besides using a higher precision arithmetic, the instability effect can be reduced by a tolerance switch: if D_i is far enough from the singularity s, the integrand function K_2 is a well-behaved function and the corresponding modified moments can be efficiently evaluated with a quadrature rule for regular integrals.

Singular integrals (16) consist of more involved computational steps than the regular integrals (15), mainly due to prerequisite computation of the modified moments. The amount of precomputed values can be greatly reduced if we consider a uniform mesh and all the shape functions $B_{i,d}^{(T)}$ are obtained by shifting one instance. In that setting the value $\mu_{k,p}^{(i)}(s)$ in (19) depends only on the relative position of s with respect to the B-spline factor:

- if $D_{i'} = D_i + s'$, then $\mu_{k,p}^{(i')}(s') = \mu_{k,p}^{(i)}(s)$,
- if $B_{k+k',p}^{(\tau)}(\cdot + s') = B_{k,p}^{(\tau)}(\cdot)$ for the two basis function constructed on D_i, then $\mu_{k+k',p}^{(i)}(s + s') = \mu_{k,p}^{(i)}(s)$.

Moreover, we do not need to completely compute new modified moments when switching to a finer mesh. If $B_{k',p}^{(\tau_h/\ell)}((t-\zeta)/\ell) = B_{k,p}^{(\tau_h)}(t)$ for some $\zeta \in \mathbb{R}$ and $\ell s + \zeta \in [a, b]$, and $D_{i'} = (D_i - \zeta)/\ell$, then the moment $\mu_{k',p}^{(i')}(s)$ for the finer mesh can be obtained from $\mu_{k,p}^{(i)}(\ell s + \zeta)$ on a coarser mesh by the expression

$$\mu_{k',p}^{(i')}(s) = \int_{D_{i'}} \log|s-t|\, B_{k',p}^{(\tau_h/\ell)}(t)\, dt$$

$$= \frac{1}{\ell}\int_{D_i} \log\left|\frac{\ell s + \zeta - t}{\ell}\right| B_{k',p}^{(\tau_h/\ell)}\left(\frac{t-\zeta}{\ell}\right) dt = \frac{1}{\ell}\,\mu_{k,p}^{(i)}(\ell s + \zeta) - \frac{\log \ell}{\ell}\,\frac{|\operatorname{supp} B_{k,p}^{(\tau_h)}|}{p+1}.$$

A similar approach can be applied when δ in (11) for closed curves is considered. In this case we are left to derive the expressions for the remaining two terms in the kernel

$$\log \delta(s,t) = \log\left(|s-t|\frac{|(s-t)^2-\gamma^2|}{\gamma^2}\right) = \log|s-t| + \log\frac{|(s-t)+\gamma|}{\gamma} + \log\frac{|(s-t)-\gamma|}{\gamma}.$$

If $B_{k',p}^{(\boldsymbol{\tau}_h/\ell)}((t-\zeta)/\ell) = B_{k,p}^{(\boldsymbol{\tau}_h)}(t)$ for some $\zeta \in \mathbb{R}$ and $\ell s \pm (\ell-1)\gamma + \zeta \in [a,b]$, and $D_{i'} = (D_i - \zeta)/\ell$, then

$$\begin{aligned}\int_{D_{i'}} \log\frac{|(s-t)\pm\gamma|}{\gamma}\, B_{k',p}^{(\boldsymbol{\tau}_h/\ell)}(t)\,dt &= \frac{1}{\ell}\int_{D_i} \log\frac{|(\ell s+\zeta-t)\pm\ell\gamma|}{\ell\gamma}\, B_{k',p}^{(\boldsymbol{\tau}_h/\ell)}\left(\frac{t-\zeta}{\ell}\right)dt \\ &= \frac{1}{\ell}\int_{D_i} \log\frac{|(\ell s\pm(\ell-1)\gamma+\zeta-t)\pm\gamma|}{\gamma}\, B_{k,p}^{(\boldsymbol{\tau}_h)}(t)\,dt \\ &\quad - \frac{\log\ell}{\ell}\,\frac{\left|\operatorname{supp} B_{k,p}^{(\boldsymbol{\tau}_h)}\right|}{p+1}.\end{aligned}$$

The asymptotic accuracy of a quadrature rule is affected by the accuracy of the approximated function g in $I_{B_i}[g]$ and $I_{w_i^s}[g]$. The quasi-interpolation operator in [18] is a projector, i.e., $\sigma_g = g$ for $g = B_{i,d}^{(\boldsymbol{T})}$, if $B_{i,d}^{(\boldsymbol{T})} \in \hat{S}_{\boldsymbol{\tau}}$, see [18] for details. The projector property implies $p \geq d$ and $\boldsymbol{T}$ to be a subset of $\boldsymbol{\tau}$ on D_i. Hence, $I_{B_i}[C]$ and $I_{w_i^s}[C]$ can be computed exactly by `procedure 1` for any constant C, when the projector operator is used. If the rule is exact only for constant functions, we can expect the asymptotic accuracy $O(h)$ for the QI operator, and the overall accuracy $O(h^2)$ for the corresponding quadrature scheme. More precisely, one additional order is obtained due to reduction of the integration domain in the h-refinement procedure.

To obtain higher approximation order for `procedure 1`, QI splines with higher degree should be employed. Furthermore the knot vector $\boldsymbol{\tau}$ should have multiple knots to satisfy $B_{i,d}^{(\boldsymbol{T})} \in \hat{S}_{\boldsymbol{\tau}}$ and a suitable generalization of the operator [18] should be developed. Another possible limitation of such quasi-interpolation operator is that the construction of the approximant involves also derivatives of g, which might not be available for all the considered integrals (e.g., in the outer integrals of A and in $\boldsymbol{\beta}$).

The derivative free quasi-interpolant variant [19] bypasses the latter limitation and it has the same approximation order as the former operator. However, the latter operator is not a projector and hence it is not applicable in `procedure 1` to obtain exact values of $I_{B_i}[C]$ and $I_{w_i^s}[C]$. On the other hand, the derivative free variant is a preferable choice for the `procedure 2`, where the projector is not needed since the quasi-interpolant is constructed only for g.

4 Accuracy of the Quadrature Rules for the Boundary Integrals

To measure the accuracy of the derived spline quasi-interpolation quadrature schemes `procedure 1` and `procedure 2`, from now on denoted by QI_1 and QI_2, we perform numerical tests for different types of integrals. The tests comprise of regular

(15) and singular integrals (16), that appear in the system matrix A and in the right-hand side vector $\boldsymbol{\beta}$. The quadratures are tested against some other quadrature rules, suitable for the boundary integrals.

The theoretical accuracy of QI_1 and QI_2 with respect to n is studied in [7]. In [11] the analysis on the convergence of QI_2 is studied, when h-refinement is performed, the amount of quadrature nodes is kept fixed and the regular part of the integrand is sufficiently smooth. We recall that the derived convergence order for QI_2 for regular and singular integrals is $O(h^{p+2})$ and $O(h^{p+2}|\log h|)$, respectively.

4.1 Perturbed System and Strang's Lemma

In this section we summarize results to estimate the needed asymptotic accuracy of the quadrature rules in BEM, when performing h-refinement on the discrete spaces, which are defined on a nested sequence of knot vectors $\boldsymbol{T}_h$. To keep the results concise, we derive only the relevant steps to obtain the error bounds for the matrix and the right-hand side entries. We refer to [25] for a more general analysis.

For a sufficiently smooth boundary and a sufficiently regular exact solution $\phi \in H^{d+1}(\Gamma)$ there exists a constant $C > 0$ such that the following error estimate on the approximate solution $\phi_h \in S_h$ holds,

$$\|\phi - \phi_h\|_{L^2(\Gamma)} \le C\, h^{d+1} \|\phi\|_{H^{d+1}(\Gamma)}, \tag{23}$$

with S_h defined in (8). For our purposes we assume that S_h is constructed on a uniformly spaced knot vector with simple knots and geometry map $\boldsymbol{F}$ is sufficiently regular with $0 < \boldsymbol{J}_{\min} \le \boldsymbol{J}(s) \le \boldsymbol{J}_{\max} < \infty$.

Mainly due to the truncation error of the quadrature rules, the Galerkin solution is usually not computed exactly and a notion of a perturbed Galerkin method needs to be introduced. Sufficiently accurate quadrature rules need to be applied at the discretization step size h, so that the computed solution $\tilde{\phi}_h$ maintains the optimal convergence order,

$$\|\phi - \tilde{\phi}_h\|_{L^2(\Gamma)} \le \tilde{C}\, h^{d+1} \|\phi\|_{H^{d+1}(\Gamma)}. \tag{24}$$

As a result of the applied quadrature rules, we denote by $\tilde{\mathcal{A}}_h$ and $\tilde{\mathcal{F}}_h$ the perturbed functionals of $\mathcal{A}$ and $\mathcal{F}$ in (7). Let us assume the perturbed Galerkin method to be stable, i.e., it satisfies the discrete inf-sup conditions. Then there exists $\gamma > 0$ and sufficiently small $h \le h_0$ such that

$$\gamma \le \inf_{\xi_h \in S_h \setminus \{0\}} \sup_{\psi_h \in S_h \setminus \{0\}} \frac{|\tilde{\mathcal{A}}_h(\xi_h, \psi_h)|}{\|\xi_h\|_{L^2(\Gamma)}\, \|\psi_h\|_{L^2(\Gamma)}},$$

$$\gamma \le \inf_{\psi_h \in S_h \setminus \{0\}} \sup_{\xi_h \in S_h \setminus \{0\}} \frac{|\tilde{\mathcal{A}}_h(\xi_h, \psi_h)|}{\|\xi_h\|_{L^2(\Gamma)}\, \|\psi_h\|_{L^2(\Gamma)}}.$$

The following estimate from Strang's first lemma (see [25, Sect. 4.2.4]) is obtained

$$\begin{aligned}\|\phi - \tilde{\phi}_h\|_{L^2(\Gamma)} &\le \|\phi - \phi_h\|_{L^2(\Gamma)} + \|\phi_h - \tilde{\phi}_h\|_{L^2(\Gamma)}\\ &\le \|\phi - \phi_h\|_{L^2(\Gamma)} + \gamma^{-1} \sup_{\psi_h \in S_h\setminus\{0\}} \frac{|\tilde{\mathcal{A}}_h(\phi_h - \tilde{\phi}_h, \psi_h)|}{\|\psi_h\|_{L^2(\Gamma)}}\\ &\le \|\phi - \phi_h\|_{L^2(\Gamma)} + \gamma^{-1} \Bigg(\sup_{\psi_h \in S_h\setminus\{0\}} \frac{|\mathcal{A}(\phi_h, \psi_h) - \tilde{\mathcal{A}}_h(\phi_h, \psi_h)|}{\|\psi_h\|_{L^2(\Gamma)}}\\ &\qquad + \sup_{\psi_h \in S_h\setminus\{0\}} \frac{|\mathcal{F}(\psi_h) - \tilde{\mathcal{F}}_h(\psi_h)|}{\|\psi_h\|_{L^2(\Gamma)}}\Bigg). \end{aligned} \tag{25}$$

The first term in the right-hand side of estimate (25) is bounded by (23). To satisfy (24) the remaining two consistency error terms in (25) should be sufficiently small. Thus the functionals $\tilde{\mathcal{A}}_h$ and $\tilde{\mathcal{F}}_h$ must be sufficiently good approximations of $\mathcal{A}$ and $\mathcal{F}$, respectively.

The estimate (25) helps us to bound the needed accuracy of the system matrix A_h and the right-hand side vector $\boldsymbol{\beta}_h$ at the discretization step size h. Let $\tilde{A}_{h,ij}$ and $\tilde{\beta}_{h,i}$ be the computed entries of $A_{h,ij}$ and $\beta_{h,i}$ defined in (12), (13) and (14). Then

$$\begin{aligned} A_{h,i,j} &:= \mathcal{A}\left(B^{(\boldsymbol{T})}_{j,d} \circ \mathbf{F}^{-1},\, B^{(\boldsymbol{T})}_{i,d} \circ \mathbf{F}^{-1}\right), & \beta_{h,i} &:= \mathcal{F}\left(B^{(\boldsymbol{T})}_{i,d} \circ \mathbf{F}^{-1}\right),\\ \tilde{A}_{h,i,j} &:= \tilde{\mathcal{A}}_h\left(B^{(\boldsymbol{T})}_{j,d} \circ \mathbf{F}^{-1},\, B^{(\boldsymbol{T})}_{i,d} \circ \mathbf{F}^{-1}\right), & \tilde{\beta}_{h,i} &:= \tilde{\mathcal{F}}_h\left(B^{(\boldsymbol{T})}_{i,d} \circ \mathbf{F}^{-1}\right). \end{aligned}$$

Here, we dropped the subscript h in the knot vector $\boldsymbol{T}$ to simplify the notation, as it is done also for the following two vectors $\boldsymbol{\alpha}, \boldsymbol{\zeta}$. Let ϕ_h and ψ_h be written in B-form in the basis of S_h,

$$\phi_h = \sum_{j=1}^{N} \alpha_j\, B^{(\mathbf{T})}_{j,d} \circ \mathbf{F}^{-1}, \quad \psi_h = \sum_{i=1}^{N} \zeta_i\, B^{(\mathbf{T})}_{i,d} \circ \mathbf{F}^{-1}.$$

Upper bounds for the vector norms $\|\boldsymbol{\alpha}\|_2 = \|(\alpha_j)_j\|_2$ and $\|\boldsymbol{\zeta}\|_2 = \|(\zeta_j)_j\|_2$ in terms of the function norms $\|\phi_h\|_{L^2(\Gamma)}$ and $\|\psi_h\|_{L^2(\Gamma)}$, respectively, are derived by the following steps. From the stability of the so-called p-norm B-splines (see [17, Sect. 9.3]) we obtain the following bounds for the standard B-splines in the second norm,

$$K^{-1}\, \|\boldsymbol{\alpha}\|_2 \le h^{-1/2} \left\| \sum_{j=1}^{N} \alpha_j\, B^{(\mathbf{T})}_{j,d} \right\|_{L^2([a,b])} \le \|\boldsymbol{\alpha}\|_2, \tag{26}$$

where $K > 0$ depends only on d. The estimate (26) helps us to bound $\|\boldsymbol{\alpha}\|_2$ with respect to $\|\phi_h\|_{L^2(\Gamma)}$ and $J_{\min}$,

$$
\begin{aligned}
K^{-1}\, J_{\min}^{1/2}\, \|\boldsymbol{\alpha}\|_2 &\le J_{\min}^{1/2}\, h^{-1/2} \left\| \sum_{j=1}^{N} \alpha_j\, B_{j,d}^{(\mathrm{T})} \right\|_{L^2([a,b])} \\
&\le h^{-1/2} \sqrt{\int_a^b \left(\sum_{j=1}^{N} \alpha_j\, B_{j,d}^{(\mathrm{T})}(s) \right)^2 J(s)\, ds} \\
&= h^{-1/2}\, \|\phi_h\|_{L^2(\Gamma)}.
\end{aligned} \tag{27}
$$

A similar estimates holds true for $\|\boldsymbol{\zeta}\|_2 = \|(\zeta_j)_j\|_2$ and $\|\psi_h\|_{L^2(\Gamma)}$.

By applying the bound (27) for $\|\boldsymbol{\alpha}\|_2$, and an analogous one for $\|\boldsymbol{\zeta}\|_2$, we finally obtain an estimate on the first consistency error term in (25),

$$
\begin{aligned}
\frac{|\mathcal{A}(\phi_h, \psi_h) - \tilde{\mathcal{A}}_h(\phi_h, \psi_h)|}{\|\psi_h\|_{L^2(\Gamma)}} &= \frac{|\boldsymbol{\zeta}^\top (A_h - \tilde{A}_h)\, \boldsymbol{\alpha}|}{\|\psi_h\|_{L^2(\Gamma)}} \\
&\le \frac{\|A_h - \tilde{A}_h\|_2\, \|\boldsymbol{\zeta}\|_2\, \|\boldsymbol{\alpha}\|_2}{\|\psi_h\|_{L^2(\Gamma)}} \\
&\le K^2\, J_{\min}^{-1}\, h^{-1}\, \|\phi_h\|_{L^2(\Gamma)}\, \|A_h - \tilde{A}_h\|_2 \\
&\le K^2\, J_{\min}^{-1}\, C'\, h^{-2}\, \|\phi_h\|_{L^2(\Gamma)}\, \max_{i,j} |A_{h,i,j} - \tilde{A}_{h,i,j}|.
\end{aligned}
$$

The last inequality is obtained by applying the matrix norm inequality between the induced second norm and the elementwise max norm, and the fact that there exists sufficiently big $C' > 0$ such that $N \le C'\, h^{-1}$. A similar estimate applies for the second consistency error term,

$$
\begin{aligned}
\frac{|\mathcal{F}(\psi_h) - \tilde{\mathcal{F}}_h(\psi_h)|}{\|\psi_h\|_{L^2(\Gamma)}} &= \frac{|\boldsymbol{\zeta}^\top (\boldsymbol{\beta}_h - \tilde{\boldsymbol{\beta}}_h)|}{\|\psi_h\|_{L^2(\Gamma)}} \\
&\le \frac{\|\boldsymbol{\beta}_h - \tilde{\boldsymbol{\beta}}_h\|_2\, \|\boldsymbol{\zeta}\|_2}{\|\psi_h\|_{L^2(\Gamma)}} \\
&\le K\, J_{\min}^{-1/2}\, h^{-1/2}\, \|\boldsymbol{\beta}_h - \tilde{\boldsymbol{\beta}}_h\|_2 \\
&\le K\, J_{\min}^{-1/2}\, C'^{1/2}\, h^{-1}\, \max_i |\beta_{h,i} - \tilde{\beta}_{h,i}|.
\end{aligned}
$$

The derived estimates and $\|\phi_h\|_{L^2(\Gamma)} \le \|\phi_h\|_{H^{d+1}(\Gamma)}$ give us the following bounds on the accuracy of $\tilde{A}_h$ and $\tilde{\boldsymbol{\beta}}_h$,

$$
\max_{i,j} |A_{h,i,j} - \tilde{A}_{h,i,j}| \le \tilde{C}\, h^{d+3}, \quad \max_i |\beta_{h,i} - \tilde{\beta}_{h,i}| \le \tilde{C}\, h^{d+2}, \tag{28}
$$

which imply the optimal convergence estimate (24) of the perturbed solution $\tilde{\phi}_h$.

4.2 Numerical Tests

In this subsection we test QI_1 and QI_2 and compare them with other quadrature rules. For each of the rules the tests are repeated twice, with lower and higher amount of quadrature nodes; let the amount of quadrature nodes on the support of a univariate B-spline for all the considered integration schemes be denoted by n_s (clearly $n_s = n + 1$ for QI_1 and QI_2). In all cases we consider uniform meshes with knot vectors $\mathbf{T}_h$. The spline degree in S_h is fixed to $d = 2$ and $p = 2, 3$ for the quasi-interpolation spaces. The exact integral values are obtained using the integration solver in Wolfram Mathematica.

4.2.1 Regular Integrals I_{B_i}

In the first test we employ the quadratures on a regular integral $I_{B_i}[g]$, defined in (15). This type of integrals appears, for example, in the right-hand side $\boldsymbol{\beta}$. Let $g(t) = 3\sin(\pi(t+1))\cos(t+1)$. In the test we also include a recently developed B-spline weighted quadrature rule (BWR) [1, 8], where the integrand B-spline is thought as a *weight* function. The exactness for these rules is imposed on the chosen test spline space or on a refinement of it. The weights of the quadrature rule are computed solving a local band system and the quadrature nodes are chosen a priori such that the Schoenberg-Whitney's conditions hold. In the test we also include the very common Gauss-Legendre quadrature rule (StdG) [10], which is usually considered as the optimal possible for polynomials for the given amount of nodes. The main drawback of this rule is that with an increasing d it is not effective anymore for B-splines since the required n_s increases quadratically with d and not linearly.

For every $h = 1/5,\ 1/10,\ 1/20,\ 1/40$ of the uniformly spaced meshes we compute the maximum error of a quadrature scheme,

$$\max_i \left| I_{B_i}[g] - \tilde{I}_{B_i}[g] \right|,$$

where $\tilde{I}_{B_i}[g]$ is the value of a numerically computed integral. Figure 1 reveals the optimal convergence order $O(h^4)$ for BWR, while for QI_2 we observe super convergence $O(h^5)$. Both of the schemes provide the optimal accuracy in the context of the perturbed Galerkin method (see (28) in Sect. 4.1). The quadrature QI_1 is steadily converging but with a reduced order $O(h^2)$, as discussed in Sect. 3.2. As expected, the accuracy of all the rules is improved if we increase the amount of quadrature nodes (Fig. 1(b)). For QI_2 we can easily increase the quasi-interpolation spline degree to $p = 3$; this increases the accuracy of the rule (see Sect. 4 and [11] for details). Using a low number of nodes $n_s = 6$ for StdG does not output the optimal convergence order. In a more common setting with more nodes, $n_s = 9 = (d+1)^2$, the rule gives the optimal order and the lowest error among the considered methods.

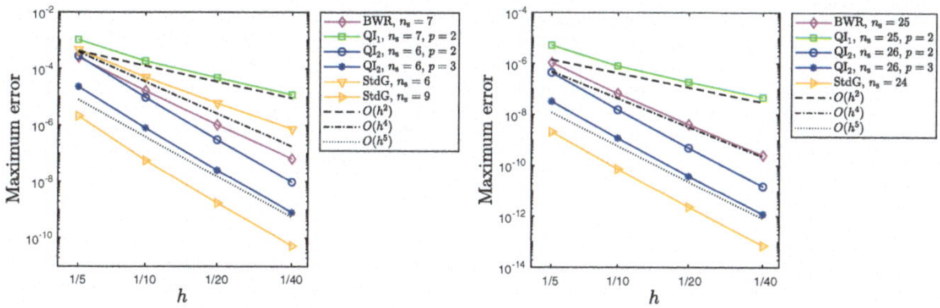

(a) Lower amount of nodes n_s (b) Higher amount of nodes n_s

Fig. 1 Error convergence plots with respect to the mesh size h for a type of integral $I_{B_i}[g]$ in Sect. 4.2.1

4.2.2 Regular and Singular Integrals $I_{w_i^s}$

In the second example we measure the accuracy of the integrals that include a term $K_2(s, t) = \log |s - t|$. The integrals are regular and singular integrals $I_{w^s_{h,i}}$, with the factor $w^s_{h.i} := K_2(s, \cdot)\, B^{(\mathbf{T}_h)}_{i,d}(\cdot)$, that appear as the inner integrals in $A^{(2)}_{i,j}$. For the test case we consider $g(t) = \sqrt{1 + 4t^2}$. The quadratic B-splines $B^{(\boldsymbol{T}_h)}_{i,d}$ are constructed on the interval $[-1, 1]$ with uniform open knot vectors $\mathbf{T}_h$ for the following mesh sizes $h = 1/5, 1/10, 1/20, 1/40$. The parameter s is restricted to a priori chosen discrete values. Specifically, $s \in \boldsymbol{T}_{h/2}$, i.e., s takes the values of all knots in the knot vector $\mathbf{T}_h$ and all the knot midpoints.

For every mesh size h we measure the maximum error

$$\max_{i,s} |I_{w^s_{h,i}} - \tilde{I}_{w^s_{h,i}}|,$$

where the computed value of the integral obtained by a quadrature rule is denoted by $\tilde{I}_{w^s_{h,i}}$.

In the test we also include other suitable quadrature rules available in the literature. The hybrid Gauss-trapezoidal quadrature rules, sometimes called just Alpert rules (Alpert), are a class of quadratures for regular and singular functions, that comprise of special quadrature nodes and weights near the (regular or singular) edges of the integration domain [5]. The construction exploits a generalization of the Euler-Maclaurin summation formula. A common technique to accurately evaluate weakly singular integrals in BEM is the Telles transformation (Telles), which consists of applying a cubic polynomial coordinate transformation to smooth out the singularity and then applying the standard Gaussian quadrature rule to evaluate the regularised integrals [29]. The singular weighted rule (SWR) [1] is based on the precomputed modified moments, like QI_1 and QI_2, but the construction of the weights requires to solve a global linear system. Finally, we consider a weighted Gaussian rule (WG) for polynomials [10] with precomputed optimal weights and nodes for different mesh sizes and parameters s. Due to high complexity to construct the rule, we employ the

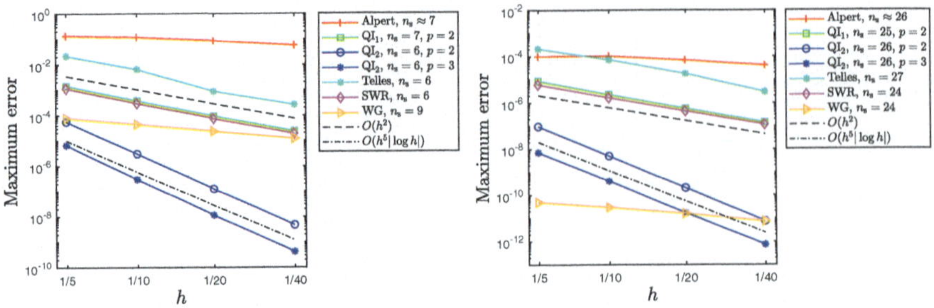

(a) Lower amount of nodes n_s (b) Higher amount of nodes n_s

Fig. 2 Error convergence plots with respect to the mesh size h for the regular and singular inner integrals $I_{w_i^s}$ in Sect. 4.2.2 for various quadrature rules available in literature

weighted rule with the weight function $\rho(\cdot) = -\log|s - \cdot|$ only for s and cells Q of supp $B_{i,d}^{(T_h)}$ that satisfy $s \in \overline{Q}$ (the rule is computed by a symbolic/high precision toolbox in Mathematica, see the implementation in [15]). When $s \notin \overline{Q}$, we employ the standard Gauss-Legendre rule ($\rho \equiv 1$).

By comparing the convergence error plots in Fig. 2(a) and (b) for a fixed h we can observe that all quadrature rules converge with the increased amount of quadrature nodes. Convergence with respect to h reveals that the procedure QI_2 is the only scheme with the optimal convergence order, $O(h^5|\log h|)$, and increasing the QI degree to $p = 3$ again naturally improves the accuracy of the rule. The procedures QI_1 and SWR behave very similarly with the suboptimal convergence order $O(h^2)$; the order can be justified by the observation that both of the rules can guarantee the exactness only if locally $J \equiv$ const. The suboptimal convergence order of WG can be attributed to the limitation of the standard Gaussian rule to properly handle nearly singular integrals. The accuracy of the rule could be improved by taking a larger domain to employ the logarithmic weighted rule but with a significant extra cost to precompute rules for additional splines $B_{i,d}^{(T_h)}$ and parameters s by solving ill-conditioned algebraic systems.

4.2.3 Outer Integrals in Matrix $A^{(2)}$

In the last test we focus on regular integrals of the type

$$A_{h,i,j}^{(2)} := -\frac{1}{2\pi} \int_{D_i} B_{i,d}^{(\mathbf{T}_h)}(s)\, I_{w_{h,i}^s}\, ds, \tag{29}$$

that appear as the outer integrals in (12) for $J \equiv 1$. The latter choice allows us to exactly evaluate the term $I_{w_{h,i}^s}$ in (29) by computing the corresponding modified moments (19). Therefore, the error of the numerical integration to compute $A_{h,i,j}^{(2)}$ is

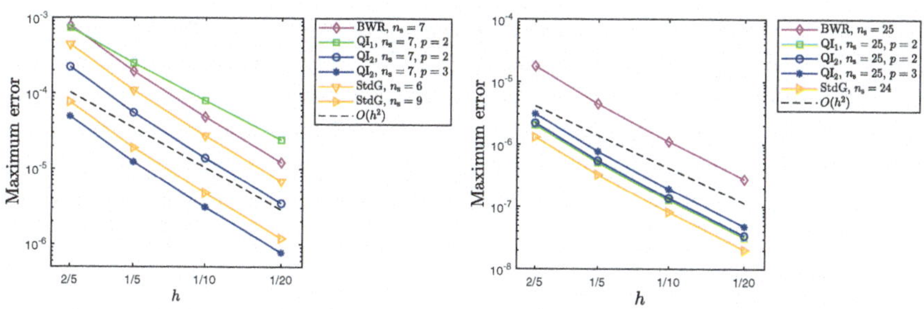

(a) Lower amount of nodes n_s (b) Higher amount of nodes n_s

Fig. 3 Error convergence plots with respect to the mesh size h for the outer integrals in $A^{(2)}$, defined in Sect. 4.2.3

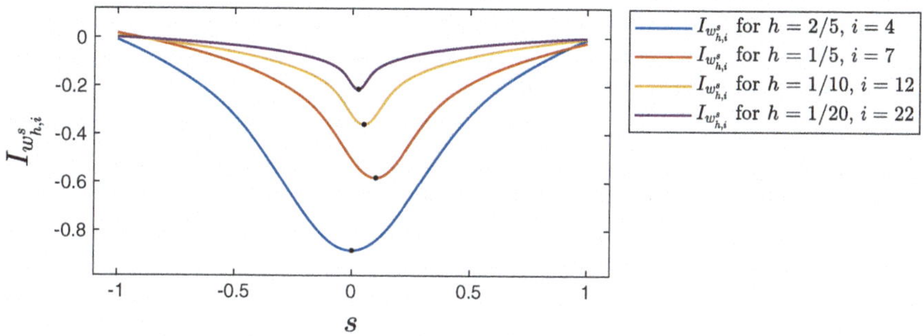

Fig. 4 Function $I_{w^s_{h,i}}$ for different h and i, from the test in Sect. 4.2.3

contributed solely by the quadrature to compute the outer integral (29). On interval $[-1, 1]$ we construct uniform meshes with $h = 2/5, 1/5, 1/10, 1/20$.

In this example we also include the other two previously tested rules for regular integrals: the B-spline weighted quadrature rule (BWR) and the standard Gauss-Legendre rule (StdG).

For each mesh size h we measure the maximum error of integrals (29) for every i and j,

$$\max_{i,j} \left| A^{(2)}_{h,i,j} - \tilde{A}^{(2)}_{h,i,j} \right|,$$

where $\tilde{A}^{(2)}_h(i, j)$ is the computed integral $A^{(2)}_h(i, j)$ with a quadrature rule. Surprisingly, a suboptimal convergence rate is obtained for all the quadratures, even though the function $I_{w^s_{h,i}}$ is well-defined and smooth. From Fig. 3 we can observe the convergence order that is slightly higher than $O(h^2)$. Again, the accuracy of the integral approximations is naturally improved, when n_s is increased. This can be observed by comparing the plot Fig. 3(a) for the lower and Fig. 3(b) for the higher amount of nodes for a fixed h.

To clarify the suboptimal convergence in the test, we plot the function $I_{w^s_{h,i}}$ as a function of s in Fig. 4 for different fixed h and i. For every mesh we plot the function only for the corresponding most central spline element, namely $i = 4, 7, 12, 22$. As we can see from the plot, the function $I_{w^s_{h,i}}$ is a smooth function of s but the highest curvature actually increases with smaller h (depicted as dots in the figure). Since the derivatives of $I_{w^s_{h,i}}$ are not bounded when $h \to 0$, none of the considered quadrature schemes can efficiently approximate this type of integrals for a fixed amount of nodes.

5 Numerical Simulation with (Galerkin) BEM

In this section we test the boundary element model to numerically solve three Laplace boundary value problems. For all the examples we evaluate the governing integrals using the presented QI_1 and QI_2 quadrature schemes and two of the presented alternatives: Alpert rules (Alpert), which is combined with the standard Gaussian rule for the outer integrals of a system matrix, and the weighted quadrature rules (WR), which comprises of the B-spline weighted rule for the outer integrals and the singular weighted rule for the inner integrals. The amount of quadrature nodes on the support of a univariate B-spline is denoted by n_{s}. We note that the distribution of the nodes for Alpert depends on the proximity to a singularity, hence n_{s} is not strictly the same for every spline. In all cases we construct several successive approximate solutions of the problem by performing a dyadic h-refinement procedure on uniform meshes. We measure the relative error of an approximated solution against the exact one in L^2 norm with respect to degrees of freedom (DoF). The first numerical example is an exterior problem to an open curve. In the next two examples we employ the direct formulation to model interior problems to closed curves.

5.1 *Exterior Dirichlet Problem to Arc of Parabola*

In this test we focus on the exterior problem described in [1], using the BIE (9). The Dirichlet BVP is defined in the exterior to an arc of parabola $\Gamma \in \mathbb{R}^2$, parametrized by a quadratic B-spline curve ($d = 2$). The transformation map $\mathbf{F}(s) = [s, 1 - s^2]^\top$, $s \in [-1, 1]$, is described in terms of B-spline basis with the following knot vector $\mathbf{T}$ and set of control points D,

$$\mathbf{T} = (-1, -1, -1, 1, 1, 1), \quad D = \begin{bmatrix} -1 & 0 & 1 \\ 0 & 2 & 0 \end{bmatrix}.$$

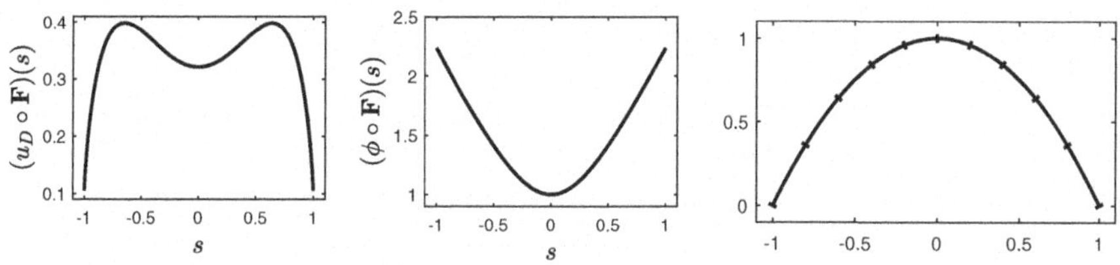

(a) Dirichlet datum (b) Exact solution (c) Initial mesh

Fig. 5 Parabola test in Sect. 5.1: The Dirichlet datum, the exact solution and the initial mesh

The Dirichlet datum u_D and the exact solution ϕ are

$$(u_D \circ \mathbf{F})(s)$$
$$= \frac{-(7-9s+4s^3)\log(2+2s+s^2)-(7+9s-4s^3)\log(2-2s+s^2)}{12\pi} + \frac{14+24s^2}{9\pi}$$
$$+ \frac{-(7+3s+4s^3)\log(1+s)-(7-3s-4s^3)\log(1-s)-(-1+12s^2)\arctan\left(\frac{2}{s^2}\right)}{6\pi},$$
$$(\phi \circ \mathbf{F})(s) = \sqrt{1+4s^2}.$$

The Dirichlet datum, the exact solution, and the initial mesh in the physical domain are depicted in Fig. 5. Observe that, although the function $u_D \circ \mathbf{F}$ is well defined for $s \in (-1, 1)$, its derivative is not bounded when $s \to \pm 1$ and can represent an additional limitation for the quadrature QI_1, which needs also the derivative information of the integrand.

The convergence orders of the approximate solutions are shown in Table 1 for the quasi-interpolant degree $p = 2$. Procedure QI_1 has a reduced accuracy near both the edges of the parametric domain. To get the optimal convergence $O(h^{d+1})$ for all refinement steps, a relative high amount of quadrature nodes is needed, $n_s = 49$. On the other hand, it is sufficient to use a small amount of quadrature nodes for the other schemes, $n_s = 5$ for the procedure QI_2, approximately $n_s = 6$ for Alpert and $n_s = 7$ for WR.

5.2 *Interior Dirichlet Problem to a Circle*

To verify the correctness of the model using the direct formulation (10), we consider a simple boundary value problem. The domain Ω is a disk with radius $1/2$ and boundary Γ parametrized by the map $\mathbf{F}(s) = 1/2\,[\cos(\pi s),\ \sin(\pi s)]^\top$ for $s \in [-1, 1]$.

For the chosen exact solution $u = x, x \in \Omega$, of the Laplacian, the Dirichlet datum is $(u_D \circ \mathbf{F})(s) = 1/2\cos(\pi s)$ and the exact flux is $(\phi \circ \mathbf{F})(s) = \cos(\pi s)$.

Table 1 Parabola test in Sect. 5.1: Errors and convergence orders of the approximated solutions for different quadrature schemes

	$QI_1, n_s = 49$		$QI_2, n_s = 5$		Alpert, $n_s \approx 6$		WR, $n_s = 7$	
DoF	Error	Conv.	Error	Conv.	Error	Conv.	Error	Conv.
12	$1.55 \cdot 10^{-4}$		$1.73 \cdot 10^{-4}$		$1.54 \cdot 10^{-4}$		$1.56 \cdot 10^{-4}$	
22	$1.66 \cdot 10^{-5}$	3.69	$1.74 \cdot 10^{-5}$	3.79	$1.66 \cdot 10^{-5}$	3.68	$1.66 \cdot 10^{-5}$	3.70
42	$2.00 \cdot 10^{-6}$	3.27	$2.03 \cdot 10^{-6}$	3.33	$2.01 \cdot 10^{-6}$	3.26	$1.99 \cdot 10^{-6}$	3.27
82	$2.44 \cdot 10^{-7}$	3.14	$2.48 \cdot 10^{-7}$	3.14	$2.53 \cdot 10^{-7}$	3.10	$2.47 \cdot 10^{-7}$	3.12
162	$3.06 \cdot 10^{-8}$	3.05	$3.08 \cdot 10^{-8}$	3.06	$3.24 \cdot 10^{-8}$	3.02	$3.08 \cdot 10^{-8}$	3.06
322	$3.95 \cdot 10^{-9}$	2.98	$3.85 \cdot 10^{-9}$	3.03	$4.37 \cdot 10^{-9}$	2.92	$3.85 \cdot 10^{-9}$	3.03

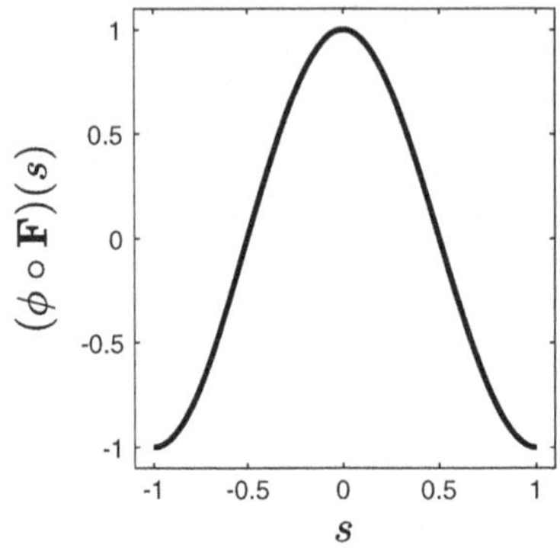

(a) Exact solution

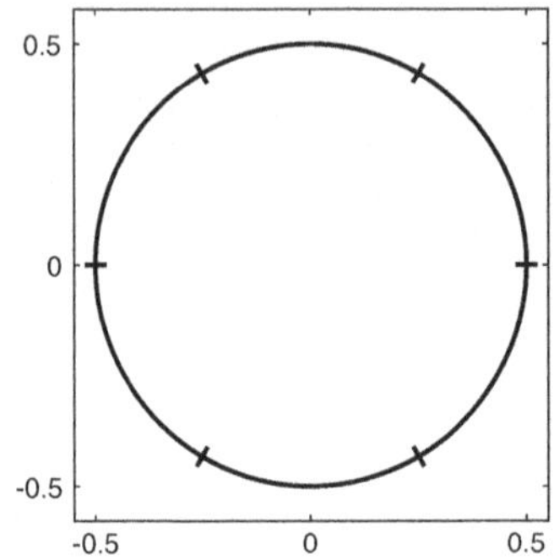

(b) Initial mesh

Fig. 6 Circle test in Sect. 5.2: The exact solution and the initial mesh

The approximate solution is sought in the space of cubic B-splines ($d = 3$) with the initial equally spaced extended knot vector

$$\mathbf{T} = (-2, -5/3, -4/3, \ldots, 4/3, 5/3, 2).$$

The exact solution and the initial mesh in the physical domain are depicted in Fig. 6.

In Table 2 we report errors and convergence rates for the approximate solutions for different amount of DoF. For QI_1 we need to take $p = d$ to satisfy the projector property (see Sect. 3.2), whereas for QI_2 it is enough to consider $p = 2$. To obtain the optimal convergence order 4 for all the refinement steps, we need to considerably increase the amount of quadrature nodes for quadrature QI_1 ($n_s = 25$), while for QI_2 we can maintain a small amount of nodes ($n_s = 5$). Adequate results are provided also by WR ($n_s = 13$), while for Alpert the amount of nodes for the singular rule needs to be relatively high ($n_s \approx 52$).

Table 2 Circle test in Sect. 5.2: Errors and convergence orders of the approximated solutions using different quadrature schemes

	QI_1, $n_s = 25$		QI_2, $n_s = 5$		Alpert, $n_s \approx 52$		WR, $n_s = 13$	
DoF	Error	Conv.	Error	Conv.	Error	Conv.	Error	Conv.
6	$1.66 \cdot 10^{-3}$		$1.73 \cdot 10^{-3}$		$1.66 \cdot 10^{-3}$		$1.66 \cdot 10^{-3}$	
12	$7.69 \cdot 10^{-5}$	4.43	$8.26 \cdot 10^{-5}$	4.39	$7.69 \cdot 10^{-5}$	4.43	$7.69 \cdot 10^{-5}$	4.43
24	$4.40 \cdot 10^{-6}$	4.13	$4.67 \cdot 10^{-6}$	4.14	$4.40 \cdot 10^{-6}$	4.13	$4.40 \cdot 10^{-6}$	4.13
48	$2.69 \cdot 10^{-7}$	4.03	$2.78 \cdot 10^{-7}$	4.07	$2.69 \cdot 10^{-7}$	4.03	$2.69 \cdot 10^{-7}$	4.03
96	$1.67 \cdot 10^{-8}$	4.01	$1.70 \cdot 10^{-8}$	4.03	$1.67 \cdot 10^{-8}$	4.01	$1.67 \cdot 10^{-8}$	4.01
192	$1.05 \cdot 10^{-9}$	4.00	$1.05 \cdot 10^{-9}$	4.02	$1.09 \cdot 10^{-9}$	3.94	$1.05 \cdot 10^{-9}$	4.00

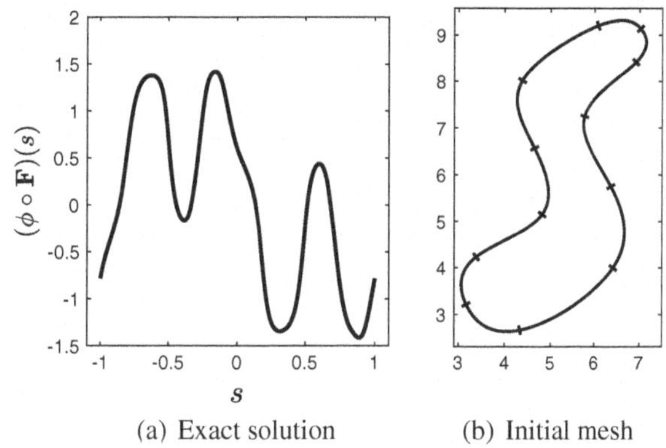

Fig. 7 S curve test in Sect. 5.3: The exact solution and the initial mesh

5.3 *Interior Dirichlet Problem to S Curve*

In the last numerical example we consider a problem with a more involved geometry, a domain described by the closed S curve [1]. The curve is parametrized by cubic B-splines with the knot vector $\mathbf{T}$ and set of control points D,

$$\mathbf{T} = (-9/6, -9/6, -9/6, -9/6, -8/6, -7/6, \ldots, 7/6, 8/6, 9/6, 9/6, 9/6, 9/6),$$

$$D = \begin{bmatrix} 3 & 4 & 7 & 6.5 & 5.2 & 7.3 & 7.1 & 6.4 & 3.8 & 4.7 & 5.3 & 3 & 3 & 4 & 7 \\ 3.2 & 2.2 & 4 & 5.8 & 7.3 & 8.5 & 9.2 & 9.5 & 8 & 6.6 & 5 & 4.3 & 3.2 & 2.2 & 4 \end{bmatrix}.$$

The boundary Dirichlet datum is set to $(u_D \circ \mathbf{F})(s) = F_1(s) + F_2(s)$ and the exact solution reads $(\phi \circ \mathbf{F})(s) = (-F_1'(s) + F_2'(s))/\|\mathbf{F}'(s)\|_2$. The exact solution and the initial mesh in the physical domain are depicted in Fig. 7.

In Table 3 we report errors for the approximate solutions when C^2 cubic test functions ($d = 3$) are used. Again, for QI_1 we set $p = 3$ and $p = 2$ for QI_2. Since the exact solution $\phi \circ \mathbf{F}$ is only C^1 regular, a reduced order of convergence for the approximate solution is expected. This is confirmed by our experiments where the

Table 3 S curve test in Sect. 5.3: Error and convergence orders for C^2 cubic test functions using different quadrature schemes

	$QI_1, n_s = 25$		$QI_2, n_s = 7$		Alpert, $n_s \approx 16$		WR, $n_s = 17$	
DoF	Error	Conv.	Error	Conv.	Error	Conv.	Error	Conv.
12	$1.20 \cdot 10^{-1}$		$1.30 \cdot 10^{-1}$		$1.20 \cdot 10^{-1}$		$1.21 \cdot 10^{-1}$	
24	$3.26 \cdot 10^{-2}$	1.88	$3.50 \cdot 10^{-2}$	1.89	$3.26 \cdot 10^{-2}$	1.88	$3.27 \cdot 10^{-2}$	1.89
48	$4.46 \cdot 10^{-3}$	2.87	$4.52 \cdot 10^{-3}$	2.95	$4.46 \cdot 10^{-3}$	2.87	$4.47 \cdot 10^{-3}$	2.87
96	$6.30 \cdot 10^{-4}$	2.82	$6.46 \cdot 10^{-4}$	2.81	$6.30 \cdot 10^{-4}$	2.82	$6.40 \cdot 10^{-4}$	2.80
192	$1.05 \cdot 10^{-4}$	2.59	$1.07 \cdot 10^{-4}$	2.59	$1.05 \cdot 10^{-4}$	2.59	$1.05 \cdot 10^{-4}$	2.61
384	$1.82 \cdot 10^{-5}$	2.52	$1.86 \cdot 10^{-5}$	2.53	$1.82 \cdot 10^{-5}$	2.52	$1.84 \cdot 10^{-5}$	2.51

Table 4 S curve test in Sect. 5.3: Error and convergence orders for C^1 quadratic test functions (left) and for cubic test functions with C^1 smoothness at the initial knots, and C^2 regularity at the inserted knots (right)

	$QI_1, n_s = 13$		$QI_2, n_s = 7$	
DoF	Error	Conv.	Error	Conv.
12	$1.24 \cdot 10^{-1}$		$1.26 \cdot 10^{-1}$	
24	$2.76 \cdot 10^{-2}$	2.17	$2.79 \cdot 10^{-2}$	2.17
48	$2.96 \cdot 10^{-3}$	3.22	$2.98 \cdot 10^{-3}$	3.23
96	$2.55 \cdot 10^{-4}$	3.53	$2.58 \cdot 10^{-4}$	3.53
192	$2.50 \cdot 10^{-5}$	3.35	$2.53 \cdot 10^{-5}$	3.35
384	$2.92 \cdot 10^{-6}$	3.10	$2.84 \cdot 10^{-6}$	3.16

	$QI_2, n_s = 7$		$QI_2, n_s = 13$	
DoF	Error	Conv.	Error	Conv.
24	$2.92 \cdot 10^{-2}$		$2.65 \cdot 10^{-2}$	
36	$7.12 \cdot 10^{-3}$	3.48	$6.67 \cdot 10^{-3}$	3.40
60	$7.28 \cdot 10^{-4}$	4.46	$6.69 \cdot 10^{-4}$	4.50
108	$4.64 \cdot 10^{-5}$	4.68	$3.08 \cdot 10^{-5}$	5.23
204	$5.51 \cdot 10^{-6}$	3.35	$1.73 \cdot 10^{-6}$	4.53
396	$8.99 \cdot 10^{-7}$	2.73	$1.31 \cdot 10^{-7}$	3.88

average convergence order drops to approximately 2.5. Only QI_2 scheme outputs accurate enough results with a relative small amount of quadrature nodes ($n_s = 7$).

In Table 4 (left) we can observe an improved convergence order 3 if we employ C^1 quadratic test functions ($d = 2$) since the test functions have the same regularity as the exact solution. Here $p = 2$ for both the quadrature rules. Of course, in this setting the space to describe the geometry is not a subspace of the test space S_h.

To conclude this test, we consider also a case with cubic test functions ($d = 3$) that are C^1 smooth on the initial mesh by using double knots. More precisely, the discretization space S_h consists of basis functions that are C^1 regular at the initial knots, and C^2 continuous at the inserted knots, obtained by dyadic refinements. In this setting, the space to describe the geometry is a subspace of S_h for every step size h. Note that, in this case for the periodic compatibility $\rho = d - m + 1 = 2$ (see Sect. 2.2). The results for the errors and convergence orders are reported in Table 4 (right) for QI_2 with $p = 2$, $n_s = 7$ and $n_s = 13$. A higher amount of quadrature nodes is necessary to recover the optimal order 4 for the approximate solution. The quadrature scheme QI_1 is not considered in this test since every basis function $B_{i,d}^{(T)} \in S_h$ should belong to the quasi-interpolation space $\hat{S}_\tau$ and the involved quasi-interpolation operator cannot handle multiple knots.

6 Conclusion

A study of two recently introduced spline quasi-interpolation quadrature schemes is performed in the context of boundary integral equations in Galerkin IgA-BEM. A comparison of the accuracy of the schemes for singular integrals was already done in [7]. The analysis with respect to the amount of employed quadrature nodes revealed the optimal order of convergence for both approaches.

In the present paper, numerical tests show a notable difference between the two schemes. For a fixed amount of quadrature nodes the accuracy of the considered integrals is examined, when performing h-refinement of the approximation space. The observed rate of convergence is optimal only for the second scheme. Among all the tested integration methods in the numerical simulations for the 2D Laplace problems, the second QI-based procedure utilizes the least amount of quadrature nodes. For example, from the obtained tests we can infer that for B-spline shape functions of odd degree and maximum regularity, suitable nodes can be chosen as the breakpoints of the knot vector. Regarding the first procedure, the amount of locally employed nodes should be increased throughout the h-refinement steps to recover the optimal order.

In future work, we would like to investigate quadrature schemes for integrals of higher order singularities for more complex differential problems and compare them with other state-of-the-art approaches for BEM. Moreover, a proper treatment of the outer integrals of the system matrix would contribute to better efficiency of the overall scheme. Finally, a valuable contribution would be to derive stable formulae for the modified moments to simplify the construction of the proposed methods.

Acknowledgements This work was partially supported by the MIUR "Futuro in Ricerca" programme through the project DREAMS (RBFR13FBI3). The authors are members of the INdAM Research group GNCS. The INdAM support through GNCS and Finanziamenti Premiali SUNRISE is gratefully acknowledged.

References

1. Aimi, A., Calabrò, F., Diligenti, M., Sampoli, M.L., Sangalli, G., Sestini, A.: Efficient assembly based on B-spline tailored quadrature rules for the IgA-SGBEM. Comput. Methods Appl. Mech. Eng. **331**, 327–342 (2018)
2. Aimi, A., Diligenti, M.: Numerical integration in 3D galerkin BEM solution of HBIEs. Comput. Mech. **28**, 233–249 (2002)
3. Aimi, A., Diligenti, M., Sampoli, M.L., Sestini, A.: Isogeometric analysis and symmetric Galerkin BEM: A 2D numerical study. Appl. Math. Comput. **272**, 173–186 (2016)
4. Aimi, A., Diligenti, M., Sampoli, M.L., Sestini, A.: Non-polynomial spline alternatives in isogeometric symmetric Galerkin BEM. Appl. Numer. Math. **116**, 10–23 (2017)
5. Alpert, B.K.: Hybrid Gauss-trapezoidal quadrature rules. SIAM J. Sci. Comput. **20**, 1551–1584 (1999)
6. de Boor, C.: A Practical Guide to Splines. Applied Mathematical Sciences, vol. 27, revised edn. Springer, New York (2001)

7. Calabrò, F., Falini, A., Sampoli, M.L., Sestini, A.: Efficient quadrature rules based on spline quasi-interpolation for application to IgA-BEMs. J. Comput. Appl. Math. **338**, 153–167 (2018)
8. Calabrò, F., Sangalli, G., Tani, M.: Fast formation of isogeometric Galerkin matrices by weighted quadrature. Comput. Methods Appl. Mech. Eng. **316**, 606–622 (2017)
9. Costabel, M.: Principles of Boundary Element Methods. Techn. Hochsch, Fachbereich Mathematik (1986)
10. Davis, P.J., Rabinowitz, P.: Methods of Numerical Integration. Computer Science and Applied Mathematics. Academic Press, New York (1984)
11. Falini, A., Giannelli, C., Kanduč, T., Sampoli, M.L., Sestini, A.: An adaptive IGA-BEM with hierarchical B-splines based on quasi-interpolation quadrature schemes. Int. J. Numer. Meth. Eng. **117**, 1038–1058 (2019)
12. Farin, G., Hansford, D.: The Essentials of CAGD. A. K. Peters/CRC Press, New York, USA (2000)
13. Feischl, M., Gantner, G., Haberl, A., Praetorius, D.: Adaptive 2D IGA boundary element methods. Eng. Anal. Bound. Elem. **62**, 141–153 (2016)
14. Feischl, M., Gantner, G., Praetorius, D.: Reliable and efficient a posteriori error estimation for adaptive IGA boundary element methods for weakly-singular integral equations. Comput. Methods Appl. Mech. Eng. **290**, 362–386 (2015)
15. Fukuda, H., Katuya, M., Alt, E.O., Matveenko, A.V.: Gaussian quadrature rule for arbitrary weight function and interval. Comput. Phys. Commun. **167**, 143–150 (2005)
16. Heltai, L., Arroyo, M., DeSimone, A.: Nonsingular isogeometric boundary element method for stokes flows in 3D. Comput. Methods Appl. Mech. Eng. **268**, 514–539 (2014)
17. Lyche, T., Mørken, K.: Spline Methods Draft. University of Oslo (2008)
18. Mazzia, F., Sestini, A.: The BS class of Hermite spline quasi-interpolants on nonuniform knot distributions. BIT Numer. Math. **49**, 611–628 (2009)
19. Mazzia, F., Sestini, A.: Quadrature formulas descending from BS Hermite spline quasi-interpolation. J. Comput. Appl. Math. **236**, 4105–4118 (2012)
20. Mørken, K.: Some identities for products and degree raising of splines. Constr. Approx. **7**, 195–208 (1991)
21. Nguyen, B.H., Tran, H.D., Anitescu, C., Zhuang, X., Rabczuk, T.: An isogeometric symmetric Galerkin boundary element method for two-dimensional crack problems. Comput. Methods Appl. Mech. Eng. **306**, 252–275 (2016)
22. Peng, X., Atroshchenko, E., Kerfriden, P., Bordas, S.P.A.: Isogeometric boundary element methods for three dimensional static fracture and fatigue crack growth. Comput. Methods Appl. Mech. Eng. **316**, 151–185 (2017)
23. Sablonnière, P.: A quadrature formula associated with a univariate spline quasi interpolant. BIT Numer. Math. **47**, 825–837 (2007)
24. Sablonnière, P., Sbibih, D., Taharichi, M.: Error estimate and extrapolation of a quadrature formula derived from a quartic spline quasi-interpolant. BIT Numer. Math. **50**, 843–862 (2010)
25. Sauter, S.A., Schwab, C.: Boundary Element Methods. Springer Series in Computational Mathematics, vol. 39. Springer, Berlin, Heidelberg (2011)
26. Schumaker, L.L.: Spline Functions: Basic Theory, 3rd edn. Cambridge University Press, Cambridge (2007)
27. Simpson, R.N., Bordas, S.P.A., Trevelyan, J., Rabczuk, T.: A two-dimensional isogeometric boundary element method for elastostatic analysis. Comput. Methods Appl. Mech. Eng. **209–212**, 87–100 (2012)
28. Taus, M., Rodin, G.J., Hughes, T.J.R.: Isogeometric analysis of boundary integral equations: High-order collocation methods for the singular and hyper-singular equations. Math. Model. Methods Appl. Sci. **26**, 1447–1480 (2016)

29. Telles, J.C.F.: A self-adaptive co-ordinate transformation for efficient numerical evaluation of general boundary element integrals. Int. J. Numer. Meth. Eng. **24**, 959–973 (1987)
30. Wendland, W.L.: On some mathematical aspects of boundary element methods for elliptic problems. In: Whiteman, J.R. (ed.) The Mathematics of Finite Elements and Applications, MAFELAP 1984, vol. 5, pp. 193–227. Academic Press Ltd., London (1985)

New Developments in Theory, Algorithms, and Applications for Pythagorean–Hodograph Curves

Rida T. Farouki, Carlotta Giannelli and Alessandra Sestini

Abstract The past decade has witnessed sustained interest in elucidating the basic theory of Pythagorean–hodograph (PH) curves, developing construction algorithms, formulating generalizations, and investigating applications. The rapid pace of this activity, encompassing diverse lines of inquiry, makes it desirable at this point to take a broad perspective of these recent developments, and assess their relationships. In the present article, we aim to address this need by categorizing recent results into a number of broad themes—extensions and specializations of the basic polynomial PH curves; rational orthonormal frames along spatial PH curves; construction and analysis algorithms for PH curves; surface design based on PH curves; and the use of PH curves in practical applications.

1 Introduction

The incorporation of Pythagorean–hodograph (PH) structures in parametric curves facilitates exact computation of many geometrical properties that otherwise require numerical approximations. The significance of this feature in diverse applications (such as path planning and motion control for robotics, manufacturing, computer animation, and unmanned aerial vehicles) has spurred intensive study of the theory, algorithms, and applications for PH curves since their inception [85] in 1990.

R. T. Farouki (✉)
Department of Mechanical and Aerospace Engineering, University of California, Davis, CA, USA
e-mail: farouki@ucdavis.edu

C. Giannelli · A. Sestini
Dipartimento di Matematica e Informatica "U. Dini", Università di Firenze, Florence, Italy
e-mail: carlotta.giannelli@unifi.it

A. Sestini
e-mail: alessandra.sestini@unifi.it

C. Giannelli and H. Speleers (eds.), *Advanced Methods for Geometric Modeling and Numerical Simulation*, Springer INdAM Series 35,
https://doi.org/10.1007/978-3-030-27331-6_7

The current state of development of theory and algorithms for PH curves has been assessed on several earlier occasions. The *Handbook of Computer Aided Geometric Design* includes a chapter [41] that offers a synopsis of the early phase of research on PH curves, up to about 2000. Subsequently the book *Pythagorean–Hodograph Curves: Algebra and Geometry Inseparable* [43], published in 2008, presented a comprehensive assessment of the field, and a special issue of the journal *Computer Aided Geometric Design* on "Pythagorean–hodograph curves and related topics" appeared [72] in the same year. Further new developments concerning PH curves were briefly described in a paper [132] that appeared a few years ago.

Recent years have witnessed a sustained and growing interest in research on PH curves, yielding a wealth of new paradigms and applications. In fact, it has become quite difficult to keep track of the volume and diversity of these new developments. Consequently, the intent of this article is to present a coherent overview of the new results since the appearance of the book [43], emphasizing novel concepts and the relationships between different avenues of investigation.

The plan for the remainder of this article is as follows. Following a brief review of the definitions and distinctive properties of PH curves in Sect. 2, we describe a number of specializations and generalizations of the basic polynomial PH curves in Sect. 3—including helical polynomial curves; rational PH space curves; PH curves based upon sparse polynomial bases, mixed polynomial–trigonometric bases, and spline bases; the Pythagorean hodograph condition under the Minkowski (rather than Euclidean) metric, and surfaces that possess Pythagorean normals.

As described in Sect. 4, the construction of rational orthonormal frames on PH curves has been an active research focus in recent years. Key advances include the identification of low–degree PH curves with *rational* adapted rotation–minimizing frames (in which the normal–plane vectors do not rotate about the tangent); the construction of rational minimal twist frames, which have the least possible rotation of the normal–plane vectors, consistent with prescribed boundary conditions; frames with no rotation of the osculating–plane vectors about the binormal, and frames that are rotation–minimizing with respect to the curve polar vector.

On account of their non–linear nature, the construction and analysis of PH curves is more challenging than for standard linear models. Nevertheless, as described in Sect. 5, there has been considerable recent progress. This includes algorithms for constructing PH curves satisfying various geometrical constraints (including arc length), and procedures for identification and "reverse engineering" of PH curves. Section 6 then addresses the use of PH curves for surface constructions, including surface patches bounded by lines of curvature, swept surfaces based on differential or integral properties of the sweep curve, and surface patches with PH isoparametric curves. Finally, some applications of PH curves in spatial kinematics and real–time motion control are discussed in Sect. 7, while Sect. 8 summarizes the results described herein, and assesses the prospects for further advancements.

2 Fundamentals of Pythagorean–Hodograph Curves

PH curves were originally conceived [85] as planar polynomial parametric curves $\mathbf{r}(\xi) = (x(\xi), y(\xi))$ with hodographs (derivatives) $\mathbf{r}'(\xi) = (x'(\xi), y'(\xi))$ satisfying the Pythagorean condition

$$x'^2(\xi) + y'^2(\xi) = \sigma^2(\xi) \tag{1}$$

for some polynomial $\sigma(\xi)$, which defines the *parametric speed* of $\mathbf{r}(\xi)$, i.e., the derivative $\mathrm{d}s/\mathrm{d}\xi$ of arc length s with respect to the curve parameter ξ. The fact that $\sigma(\xi)$ is a polynomial (rather than the square–root of a polynomial) means that its integral—the arc length function $s(\xi)$—is likewise a polynomial, and the unit tangent $\mathbf{t}(\xi)$, normal $\mathbf{n}(\xi)$, curvature $\kappa(\xi)$, and the *offset curves* $\mathbf{r}_d(\xi) = \mathbf{r}(\xi) + d\,\mathbf{n}(\xi)$ at each distance d from $\mathbf{r}(\xi)$ are all rational in ξ.

Spatial polynomial PH curves $\mathbf{r}(\xi) = (x(\xi), y(\xi), z(\xi))$ satisfying the extension of (1) to $\mathbb{R}^3$, namely

$$x'^2(\xi) + y'^2(\xi) + z'^2(\xi) = \sigma^2(\xi) \tag{2}$$

for some polynomial $\sigma(\xi)$, were first considered in [87]. The simplest (non–trivial) planar and spatial PH curves are the cubics, which admit characterizations in terms of fairly simple geometrical constraints on their Bézier control polygons. However, attempts to characterize higher–order PH curves in terms of their control polygons often lead to cumbersome, non–intuitive constraints [37, 39, 207, 213] of dubious value in practical use. Consequently, PH curves are usually specified as interpolants to given discrete data (end points and tangents, total arc length, etc.).

The construction of PH curve interpolants is greatly facilitated by the adoption of a suitable algebraic model. For planar curves, this is based on using the complex representation [39]—i.e., on identifying points (x, y) in the Euclidean plane with complex values $x + \mathrm{i}\,y$. Setting

$$\mathbf{r}'(\xi) = x'(\xi) + \mathrm{i}\,y'(\xi) = \mathbf{w}^2(\xi) \tag{3}$$

for some complex polynomial $\mathbf{w}(\xi) = u(\xi) + \mathrm{i}\,v(\xi)$, the Pythagorean condition (1) is satisfied with

$$x'(\xi) = u^2(\xi) - v^2(\xi), \quad y'(\xi) = 2\,u(\xi)v(\xi), \quad \sigma(\xi) = u^2(\xi) + v^2(\xi), \tag{4}$$

and once the coefficients of $\mathbf{w}(\xi)$ have been computed by satisfying the interpolation conditions, the planar PH curve $\mathbf{r}(\xi)$ can be obtained by integration of $\mathbf{r}'(\xi)$. The parametric speed and cumulative arc length of $\mathbf{r}(\xi)$ are specified by the polynomial functions

$$\sigma(\xi) = |\mathbf{w}(\xi)|^2, \quad s(\xi) = \int_0^\xi \sigma(u)\,\mathrm{d}u$$

in the curve parameter ξ, while the unit tangent and normal vectors and the curvature are defined by the rational expressions

$$\mathbf{t}(\xi) = \frac{\mathbf{w}^2(\xi)}{|\mathbf{w}(\xi)|^2}, \quad \mathbf{n}(\xi) = -\mathrm{i}\,\frac{\mathbf{w}^2(\xi)}{|\mathbf{w}(\xi)|^2}, \quad \kappa(\xi) = 2\,\frac{\mathrm{Im}(\overline{\mathbf{w}}(\xi)\mathbf{w}'(\xi))}{|\mathbf{w}(\xi)|^4}.$$

The initial study [87] of spatial PH curves was based upon a sufficient, but not necessary, characterization for the solutions of equation (2). To obtain a formulation compatible with arbitrary orientations of the x, y, z coordinate frame, the quaternion algebra is usually invoked [30, 50]. Namely, if $\mathcal{A}(\xi) = u(\xi) + v(\xi)\,\mathbf{i} + p(\xi)\,\mathbf{j} + q(\xi)\,\mathbf{k}$ is any quaternion polynomial, the product[1]

$$\mathbf{r}'(\xi) = x'(\xi)\,\mathbf{i} + y'(\xi)\,\mathbf{j} + z'(\xi)\,\mathbf{k} = \mathcal{A}(\xi)\,\mathbf{i}\,\mathcal{A}^*(\xi), \tag{5}$$

where $\mathcal{A}^*(\xi) = u(\xi) - v(\xi)\,\mathbf{i} - p(\xi)\,\mathbf{j} - q(\xi)\,\mathbf{k}$ is the conjugate of $\mathcal{A}(\xi)$, yields

$$\begin{aligned}
x'(\xi) &= u^2(\xi) + v^2(\xi) - p^2(\xi) - q^2(\xi)\,, \\
y'(\xi) &= 2\,[\,u(\xi)q(\xi) + v(\xi)p(\xi)\,]\,, \\
z'(\xi) &= 2\,[\,v(\xi)q(\xi) - u(\xi)p(\xi)\,]\,, \\
\sigma(\xi) &= u^2(\xi) + v^2(\xi) + p^2(\xi) + q^2(\xi)\,,
\end{aligned} \tag{6}$$

which satisfies the PH condition (2) in $\mathbb{R}^3$. The map $(u, v, p, q) \rightarrow (x', y', z', \sigma)$ has also been interpreted [33] as a *generalized stereographic projection*. The parametric speed and cumulative arc length of the PH curve $\mathbf{r}(\xi)$ obtained by integrating $\mathbf{r}'(\xi)$ are specified by the polynomials

$$\sigma(\xi) = |\mathcal{A}(\xi)|^2, \quad s(\xi) = \int_0^\xi \sigma(u)\,\mathrm{d}u\,,$$

and the rational orthonormal frame defined along $\mathbf{r}(\xi)$ by

$$(\mathbf{e}_1(\xi), \mathbf{e}_2(\xi), \mathbf{e}_3(\xi)) = \frac{(\mathcal{A}(\xi)\,\mathbf{i}\,\mathcal{A}^*(\xi), \mathcal{A}(\xi)\,\mathbf{j}\,\mathcal{A}^*(\xi), \mathcal{A}(\xi)\,\mathbf{k}\,\mathcal{A}^*(\xi))}{|\mathcal{A}(\xi)|^2} \tag{7}$$

is known as the *Euler–Rodrigues frame* (ERF)—$\mathbf{e}_1(\xi)$ coincides with the curve tangent, while $\mathbf{e}_2(\xi), \mathbf{e}_3(\xi)$ span the curve normal plane at each point [26, 116].

An alternative approach [30] is to construct spatial PH curves from a pair of complex polynomials $\boldsymbol{\alpha}(\xi) = u(\xi) + \mathrm{i}\,v(\xi), \boldsymbol{\beta}(\xi) = q(\xi) + \mathrm{i}\,p(\xi)$ through the Hopf map expression

$$\mathbf{r}'(\xi) = (|\boldsymbol{\alpha}(\xi)|^2 - |\boldsymbol{\beta}(\xi)|^2, 2\,\mathrm{Re}(\boldsymbol{\alpha}(\xi)\overline{\boldsymbol{\beta}}(\xi)), 2\,\mathrm{Im}(\boldsymbol{\alpha}(\xi)\overline{\boldsymbol{\beta}}(\xi)))\,. \tag{8}$$

[1]The use of the basis vector $\mathbf{i}$ in the product (5) is only conventional—any other unit vector $\mathbf{u}$ may be used instead, corresponding to a change of coordinates.

Setting $\mathcal{A}(\xi) = \boldsymbol{\alpha}(\xi) + \mathbf{k}\,\boldsymbol{\beta}(\xi)$, this is seen to be equivalent to the quaternion form when the imaginary unit i is identified with the quaternion basis element $\mathbf{i}$.

The PH forms (3) and (5) may be generalized by multiplying them with a scalar polynomial $f(\xi)$, but the resulting hodographs are no longer *primitive*—i.e., their components are not relatively prime, and a real root of $f(\xi)$ will incur a *cusp* (i.e., sudden tangent reversal) on $\mathbf{r}(\xi)$. Without the factor $f(\xi)$, the planar PH form (3) always generates a primitive hodograph when $\gcd(u(\xi), v(\xi))$ is a constant. For the spatial form (5), however, the situation is more subtle: $\gcd(u(\xi), v(\xi), p(\xi), q(\xi)) =$ constant does not ensure a primitive hodograph, since the real polynomial

$$h(\xi) = |\gcd(u(\xi) + \mathrm{i}\,v(\xi), p(\xi) - \mathrm{i}\,q(\xi))\,|^2,$$

if non–constant, is a common factor of $x'(\xi), y'(\xi), z'(\xi)$.

The complex polynomial $\mathbf{w}(\xi)$ in (3) and quaternion polynomial $\mathcal{A}(\xi)$ in (5) are called the *pre–images* of planar and spatial Pythagorean hodographs. For a planar $\mathbf{r}'(\xi)$ there are two pre–images $\mathbf{w}(\xi)$, that differ only in sign. For a spatial $\mathbf{r}'(\xi)$, however, there exists a one–parameter family of pre–images, since $\tilde{\mathcal{A}}(\xi)\,\mathbf{i}\,\tilde{\mathcal{A}}^*(\xi) = \mathcal{A}(\xi)\,\mathbf{i}\,\mathcal{A}^*(\xi)$ when $\tilde{\mathcal{A}}(\xi) = \mathcal{A}(\xi)\,(\cos\theta + \sin\theta\,\mathbf{i})$ for any θ. Thus, the pre–image is actually a "ringed surface" in 4–dimensional space [50, 68].

A more comprehensive treatment of the definitions, properties, construction, and applications of planar and spatial polynomial PH curves may be found in [43], and the extension of the Pythagorean hodograph structure to Euclidean spaces of higher dimension is discussed in [183].

3 Specializations and Generalizations of Polynomial PH Curves

In their original formulation [85], the PH curves were investigated from an algebraic perspective as plane polynomial curves with derivatives that satisfy the Pythagorean condition (1). The extension to rational planar PH curves was developed in [179] from a more geometrical perspective, and the introduction [30] of the quaternion representation has established a firm foundation for the theory of spatial PH curves. As described below, many novel specializations and generalizations of these basic PH curve formulations have emerged during the past decade.

3.1 Helical Curves and Double PH Curves

A *helical curve* is characterized by the fact that its unit tangent vector $\mathbf{t}(\xi)$ maintains a constant angle ψ relative to a fixed unit vector $\mathbf{a}$. For a polynomial curve $\mathbf{r}(\xi)$, this condition is equivalent to

$$\mathbf{a}\cdot\mathbf{r}'(\xi) \,=\, \cos\psi\,|\mathbf{r}'(\xi)|\,,$$

and can only be satisfied if $\mathbf{r}(\xi)$ is a PH curve [70], for which $|\mathbf{r}'(\xi)|$ is a polynomial. An alternative characterization is that helical curves have a constant ratio of their curvature and torsion—namely, $\kappa(\xi)/\tau(\xi) = \tan\psi$. The *double DPH* (DPH) *curves* are a special class of PH curves, for which $|\,\mathbf{r}'(\xi)\times\mathbf{r}''(\xi)\,|$—in addition to $|\mathbf{r}'(\xi)|$—is a polynomial. Now every spatial PH curve satisfies [61] the property

$$|\,\mathbf{r}'(\xi)\times\mathbf{r}''(\xi)\,|^2 \,=\, \sigma^2(\xi)\,\rho(\xi)\,,$$

where $\rho(\xi)$ is the polynomial defined in terms of the components $u(\xi)$, $v(\xi)$, $p(\xi)$, $q(\xi)$ of the quaternion polynomial $\mathcal{A}(\xi)$ in (5) as

$$\rho \,=\, 4\,[\,(up'-u'p+vq'-v'q)^2+(uq'-u'q-vp'+v'p)^2\,]\,.$$

This may be more compactly expressed using the complex polynomials $\boldsymbol{\alpha}(\xi)$, $\boldsymbol{\beta}(\xi)$ in the Hopf map form (8) as $\rho = |\,\boldsymbol{\alpha}\,\boldsymbol{\beta}'-\boldsymbol{\alpha}'\boldsymbol{\beta}\,|^2$. Consequently, the DPH curves are those PH curves for which $\rho(\xi)$ is a perfect square, i.e., $\rho(\xi)=\omega^2(\xi)$ for some polynomial $\omega(\xi)$, and they coincide with the set of polynomial curves for which the Frenet frame vectors and the curvature and torsion functions have a rational dependence [204] on the curve parameter ξ, namely

$$\mathbf{t} \,=\, \frac{\mathbf{r}'}{\sigma}\,,\quad \mathbf{n} \,=\, \frac{\sigma\,\mathbf{r}''-\sigma'\mathbf{r}'}{\sigma\,\omega}\,,\quad \mathbf{b} \,=\, \frac{\mathbf{r}'\times\mathbf{r}''}{\sigma\,\omega}\,,\quad \kappa \,=\, \frac{\omega}{\sigma^2}\,,\quad \tau \,=\, \frac{(\mathbf{r}'\times\mathbf{r}'')\cdot\mathbf{r}'''}{\sigma^2\omega^2}\,.$$

It was noted already in [87] that every spatial PH cubic is a helical curve, but not all higher–order spatial PH curves are helical. The *tangent indicatrix* of a helical curve $\mathbf{r}(\xi)$—i.e., the locus traced by the unit tangent vector $\mathbf{t}(\xi)$—is a (small) circle on the unit sphere. Two types of helical PH space quintics were identified in [70], based on the behavior of the tangent indicatrix: a *monotone–helical* PH quintic exhibits a consistent sense of rotation of $\mathbf{t}(\xi)$ about the axis vector $\mathbf{a}$, while a *general helical* PH quintic may exhibit reversals in the sense of rotation of $\mathbf{t}(\xi)$.

It has been shown in [156] that a spatial PH curve of arbitrary degree is helical if and only if the rational plane curve defined in terms of the complex polynomials $\boldsymbol{\alpha}(\xi)$, $\boldsymbol{\beta}(\xi)$ in the Hopf map form (8) by $\mathbf{z}(\xi)=\boldsymbol{\alpha}(\xi)/\boldsymbol{\beta}(\xi)$ defines a straight line or a circle, and this characterization was also employed in [156] to solve the first–order Hermite interpolation problem with helical PH quintic space curves.

It is known [12] that the cubic and quintic DPH curves coincide exactly with the helical PH curves of the same degree, but DPH curves of degree 7 exist that are not helical. More precisely, all helical polynomial curves must be DPH curves, although there are non–helical DPH curves of degree ≥ 7. A comprehensive treatment of the theory of helical polynomial curves and DPH curves, emphasizing the relationship between the quaternion and Hopf map models, and the enumeration of all curve types (helical and non–helical) up to degree 7, was presented in [61, 62].

Geometric Hermite interpolation with monotone helical PH quintic curves was investigated in [103], while an algorithm to construct cubic helical PH splines with Frenet frame continuity was presented in [104]. More recently, the construction of helical PH curves that satisfy initial and final positions and motion directions was investigated in [48], motivated by the problem of defining paths for unmanned aerial vehicles with limited climb rates to achieve a desired change in altitude.

Finally, DPH curves also arise in the theory of *Salkowski curves*—a family of space curves characterized by constant curvature and non–constant torsion [157]. Namely, with suitable constants and re–parameterizations, certain Salkowski curves correspond to rational DPH curves. These properties have been exploited in [157] to construct closed non–planar paths of constant curvature, consisting of segments of Salkowski curves and circular helices.

3.2 Rational Pythagorean–Hodograph Curves

Since the integration of a rational Pythagorean hodograph does not necessarily yield a rational curve, alternative approaches are required to construct rational PH curves. The planar case was first investigated in [94, 179, 180] using the *dual* representation, in which a plane curve is interpreted as the envelope of a one–parameter family of tangent lines rather than a point locus. To generate a rational planar PH curve, the tangent lines are specified by a family of rational unit tangent vectors and a rational *support function*, defining the normal distance of the tangent lines from the origin.

A different approach is required for the construction of spatial rational PH curves, since planes (rather than lines) are dual to points in $\mathbb{R}^3$. Such an approach was briefly outlined in [178] and developed in greater detail in [91]. The method begins by specifying a field of rational unit tangent vectors $\mathbf{t}(\xi)$ for the desired rational PH curve. Together with its derivative $\mathbf{t}'(\xi)$, this field defines the orientation of the curve osculating planes, which are uniquely determined by also fixing a rational support function $f(\xi)$ specifying their normal distances from the origin. The envelope of the osculating planes is a developable ruled surface, and the rational PH space curve $\mathbf{r}(\xi)$ is identified as its *edge of regression* (or *cuspidal edge*), which has the closed–form expression

$$\mathbf{r}(\xi) = \frac{f(\xi)\,\mathbf{u}'(\xi)\times\mathbf{u}''(\xi) + f'(\xi)\,\mathbf{u}''(\xi)\times\mathbf{u}(\xi) + f''(\xi)\,\mathbf{u}(\xi)\times\mathbf{u}'(\xi)}{\mathbf{u}(\xi)\cdot[\,\mathbf{u}'(\xi)\times\mathbf{u}''(\xi)\,]},$$

where $\mathbf{u}(\xi) = \mathbf{t}(\xi)\times\mathbf{t}'(\xi)$. As noted in [91], the interpolation of G^1 or G^2 data with these spatial rational PH curves incurs only linear equations, rather than the non–linear equations that occur with polynomial PH curves. Note, however, that rational PH curves do not (in general) have rational arc lengths [45].

This approach to rational spatial PH curves has been further developed in [134], employing a quaternion model to formulate a polynomial dual representation that facilitates the enumeration of curves of low *class* (i.e., degree of the dual form). With this model, G^1 Hermite interpolation with rational PH space curves was discussed in [138] using rational cubics, and in [139] using curves of class 4.

3.3 Algebraic–Trigonometric PH Curves

The extension of Pythagorean hodograph structures to more general function spaces is exemplified by the recent study [182] of *algebraic–trigonometric Pythagorean–hodograph* (ATPH) *curves*. The starting point for this generalization is the choice of a suitable normalized B–basis for the space of trigonometric or mixed algebraic–trigonometric polynomials. In this context, Bernstein–like bases can be defined that incorporate a tension parameter, which can be used to modify the curve shape. As a consequence, algebraic–trigonometric Bézier curves share properties analogous to their classical polynomial counterparts. It should be noted that PH curves were already extended to complete Chebyshev systems in [135] to define Pythagorean–hodograph cycloidal curves (see also Sect. 3.6).

The main focus of [182] is to formulate ATPH analogs of the well–known planar polynomial quintic PH curves, by integrating the expression $\mathbf{r}'(\xi) = \mathbf{w}^2(\xi)$, where for complex values $\mathbf{w}_0, \mathbf{w}_1, \mathbf{w}_2$ the function

$$\mathbf{w}(\xi) \,=\, \mathbf{w}_0 \tilde{B}_0^2(\xi) + \mathbf{w}_1 \tilde{B}_1^2(\xi) + \mathbf{w}_2 \tilde{B}_2^2(\xi)$$

is defined in terms of the normalized B–basis specified by

$$\tilde{B}_0^2(\xi) = \frac{\cos(\alpha - \xi) - 1}{\cos\alpha - 1}\,,$$
$$\tilde{B}_1^2(\xi) = \frac{\cos\alpha - \cos\xi - \cos(\alpha - \xi) + 1}{\cos\alpha - 1}\,,$$
$$\tilde{B}_2^2(\xi) = \frac{\cos\xi - 1}{\cos\alpha - 1}\,,$$

which specializes to the quadratic Bernstein basis in the limit $\alpha \to 0$ of the shape parameter α. The resulting curve $\mathbf{r}(\xi)$ belongs to the function space spanned by the basis $\{\, 1, \xi, \cos\xi, \sin\xi, \cos 2\xi, \sin 2\xi \,\}$ and in the context of this space has properties (arc length, offset curves, de Casteljau–like algorithm, etc.) analogous to those for the polynomial PH curves. The adaptation of the standard C^1 PH quintic Hermite interpolation algorithm [80] to these ATPH curves, and the construction of ATPH spiral segments, is also discussed in [182].

Further adaptations of established algorithms for polynomial PH curves to the context of ATPH curves include the construction of C^2 spline interpolants that have shape parameters associated with each spline segment [98], and the formulation of a quaternion representation for spatial ATPH curves, with application to solutions of the C^1 Hermite interpolation problem that incorporate a free shape parameter [181].

3.4 Minkowski Pythagorean–Hodograph Curves

The *medial axis* of a planar domain D consists of the locus of centers (x, y) of all maximal inscribed circles, that touch the domain boundary ∂D in at least two points, and the *medial axis transform* (MAT) specifies the radius r of the maximal disk that is centered at each point of the medial axis [18]. The domain boundary ∂D can, in principle, be reconstructed from its MAT as the envelope of the family of maximal disks specified by given data (x, y, r). The generalization of the PH property to the Minkowski space $\mathbb{R}^{2,1}$ of Special Relativity theory (with a metric signature $++-$) leads to the definition of *Minkowski Pythagorean–hodograph* (MPH) *curves*, which permit precise (rational) reconstructions of planar shape boundaries from MATs that are specified by MPH curves [27, 159–161].

A polynomial curve $(x(\xi), y(\xi), r(\xi))$ in the Minkowski space $\mathbb{R}^{2,1}$ is an MPH curve if its derivative components satisfy the Pythagorean relation

$$x'^2(\xi) + y'^2(\xi) - r'^2(\xi) = \sigma^2(\xi)$$

for some polynomial $\sigma(\xi)$. The solutions to this condition may be characterized [160, 161] in terms of four polynomials $u(\xi), v(\xi), p(\xi), q(\xi)$ by a modification of (6) as

$$\begin{aligned} x'(\xi) &= u^2(\xi) - v^2(\xi) + p^2(\xi) - q^2(\xi), \\ y'(\xi) &= 2\,[\,u(\xi)v(\xi) - p(\xi)q(\xi)\,], \\ r'(\xi) &= 2\,[\,u(\xi)p(\xi) - v(\xi)q(\xi)\,], \\ \sigma(\xi) &= u^2(\xi) + v^2(\xi) - p^2(\xi) - q^2(\xi). \end{aligned} \tag{9}$$

One may consider (6) and (9) as instances of the forms generated by the *generalized stereographic projection* introduced in [33].

The interpolation of G^1 Hermite data in $\mathbb{R}^{2,1}$ using MPH cubics was considered in [128]. Subsequently, C^1 Hermite interpolation by means of MPH quintics was addressed in [129] using a quaternion model and, as with the problem of C^1 Hermite interpolation by PH quintics in $\mathbb{R}^3$ [51, 56], the interpolants were found to depend on two free parameters. This method was extended in [133] to Hermite interpolants of C^2 data in $\mathbb{R}^{2,1}$ by degree 9 MPH curves, which exhibit four free parameters.

The generalization of Minkowski Pythagorean–hodograph curves to the rational case was introduced in [130]. In particular, it was shown there that any rational MPH curve $\mathbf{c}(\xi)$ in $\mathbb{R}^{2,1}$ can be obtained in terms of a rational PH curve $(x(\xi), y(\xi))$ in $\mathbb{R}^2$ and an additional rational function $r(\xi)$ in the form

$$\mathbf{c}(\xi) = \left(x(\xi) + \frac{r(\xi)y'(\xi)}{\sigma(\xi)}, y(\xi) - \frac{r(\xi)x'(\xi)}{\sigma(\xi)}, r(\xi) \right).$$

The analysis of interconnections between spatial MPH curves and their planar PH counterparts was continued [131] in the context of Hermite interpolation problems. The concept of *rational envelope curves*, as a superset of the class of MPH curves, was introduced in [16], and MPH–preserving mappings were investigated in [127].

3.5 Pythagorean–Normal and Linear Normal Surfaces

By analogy with the constructions of rational PH curves discussed in Sect. 3.2, the characterization of rational surfaces $\mathbf{s}(u, v)$ with rational unit normals $\mathbf{n}(u, v) = (\mathbf{s}_u \times \mathbf{s}_v)/|\mathbf{s}_u \times \mathbf{s}_v|$ (and hence rational offset surfaces) was considered in [179], using the dual form in which a surface is interpreted as the envelope[2] of a two–parameter family of tangent planes (see also [175]). Such surfaces are known as *Pythagorean normal* (PN) surfaces, and their representation under certain parameterizations has been further discussed in the context of convolution surfaces in [147].

A quaternion approach to polynomial PN surfaces was proposed in [137], based on identifying constraints on the coefficients of a bivariate quaternion pre–image polynomial, that are sufficient to ensure it will generate a surface with Pythagorean unit normals. These PN surfaces may be of odd or even degree, and constructions for cubic, quartic, and quintic instances were enumerated. A comprehensive unifying theory of PN surfaces based on a geometric algebra formalism has been introduced in [143], and PN surface construction methods have been addressed in [17, 148].

The *LN surfaces*, originally proposed in [117], are an alternative approach to the Pythagorean normal principle for identifying surfaces with rational offsets. These surfaces admit a field of (non–unit) normal vectors that are linearly dependent on the surface parameters, and by an appropriate re–parameterization one can show that they possess Pythagorean unit normals, and thus have rational offsets. LN surfaces have been characterized [99] as surfaces that possess odd (antisymmetric) rational support functions. It is known [10, 146, 173] that all quadratic triangular Bézier surface patches are LN surfaces. Further developments concerning the theory of LN surfaces and their construction algorithms may be found in [120, 184]. Finally, the studies [172, 174] approach the problem of surfaces with rational offsets from the more general perspective of the envelopes of two–parameter families of spheres.

3.6 Other Special Forms

The *roulettes* are curves generated kinematically as the loci traced by a point of one curve that "rolls without slipping" against a second curve. They have played a key role [86] in the history of science and technology. Some familiar examples are the

[2]Since they have just a one–parameter family of tangent planes, the *developable* surfaces require a separate treatment.

cycloid (generated by a circle rolling along a straight line) and the epicycloids and hypocycloids (generated by a circle rolling on the outside or inside of another circle, respectively). It was shown in [193] that rational epicycloids and hypocycloids have rational offset curves, and an algorithm for G^1 Hermite interpolation with segments of such curves was also presented. Certain roulettes can also be identified as curves with polynomial support functions [195]. *Pythagorean–hodograph cycloidal* (PHC) *curves* were proposed in [135], as a generalization of the PH cubics to extended complete Chebyshev systems (see also Sect. 3.3). The analysis of PHC curves in [135] includes a G^1 Hermite interpolation scheme and its asymptotic behavior.

The fact that they admit rational parameterization of their one–sided offset curves [85] is key property of planar PH curves. However, certain planar curves that are not generated by the complex form (3) may also admit rational parameterization of their (two–sided) offsets, after imposing a rational transformation of the parameter that yields an improper (i.e., doubly–traced) representation of the curve. It was shown in [152] that such curves correspond to complex hodographs defined by multiplying (3) with a complex linear factor $\mathbf{z}\,\xi + 1$ with $\mathbf{z} \neq 0$. A geometrical characterization of the cubic curves with rational two–sided offsets has been presented in [153].

Another approach to generating PH curves is through appropriate polynomial or rational mappings of other PH curves. For example, *stereographic projection* can be used [203] to map planar polynomial PH curves into rational spherical PH curves (i.e., PH curves on the unit sphere). Möbius transformations of the complex plane are employed in [150] to map planar polynomial PH cubics into rational PH curves, that can be used to solve the C^1 Hermite interpolation problem. A family of (scaled) rational mappings from $\mathbb{R}^2$ to $\mathbb{R}^3$ that preserve the Pythagorean hodograph property has been studied in [122], and used to solve the problem of Hermite interpolation with spatial PH curves by mapping of planar PH curve interpolants.

The notion of "sparse" Pythagorean–hodograph curves, defined in terms of bases that span only a subset of the polynomials of a given degree, was introduced in [2]. In particular, this study focuses on certain "cubic–like" PH curves defined for non–negative integer k in terms of the basis $\{1, \xi^{k+1}, \xi^{k+2}, \xi^{k+3}\}$ on a positive interval $\xi \in [\,a, b\,]$ which spans a *Müntz space*. As with the standard polynomial PH cubics [85], these sparse cubic–like PH curves are shown to admit simple characterizations in terms of the control polygon geometry, and they incorporate free parameters that prove useful in constructing G^1 Hermite interpolants. The "quintic–like" sparse PH curves, which may exhibit inflections, are also addressed in [2].

Finally, we note that the algebraic models used to construct PH curves (based on the complex numbers and quaternions) are simple instances of the *Clifford algebra*. Although it has only occasionally been used in the development of PH curve theory and algorithms—for example, to characterize MPH curves in $\mathbb{R}^{2,1}$ and $\mathbb{R}^{3,1}$ [129, 160] and as an alternative model for PH curves in $\mathbb{R}^3$ [170]—further exploitation of the methodology of Clifford algebra (also known as *geometric algebra* [171]) may yield new insights, algorithms, or generalizations for PH curves.

4 Rational Orthonormal Frames Along PH Curves

The specification of orientation along a spatial path $\mathbf{r}(\xi)$ by an orthonormal frame $(\mathbf{f}_1(\xi), \mathbf{f}_2(\xi), \mathbf{f}_3(\xi))$ is a fundamental problem in kinematics, and it is desirable that the frame vectors have a *rational* dependence on the curve parameter ξ. Of special interest are cases where the frame orientation is correlated with the path geometry. As noted in Sect. 2, every polynomial PH curve admits a rational *adapted* frame, called the Euler–Rodrigues frame (ERF), which consists of the tangent and mutually orthogonal vectors that span the normal plane.

The ERF serves as the point of departure for constructing other rational adapted frames: the *rotation–minimizing frame* (RMF) and *minimal twist frame* (MTF). The variation of a frame is defined by its angular velocity $\boldsymbol{\omega}$. An RMF is characterized by a zero angular velocity component ω_1 in the tangent direction, while an MTF has the least absolute value for the arc–length integral of this component, consistent with prescribed initial and final orientations. The familiar *Frenet frame* is rotation–minimizing with respect to the principal normal, rather than the tangent, and for the family of *double PH curves* (see Sect. 3.1) it is rational. Other frames considered below include the *rotation–minimizing osculating frame* (which has a zero angular velocity component in the direction of the binormal), and the *rotation–minimizing directed frame*, in which $\mathbf{f}_1(\xi) = \mathbf{r}(\xi)/|\mathbf{r}(\xi)|$ is the polar vector.

4.1 Rational Rotation–Minimizing Adapted Frames

The variation of an adapted orthonormal frame $(\mathbf{f}_1(\xi), \mathbf{f}_2(\xi), \mathbf{f}_3(\xi))$ on a space curve $\mathbf{r}(\xi)$, where $\mathbf{f}_1(\xi) = \mathbf{r}'(\xi)/|\mathbf{r}'(\xi)|$ is the tangent and $\mathbf{f}_2(\xi)$, $\mathbf{f}_3(\xi)$ span the normal plane, is specified by its angular velocity $\boldsymbol{\omega}$ through the relations

$$\frac{\mathrm{d}\mathbf{f}_1}{\mathrm{d}s} = \boldsymbol{\omega} \times \mathbf{f}_1 \,, \qquad \frac{\mathrm{d}\mathbf{f}_2}{\mathrm{d}s} = \boldsymbol{\omega} \times \mathbf{f}_2 \,, \qquad \frac{\mathrm{d}\mathbf{f}_3}{\mathrm{d}s} = \boldsymbol{\omega} \times \mathbf{f}_3 \,, \tag{10}$$

with s being arc length measured along $\mathbf{r}(\xi)$. If $\boldsymbol{\omega}$ is expressed in the form

$$\boldsymbol{\omega} = \omega_1\mathbf{f}_1 + \omega_2\mathbf{f}_2 + \omega_3\mathbf{f}_3 \,, \tag{11}$$

its components are

$$\omega_1 = \frac{\mathrm{d}\mathbf{f}_2}{\mathrm{d}s} \cdot \mathbf{f}_3 \,, \quad \omega_2 = \frac{\mathrm{d}\mathbf{f}_3}{\mathrm{d}s} \cdot \mathbf{f}_1 \,, \quad \omega_3 = \frac{\mathrm{d}\mathbf{f}_1}{\mathrm{d}s} \cdot \mathbf{f}_2 \,. \tag{12}$$

The component ω_1 specifies the rate of rotation of the normal–plane vectors $\mathbf{f}_2, \mathbf{f}_3$ about the curve tangent $\mathbf{f}_1$, and for applications such as swept surface constructions, 5–axis CNC machining, computer animation, spatial motion planning, and robotics, it is often desirable to have adapted frames satisfying $\omega_1 \equiv 0$, i.e., the normal–plane vectors have no instantaneous rotation about the curve tangent. Adapted frames with $\boldsymbol{\omega} \cdot \mathbf{t} \equiv 0$ are called *rotation–minimizing frames* (RMFs) or *Bishop frames* [15].

Since the construction of RMFs on general polynomial/rational curves requires a numerical quadrature, many methods have been proposed [69, 119, 155, 209, 210] to construct rational approximations of them. It should be noted [42] that *any* spatial PH curve admits exact RMF computation, but in general this incurs transcendental terms. However, a subset of the spatial PH curves, identified by algebraic constraints on the coefficients of their generating polynomials, admits exact *rational* RMFs. We summarize here the characterizations and properties of these *RRMF curves*, which have been subject to intensive investigation over the past decade. A more detailed review, covering results up to a few years ago, may be found in [46].

It was observed in [26] that, for the spatial PH curve generated by the quaternion polynomial $\mathcal{A}(\xi) = u(\xi) + v(\xi)\,\mathbf{i} + p(\xi)\,\mathbf{j} + q(\xi)\,\mathbf{k}$, the angular velocity component ω_1 of the ERF is given by

$$\omega_1 \,=\, \frac{\mathrm{d}\mathbf{e}_2}{\mathrm{d}s} \cdot \mathbf{e}_3 \,=\, \frac{2(uv' - u'v - pq' + p'q)}{\sigma^2}\,, \tag{13}$$

and the possibility of identifying PH curves for which $\omega_1 \equiv 0$—i.e., the ERF is itself an RMF—was considered. The simplest PH curves satisfying this condition are of degree 7, and incorporate 16 free parameters. A simpler characterization of these degree 7 RRMF curves [71] is in terms of the five constraints

$$\begin{aligned} \mathrm{scal}(\mathcal{A}_0\,\mathbf{i}\,\mathcal{A}_1^*) &= \mathrm{scal}(\mathcal{A}_0\,\mathbf{i}\,\mathcal{A}_2^*) = 0\,, \\ 3\,\mathrm{scal}(\mathcal{A}_1\,\mathbf{i}\,\mathcal{A}_2^*) &+ \mathrm{scal}(\mathcal{A}_0\,\mathbf{i}\,\mathcal{A}_3^*) = 0\,, \\ \mathrm{scal}(\mathcal{A}_1\,\mathbf{i}\,\mathcal{A}_3^*) &= \mathrm{scal}(\mathcal{A}_2\,\mathbf{i}\,\mathcal{A}_3^*) = 0\,, \end{aligned} \tag{14}$$

on the coefficients of the cubic quaternion polynomials generating spatial PH curves of degree 7, expressed in Bernstein form as

$$\mathcal{A}(\xi) \,=\, \mathcal{A}_0(1-\xi)^3 + \mathcal{A}_1 3(1-\xi)^2\xi + \mathcal{A}_2 3(1-\xi)\xi^2 + \mathcal{A}_3\xi^3\,.$$

The design of rational rotation–minimizing rigid–body motions using these curves, that satisfy initial and final positions and orientations, was considered in [71]. This involves the solution of four quadratic equations in four real unknowns, with two free parameters available for shape optimization.

The possibility of identifying lower–degree RRMF curves, for which the ERF is not rotation–minimizing, was considered in [102], based on taking $\mathbf{f}_1(\xi) = \mathbf{e}_1(\xi)$ and imposing a rational normal–plane rotation of the form

$$\begin{bmatrix} \mathbf{f}_2(\xi) \\ \mathbf{f}_3(\xi) \end{bmatrix} = \frac{1}{a^2(\xi) + b^2(\xi)} \begin{bmatrix} a^2(\xi) - b^2(\xi) & -2\,a(\xi)b(\xi) \\ 2\,a(\xi)b(\xi) & a^2(\xi) - b^2(\xi) \end{bmatrix} \begin{bmatrix} \mathbf{e}_2(\xi) \\ \mathbf{e}_3(\xi) \end{bmatrix}, \tag{15}$$

for polynomials $a(\xi), b(\xi)$ with $\gcd(a(\xi), b(\xi)) = \text{constant}$. This amounts to orienting $\mathbf{f}_2(\xi), \mathbf{f}_3(\xi)$ relative to $\mathbf{e}_2(\xi), \mathbf{e}_3(\xi)$ through the angle

$$\theta(\xi) \;=\; -2\arctan\frac{b(\xi)}{a(\xi)}\,. \tag{16}$$

If $(\mathbf{f}_1(\xi), \mathbf{f}_2(\xi), \mathbf{f}_3(\xi))$ is to be an RMF, the angular velocity component

$$\frac{\mathrm{d}\theta}{\mathrm{d}s} \;=\; -\,\frac{1}{\sigma(\xi)}\,\frac{a(\xi)b'(\xi) - a'(\xi)b(\xi)}{a^2(\xi) + b^2(\xi)}$$

incurred by the rotation (15), in the direction $\mathbf{f}_1 = \mathbf{e}_1$, must exactly cancel the ERF angular velocity component (13). Hence, the spatial PH curve with quaternion pre–image polynomial $\mathcal{A}(\xi) = u(\xi) + v(\xi)\,\mathbf{i} + p(\xi)\,\mathbf{j} + q(\xi)\,\mathbf{k}$ is an RRMF curve if and only if the condition

$$\frac{uv' - u'v - pq' + p'q}{u^2 + v^2 + p^2 + q^2} \;=\; \frac{ab' - a'b}{a^2 + b^2} \tag{17}$$

is satisfied for polynomials $a(\xi), b(\xi)$ with $\gcd(a(\xi), b(\xi)) = \text{constant}$. Based on this, it was shown that no non–planar cubic RRMF curves exist [102].

The existence of non–planar quintic RRMF curves was first demonstrated in [57], based on the Hopf map formulation in terms of two quadratic complex polynomials $\boldsymbol{\alpha}(\xi)$ and $\boldsymbol{\beta}(\xi)$. These quintic RRMF curves were identified by two constraints of degree 4 and 6 imposed on the Bernstein coefficients $\boldsymbol{\alpha}_0, \boldsymbol{\alpha}_1, \boldsymbol{\alpha}_2$ and $\boldsymbol{\beta}_0, \boldsymbol{\beta}_1, \boldsymbol{\beta}_2$ of $\boldsymbol{\alpha}(\xi), \boldsymbol{\beta}(\xi)$ and their conjugates. However, these constraints are not symmetric with respect to reversing the order of the coefficient indices 0, 1, 2 (which corresponds to the simple re–parameterization $\xi \to 1 - \xi$ of $\mathbf{r}(\xi)$).

By employing the quaternion and Hopf map forms interchangeably, a simpler characterization was subsequently identified in [44], that involves only *quadratic* constraints, and is symmetric in the coefficient indices. In terms of the quaternion form, it may be phrased as follows—a quadratic quaternion pre–image polynomial $\mathcal{A}(\xi)$ will generate a quintic RRMF curve if and only if its Bernstein coefficients $\mathcal{A}_0, \mathcal{A}_1, \mathcal{A}_2$ satisfy the vector constraint

$$\text{vect}(\mathcal{A}_2\,\mathbf{i}\,\mathcal{A}_0^*) \;=\; \mathcal{A}_1\,\mathbf{i}\,\mathcal{A}_1^*\,. \tag{18}$$

Figure 1 compares the variation of the normal–plane vectors for the familiar Frenet frame and the RMF on a spatial PH quintic. Note the strong rotation of these vectors about the tangent for the Frenet frame, and also the sudden reversal at the inflection point (where the curvature vanishes).

Condition (18) is based on the assumption that, in the satisfaction of (17), the numerators and denominators on the left or right have no common factors that can be cancelled out—i.e., $\deg(u, v, p, q) = \deg(a, b)$. This assumption also permits formulation of a criterion to identify RRMF curves of arbitrary degree [88]. Namely, introducing the two polynomials defined by

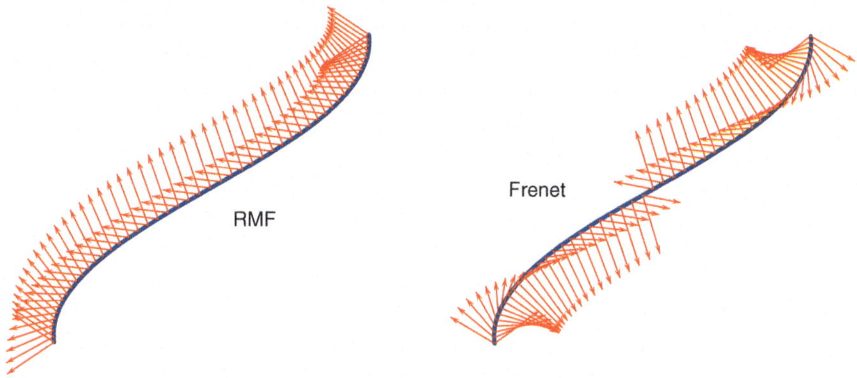

Fig. 1 A comparison of the normal–plane vectors for the rotation–minimizing frame (left) and the Frenet frame (right) along a spatial PH quintic curve with an inflection point

$$\rho = (up' - u'p + vq' - v'q)^2 + (uq' - u'q - vp' + v'p)^2 \,,$$
$$\eta = (uu' + vv' + pp' + qq')^2 + (uv' - u'v - pq' + p'q)^2 \,,$$

which satisfy

$$\rho + \eta \, = \, (u^2 + v^2 + p^2 + q^2)(u'^2 + v'^2 + p'^2 + q'^2) \,, \tag{19}$$

a spatial PH curve generated by a quaternion polynomial $\mathcal{A}(\xi)$ of any degree is an RRMF curve if and only if $\rho(\xi)$ and $\eta(\xi)$ are divisible by the parametric speed $\sigma(\xi)$. Note that, in view of (19), it is only necessary to check divisibility of either ρ or η by σ. However, the significance of this divisibility condition for the actual construction of RRMF curves has not yet been clarified.

Subsequently, RRMF curves were identified that satisfy the condition (17) with $\deg(u, v, p, q) \neq \deg(a, b)$. A family of RRMF quintics was described in [89], that satisfies (17) with $\deg(u, v, p, q) = 2$ and $\deg(a, b) = 1$, so a common quadratic factor cancels from the numerator and denominator of the rational function on the left in (17). Moreover, it has been shown in [22] that RRMF curves satisfying (17) with $\deg(u, v, p, q) < \deg(a, b)$ also exist, due to a cancellation on the right in (17). These results reveal an intricate structure to the entire space of RRMF curves.

The paper [52] formulates a characterization of the quintic RRMF curves in terms of the root structures of their quadratic quaternion pre–image polynomials $\mathcal{A}(\xi)$, for the case $\deg(u, v, p, q) = \deg(a, b) = 2$ in (17). It was shown that the roots of $\mathcal{A}(\xi)$ can be obtained by a simple algorithm, requiring computation of the unique positive root of a real cubic equation, and the degenerate (linear or planar) instances of the quintic RRMF curves were found to correspond to cases where $\mathcal{A}(\xi)$ has a double root. For non–planar RRMF quintics, the real cubic factors into linear and quadratic terms, and closed–form expressions for the roots of $\mathcal{A}(\xi)$ can be derived.

A more comprehensive analysis of the structure of quaternion polynomials $\mathcal{A}(\xi)$ of general degree, that are compatible with the satisfaction of equation (17) with the possibility of cancellation on the left or the right, was developed in [54]. This analysis shows that any quaternion polynomial $\mathcal{A}(\xi)$ generating an RRMF curve can be related to another quaternion polynomial $\mathcal{B}(\xi)$, which generates a rotation–minimizing ERF (i.e., the expression on the left in (17) vanishes identically). As noted in [26], the simplest quaternion polynomials $\mathcal{B}(\xi)$ yielding non–degenerate space curves with rotation–minimizing ERFs are cubic, and in [54] it is shown that such polynomials also exist for any degree > 3.

The paper [7] proposes an interesting approach to rational RMFs on space curves, based on the observation that a general Möbius transformation in $\mathbb{R}^3$ (which may be interpreted as a composition of inversions in planes and spheres) maps curves with rational RMFs into other curves with rational RMFs. This property was used in [7] to develop a scheme for G^1 Hermite interpolation by rational PH curves with rational RMFs, based on transformations of spatial PH cubic segments. Note that, in general, a Möbius transformation maps a polynomial PH curve into a rational PH curve of higher degree, which does not (in general) have a rational arc length, since the integration of a rational parametric speed may incur transcendental terms.

4.2 Rational Adapted Frames of Minimal Twist

The construction of an RMF on a given curve $\mathbf{r}(\xi)$ is an *initial value problem* [124]—i.e., fixing the orientation of the normal–plane vectors $\mathbf{f}_2, \mathbf{f}_3$ at any point of $\mathbf{r}(\xi)$ determines their orientation at every other point. Hence, in general, it is impossible to construct rotation–minimizing spatial motions along a given curve that satisfies prescribed initial *and* final orientations.[3] Upon completing a rotation–minimizing motion along a smooth closed path, for example, the final orientation of a rigid body will differ from its initial orientation.

Since it is often desirable to independently specify the path geometry and the initial/final orientation states for an adapted spatial rigid–body motion, the concept of a *minimal twist frame* (MTF) has been introduced in [79]. An adapted frame $(\mathbf{f}_1(\xi), \mathbf{f}_2(\xi), \mathbf{f}_3(\xi))$ along a space curve $\mathbf{r}(\xi)$, $\xi \in [\,0, 1\,]$ of arc length S is an MTF with respect to given initial and final instances

$$(\mathbf{f}_1(0), \mathbf{f}_2(0), \mathbf{f}_3(0)) \quad \text{and} \quad (\mathbf{f}_1(1), \mathbf{f}_2(1), \mathbf{f}_3(1))$$

if its angular velocity $\boldsymbol{\Omega} = \Omega_1\mathbf{f}_1 + \Omega_2\mathbf{f}_2 + \Omega_3\mathbf{f}_3$ (where $\mathbf{f}_1$ is the tangent) satisfies the following conditions:

(1) Ω_1 does not change sign for $\xi \in (0, 1)$—i.e., the normal–plane vectors $\mathbf{f}_2, \mathbf{f}_3$ maintain a *monotone* sense of rotation about $\mathbf{f}_1$;

[3]For the rotation–minimizing motions with fixed initial/final orientations in [58, 71] the curve was not defined a priori, but was rather an outcome of the construction procedure.

(2) the frame achieves the least possible absolute value of the *twist integral*,

$$T = \int_0^S \Omega_1 \, \mathrm{d}s = \int_0^1 \Omega_1(\xi)\, \sigma(\xi)\, \mathrm{d}\xi \,. \tag{20}$$

A point where Ω_1 changes sign is an *inflection* of the frame $(\mathbf{f}_1, \mathbf{f}_2, \mathbf{f}_3)$ and condition (1) excludes cancellation of clockwise and anti–clockwise rotations of $\mathbf{f}_2, \mathbf{f}_3$ about $\mathbf{f}_1$. In the absence of inflections, the values of T for different frames with the same end states can differ only by an integer multiple of 2π, and among them there is a unique value whose magnitude does not exceed π.

To construct *rational* MTFs, a rational normal–plane rotation of the ERF is used, of the form (15) employed in constructing rational RMFs. Combining $a(\xi), b(\xi)$ into the complex polynomial $\mathbf{w}(\xi) = a(\xi) + \mathrm{i}\, b(\xi)$, such that

$$\exp(\mathrm{i}\,\theta(\xi)) = \frac{\mathbf{w}^2(\xi)}{|\mathbf{w}(\xi)|^2} = \frac{a^2(\xi) - b^2(\xi) + \mathrm{i}\, 2a(\xi)b(\xi)}{a^2(\xi) + b^2(\xi)}\,, \tag{21}$$

the angular velocity component Ω_1 of the frame $(\mathbf{f}_1(\xi), \mathbf{f}_2(\xi), \mathbf{f}_3(\xi))$ becomes

$$\Omega_1(\xi) = \omega_1(\xi) + \frac{\theta'(\xi)}{\sigma(\xi)}\,, \tag{22}$$

where $\omega_1(\xi)$ is the tangent component (13) of the ERF angular velocity, and

$$\theta'(\xi) = 2\,\frac{\mathrm{Im}(\overline{\mathbf{w}}(\xi)\mathbf{w}'(\xi))}{|\mathbf{w}(\xi)|^2} = 2\,\frac{a(\xi)b'(\xi) - a'(\xi)b(\xi)}{a^2(\xi) + b^2(\xi)}\,.$$

The total twist can thus be expressed as

$$T = \int_0^S \Omega_1 \, \mathrm{d}s = \int_0^1 \omega_1\sigma + \theta' \, \mathrm{d}\xi = T_{\mathrm{ERF}} + \theta_f - \theta_i\,, \tag{23}$$

where T_{ERF} is the twist of the ERF, and $\theta_i, \theta_f \in (-\pi, +\pi\,]$ are the values of θ that specify the initial and final orientations of $\mathbf{f}_2, \mathbf{f}_3$ relative to $\mathbf{e}_2, \mathbf{e}_3$. For the PH quintics, T_{ERF} admits closed–form evaluation by partial fraction expansion of the integrand [79]. The computed value of T is replaced by the *reduced twist* $T_{\min}$, which is equal to T when $T \in (-\pi, +\pi\,]$ and is otherwise defined by adding or subtracting the integer multiple of 2π that yields a value in the domain $(-\pi, +\pi\,]$.

For an MTF, the mean value of Ω_1 is $\overline{\Omega}_1 = T_{\min}/S$. The condition $\Omega_1 \equiv \overline{\Omega}_1$ cannot be achieved by a rational MTF, but it can be closely approximated. With a quadratic polynomial $\mathbf{w}(\xi) = \mathbf{w}_0(1-\xi)^2 + \mathbf{w}_1 2(1-\xi)\xi + \mathbf{w}_2\xi^2$, choosing $\mathbf{w}_0 = \exp(\mathrm{i}\frac{1}{2}\theta_i)$ and $\mathbf{w}_2 = \gamma\, \exp(\mathrm{i}\frac{1}{2}\theta_f)$, where γ is a free real parameter, satisfies the boundary conditions $\theta(0) = \theta_i$ and $\theta(1) = \theta_f$. Moreover, setting $\Delta\theta = \theta_f - \theta_i$, the choice

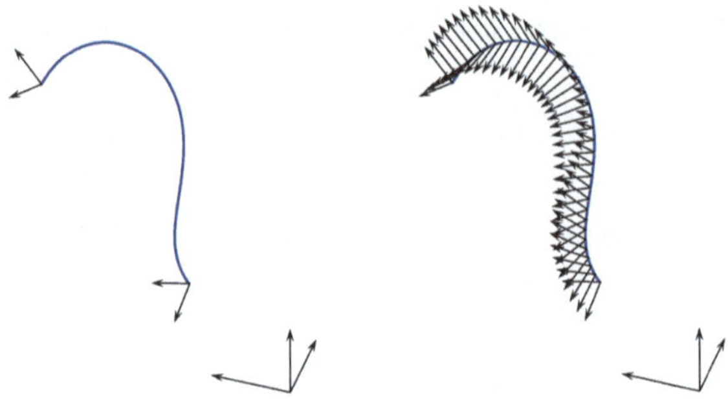

Fig. 2 Left: a spatial PH quintic with indicated initial and final normal plane vectors for a minimal–twist frame (MTF). Right: a rational MTF compatible with the prescribed boundary conditions

$$\mathbf{w}_1 = \frac{\sigma(0)\,[\,\overline{\Omega}_1 - \omega_1(0)\,]\exp(\mathrm{i}\tfrac{1}{2}\theta_f) + \gamma\,\sigma(1)\,[\,\overline{\Omega}_1 - \omega_1(1)\,]\exp(\mathrm{i}\tfrac{1}{2}\theta_i)}{4\sin\tfrac{1}{2}\Delta\theta}$$

ensures that $\Omega_1(0) = \Omega_1(1) = \overline{\Omega}_1$, so $\Omega_1(\xi)$ agrees with $\overline{\Omega}_1$ at the two end points, and the free parameter γ can be used to minimize the mean square deviation of $\Omega_1(\xi)$ about $\overline{\Omega}_1$. When the ERF angular velocity component $\omega_1(\xi)$ of $\mathbf{r}(\xi)$ exhibits relatively mild variations, the constructed MTF closely approximates the condition $\Omega_1(\xi) \equiv \overline{\Omega}_1$ (see Fig. 2). For instances where $\omega_1(\xi)$ has pronounced variations, this condition can be more accurately approximated [79] by subdividing the $[\,0, 1\,]$ parameter domain at the extrema and/or inflections of $\omega_1(\xi)$.

A method for constructing spatial PH curves with rational minimal–twist frames and prescribed end points, frames orientations, and total arc length, is described in [67]. This method employs the degree 7 PH curves with rotation–minimizing ERFs, identified by the constraints (14) on the coefficients of the pre–image polynomial. The problem can be reduced to computing the real roots of four quadratic equations in four real variables by Newton–Raphson iterations. The equations incorporate two free parameters (specifying the end derivative magnitudes) that can be exploited to optimize a suitable intrinsic measure of the curve shape.

4.3 *Rotation–Minimizing Osculating Frames*

The Frenet frame is rotation–minimizing with respect to the principal normal of a curve, while the adapted RMFs are rotation–minimizing with respect to the tangent. Thus, it seems natural to also consider the frames that are rotation–minimizing with respect to the remaining Frenet frame vector, the binormal [144]. By analogy with aeronautical terminology, these three frames may be considered to define *pitch–free*, *roll–free*, and *yaw–free* motions along a given spatial path.

Frames defining yaw–free motions, called *rotation–minimizing osculating frames* (RMOFs)—since the tangent and principal normal span the curve osculating plane, which has no instantaneous rotation about the binormal—were investigated in [60]. To identify curves with rational RMOFs, the focus of this study was on the double PH (DPH) curves described in Sect. 3.1, since they possess rational binormals. The possibility of generating a rational RMOF by means of a rational rotation of the rational tangent and principal normal vectors of a DPH curve was analyzed, and it was observed that this is not possible for space curves of degree < 7. The RMOF was also employed in [60] to construct certain ruled surfaces, whose tangent planes coincide with the osculating planes of a given space curve, and whose rulings have the least possible rotation consistent with this constraint.

An alternative approach to constructing space curves with rational RMOFs was developed in [136], based on space curves with *Pythagorean binormals* (PB curves), rather than the DPH curves. The PB curves constitute a superset of the DPH curves, since they require $|\mathbf{r}'(\xi) \times \mathbf{r}''(\xi)|$—but not $|\mathbf{r}'(\xi)|$—to be a polynomial or rational function of the curve parameter ξ. Consequently the PB curves do not, in general, admit rational representation of their cumulative arc lengths.

The construction of PB curves employs the dual representation [134], in which a space curve is defined by prescribing a one–parameter family of osculating planes, and they possess rational orthonormal frames, analogous to the ERF on spatial PH curves, comprising the binormal and orthogonal unit vectors in the osculating plane. By an appropriate rational rotation in the osculating plane, analogous to that used to construct rational rotation–minimizing adapted frames, a rational RMOF can be constructed. It was shown in [134] that such constructions are feasible for rational PB curves of degree ≥ 6, and for polynomial PB curves of degree ≥ 7.

4.4 Rotation–Minimizing Directed Frames

The notion of a rotation–minimizing *directed* frame (RMDF) was first introduced in [55], motivated by the problem of specifying the image–plane orientation for a camera executing a spatial path $\mathbf{r}(\xi)$ while imaging a static object,[4] situated at the origin. The intent is to minimize the apparent rotation of the image incurred by the camera motion. The RMDF comprises the unit polar vector $\mathbf{o}(\xi) = \mathbf{r}(\xi)/|\mathbf{r}(\xi)|$ and mutually orthogonal unit vectors $\mathbf{u}(\xi), \mathbf{v}(\xi)$ in the image plane, perpendicular to $\mathbf{o}(\xi)$, that exhibit no instantaneous rotation about $\mathbf{o}(\xi)$.

By replacing each derivative of $\mathbf{r}(\xi)$ in its Frenet–Serret apparatus [144] by a derivative of order one lower, one obtains the *directed* Frenet frame

$$\mathbf{o}(\xi) = \frac{\mathbf{r}(\xi)}{|\mathbf{r}(\xi)|}, \quad \mathbf{u}(\xi) = \frac{\mathbf{r}(\xi) \times \mathbf{r}'(\xi)}{|\mathbf{r}(\xi) \times \mathbf{r}'(\xi)|} \times \mathbf{o}(\xi), \quad \mathbf{v}(\xi) = \frac{\mathbf{r}(\xi) \times \mathbf{r}'(\xi)}{|\mathbf{r}(\xi) \times \mathbf{r}'(\xi)|},$$

[4]The assumption that the object is static and situated at the origin is not essential, since only the *relative* position of the object and camera matter.

and the *polar curvature and torsion*,

$$\lambda(\xi) = \frac{|\mathbf{r}(\xi) \times \mathbf{r}'(\xi)|}{|\mathbf{r}(\xi)|^3}, \quad \upsilon(\xi) = \frac{[\,\mathbf{r}(\xi) \times \mathbf{r}'(\xi)\,] \cdot \mathbf{r}''(\xi)}{|\mathbf{r}(\xi) \times \mathbf{r}'(\xi)|^2},$$

and hence a complete theory of *polar differential geometry*. For a *rational* RMDF, the theory of *Pythagorean–hodograph* (PH) curves can be adapted to introduce the *Pythagorean* (P) curves, characterized by solutions of the equation

$$x^2(\xi) + y^2(\xi) + z^2(\xi) = \rho^2(\xi) \tag{24}$$

for some polynomial $\rho(\xi)$, which ensures a rational polar vector $\mathbf{o}(\xi) = \mathbf{r}(\xi)/\rho(\xi)$.

The theory of P curves coincides with that of PH curves, applied to the *anti–hodograph* $\mathbf{s}(\xi) = \int \mathbf{r}(\xi)\,\mathrm{d}\xi$ (i.e., the indefinite integral) of the given curve $\mathbf{r}(\xi)$. The methods used to construct spatial PH curves, and to identify the existence of rational rotation–minimizing adapted frames on them, can thus be adapted to define spatial P curves and identify those instances that admit rational RMDFs. Algorithms that employ P curves to construct rational rotation–minimizing camera motions, which interpolate given positions and RMDF orientations, are described in [64].

5 Algorithms for Pythagorean–Hodograph Curves

To facilitate their adoption in practical applications, algorithms to construct, modify, and analyze PH curves are required. Building upon prior advances, this continues to be a very active area of research, with the development of many new capabilities. The flexibility of PH curves in practical use has been confirmed by the proof of a Weierstrass–type approximation theorem in [29]. This states that, for any C^1 curve in the Euclidean spaces $\mathbb{R}^2$ and $\mathbb{R}^3$ or Minkowski space $\mathbb{R}^{2,1}$ there exists a sequence of polynomial PH curves of increasing degree converging uniformly to that curve.

In practice, to obtain more shape flexibility, spline extensions of polynomial PH curves are typically preferred to individual PH curves of higher degree. However, to import PH splines into commercial CAD systems, it is necessary to express them[5] in the standard free–form B–spline representation (B–form). The B–form of C^2 PH quintic splines was derived in [66]: since the PH segments are quintic, but meet with C^2 continuity, triple interior knots must be used. A scheme for local modification of planar PH quintic splines was also proposed in [66], that preserves the PH structure in response to a generic control point displacement. This procedure modifies only two contiguous spline segments, and the continuity between the modified and unmodified segments must be relaxed from C^2 to C^1 to preserve the PH structure of the spline. An example is shown in Fig. 3—see [66] for complete details.

[5]To exploit the advantageous features of PH splines, one must also store the complex or quaternion pre–image polynomials of each spline segment—see Sects. 5.4 and 7.3.

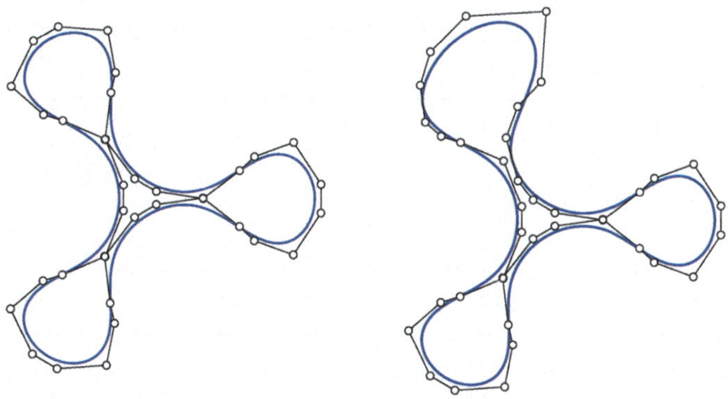

Fig. 3 A C^2 PH quintic spline (left) and the modified PH quintic spline (right), obtained after applying a displacement to one of the B-spline control point. The B-spline control polygons of the two PH quintic splines are also shown

Planar C^2 PH quintic splines have traditionally been constructed [4, 73] as the interpolants to a given sequence of points $\mathbf{p}_0, \ldots, \mathbf{p}_n$. Alternative control–polygon based methodologies for specifying planar and spatial C^2 PH spline curves were developed in [75, 169]. In [3] the B–spline form of a planar C^2 PH quintic spline is obtained by direct integration of (3), with the complex pre–image polynomial $\mathbf{z}(\xi)$ replaced by a complex spline function $\mathbf{Z}(\xi)$ specified by its knots and coefficients. Closed–form expressions for the PH B–spline control points in terms of the knots and coefficients of $\mathbf{Z}(\xi)$ are derived in [3]. Because of the non–linear dependence of the former on the latter, the main difficulty with this approach is in ensuring that the PH spline curve satisfies prescribed geometrical properties.

5.1 Construction Algorithms

Because of their non–linear nature, algorithms to construct or modify PH curves based on control polygon paradigms are, in general, impractical—see, however, [123]. Consequently, the typical approach [4, 80] to designing individual PH curve segments or PH spline curves is based upon interpolating discrete geometrical data (points, tangents, etc.). These interpolation problems typically admit a multiplicity of solutions that may differ substantially in shape, and the identification of a "good" solution is usually based on minimization of a shape measure, such as the *absolute rotation index* or the *elastic bending energy*, defined for a curve of total length S by

$$R = \int_0^S |\kappa| \, \mathrm{d}s \,, \quad E = \int_0^S \kappa^2 \, \mathrm{d}s \,, \tag{25}$$

where κ is the curvature and s is arc length [40, 80]. Note, however, that minimizing E without any constraint on S can lead to unbounded interpolants [14].

The construction of PH spline interpolants can be achieved through both local and global schemes, based on prescribed discrete data (points and possibly also unit tangent vectors). Since the minimization of a shape measure typically requires the use of numerical methods, and can be computationally quite expensive, alternative strategies to identify the "good" solution have been proposed in the literature. As outlined in Sects. 5.1.1 and 5.1.2 below, a common theme is to identify the PH solution that is in some sense "near" to the standard ordinary cubic solution for the prescribed data. This is based on the perception that cubic interpolants usually have reasonable shape (although it should be noted that, for derivative magnitudes large compared to the spacing between points, planar cubic interpolants will often exhibit loops in cases where at least one PH interpolant is free of self–intersections).

5.1.1 Local Interpolation Schemes

First order Hermite interpolation has often been used to design planar or spatial PH curves, since it allows development of local schemes based on polynomial/rational PH curves of low degree. With single segments of primitive planar polynomial PH curves, the minimum degree of practical value is five, since the planar PH curves of degree three are all segments of the Tschirnhaus cubic, which has insufficient shape flexibility to interpolate even general G^1 Hermite data [80]. A characterization of the set of G^1 Hermite data for which planar PH cubic interpolants exist was given in [20], together with an identification of the subset of these data that also guarantees the absence of a self–intersection loop in the interpolant. The use *TC–biarcs* (i.e., a pair of Tschirnhaus cubic segments that meet with tangent continuity) to solve the C^1 Hermite interpolation problem has been discussed in [11, 84].

On the other hand, it is well–known [80] that in general four distinct planar PH quintic interpolants to given C^1 Hermite data always exist, and the absolute rotation index R defined in (25) provides a robust, closed–form measure for identification of the "good" solution. The method of [80] was extended in [118] to accommodate interpolation of C^1 data together with end–point curvatures, using planar PH curves of degree 7, and it was shown that up to eight distinct interpolants exist.

An alternative topological scheme for selecting the "good" planar PH quintic interpolant was proposed in [25], based on the notion that it should be continuously deformable into the unique ordinary cubic interpolant matching the given C^1 data. More precisely, it was shown in [25] that at least one of the four PH quintic Hermite interpolants exhibits this property, which corresponds to vanishing of the *hodograph winding number* relative to the ordinary cubic. An explicit identification scheme based on this idea was developed, that does not rely upon the evaluation of fairness measures, and is still applicable when the uniqueness fails [25].

In order to obtain greater flexibility, several variants of the basic polynomial PH curve Hermite interpolation schemes have been proposed. For example, a family of polynomial PH curves with complex hodographs of the form $\mathbf{r}'(\xi) = \mathbf{k}\,(\xi - \mathbf{c})^{2n+1} + \mathbf{d}$ (where $\mathbf{k}, \mathbf{c}, \mathbf{d}$ are complex values) was proposed in [126] for solving the C^1 Hermite interpolation problem, and a polynomial re–parameterization was used to separately control the shape and parametric speed of the interpolant. A Möbius transformation was applied to planar PH cubics in [150], that yields an additional complex degree of freedom allowing the determination of four rational PH cubic interpolants to planar C^1 Hermite data. In [149], conditions on the complex coefficients of a truncated Laurent series of the form

$$\mathbf{r}(\xi) = \frac{\mathbf{a}_{-1}}{\xi - \mathbf{c}} + \mathbf{a}_0 + \mathbf{a}_1(\xi - \mathbf{c}) + \mathbf{a}_3(\xi - \mathbf{c})^3 ,$$

where $\mathbf{c}$ is a complex value, were identified to ensure that $\mathbf{r}(\xi)$ is a planar rational PH curve, and it was shown that this form generically admits four interpolants to C^1 data. The numerical examples in [149, 150] indicate that one of the constructed interpolants typically has reasonable shape, and some fairness measures may even improve on the standard PH quintic interpolants. However, in some instances the constructed PH curves are of higher degree, and the rational interpolants do not in general possess rational arc length functions [45].

The interpolation of generic spatial Hermite again cannot be achieved by cubic polynomial PH curves, even in the case when only unit end tangents are specified [119, 145, 168], unless an adjustment of the orientation of one of them is allowed [168]. Consequently, if only primitive polynomial PH curves are desired, it is again necessary to appeal to the quintics. Using the quaternion representation, it was first shown in [51] that interpolation of C^1 Hermite data with spatial PH quintics incurs a two–parameter family of interpolants,[6] rather than a finite number as in the planar case. Since the shape of the interpolants can be rather sensitive to the choice of the two free angular parameters, three criteria were proposed in [56] for their selection. These criteria ensure at least one reasonable interpolant, and also degree reduction to a PH cubic interpolant when the Hermite data are compatible with its existence. Moreover, one of them (the CC criterion) admits easy closed–form implementation, and it has been shown [188] to yield fourth–order approximation for the spatial PH quintic interpolation scheme, as with ordinary cubic curves. Prior work on the approximation order of planar and spatial PH curves may be found in [194, 196].

Interpolation of spatial G^1 Hermite data under shape constraints by PH quintic splines was addressed in [76], by an extension of the analysis previously developed in the planar case [78]. A tension parameter is associated with each spline segment, and the imposed shape constraints are concerned with preservation of convexity at the knots and the sign of the discrete torsion on each spline segment. An asymptotic analysis developed in terms of the tension parameters shows that satisfaction of the prescribed shape constraints can always be achieved for each segment by using the CC criterion to fix the two free angular parameters of that segment, requiring only mild application of the tension parameters that does not compromise the overall fairness of the interpolant, as illustrated in Fig. 4.

To achieve low–degree interpolants, local Hermite interpolation schemes based on spatial PH cubic biarcs were introduced in [8, 187]. Quintic triarcs have been

[6] This is a consequence of the fact that, under the map (5), the pre–image of a point in $\mathbb{R}^3$ is a circle in the quaternion space $\mathbb{H}$—see [68].

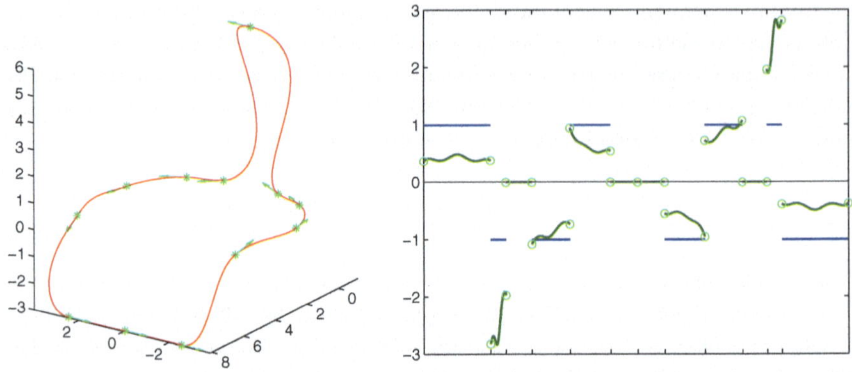

Fig. 4 An example spatial G^1 PH quintic spline constructed by the local interpolation scheme in [76]. Left: the PH spline (red) interpolating the "chair" point set and associated unit tangents (green arrows). Right: the torsion plot (green), and the sign of the discrete torsion (blue). In each spline segment the tension is set equal to 1 and the two angular parameters are fixed by the CC criterion

adopted in [9] to produce C^2 PH spline interpolants to second order Hermite data, but a six–dimensional family of solutions is obtained in the spatial case.

As an alternative to the customary first–order Hermite interpolation approach, a local Lagrange geometric interpolation scheme based on the planar PH cubics was proposed in [111], requiring as input just a sequence of four points, but achieving only C^0 smoothness at the end points. Even in this case "reasonable data" conditions are necessary to ensure the existence of an interpolant, whose computation requires the solution of a non–linear rational system in the two unknown parameter values associated with the interior data points.

The study [113] develops a scheme for geometric interpolation using PH curves in Euclidean spaces of any dimension, independent of the specific forms (3) and (5) in $\mathbb{R}^2$ and $\mathbb{R}^3$. In this scheme, the PH condition is imposed as a constraint on the curve coefficients, rather than being built in a priori by an algebraic representation. The method incurs the solution of a non–linear system of equations, and is illustrated by construction of cubic Hermite and Lagrange interpolants, using either standard polynomial root–solvers or a homotopy approach. This scheme was also applied in [114] to local G^2 interpolation with PH quintics that interpolate end points together with unit tangents and curvature values. Although multiple solutions exist, a method to identify the "good" interpolant is suggested by means of an asymptotic analysis. Prior work on the construction of higher degree PH curve Hermite interpolants may be found in, for example, [118, 197].

5.1.2 Global Interpolation Schemes

If derivative values at the data points are not available, an approximation (e.g. finite difference) scheme must be first applied in order to use a local Hermite interpolation method. In this case, the shape of the resulting spline curve may depend sensitively on the quality of the derivative estimation. Moreover, local constructions of spline curves cannot address the important issue of *global* shape quality.

Traditional spline constructions adopt a Lagrange interpolation scheme, wherein the input consists of a sequence of discrete data points that are to be interpolated at the spline nodes. This approach typically involves the solution of a global system of equations, but (for a given degree of the spline segments) yields a higher order of continuity at the nodes, and favorable global shape fairness measures. This is the approach first adopted [4, 73] for the construction of planar C^2 PH quintic splines, and subsequently extended to the spatial C^2 PH quintic splines in [77].

For the interpolation of $n + 1$ data points in the planar case, there are 2^{n+m} distinct PH interpolants, where $m \in \{-1, 0, 1\}$ depends on the adopted end conditions. The initial study [4] employed a homotopy scheme to generate the complete space of all formal C^2 PH quintic spline interpolants, but this is too expensive when only the "good" spline interpolant is required. As an efficient alternative, Newton–Raphson iterations was adopted in [73] to numerically solve the non–linear tridiagonal system of complex quadratic equations characterizing the solution space. In this context, an accurate starting approximation is essential for efficient and reliable convergence to the "good" PH spline solution, and to achieve this the nodal derivatives of the "ordinary" C^2 cubic spline interpolating the same data points are exploited.

A similar approach was adopted in [28], but a criterion based on the *absolute hodograph winding number*—an extension of the concept introduced in [25]—is employed to choose the C^2 PH quintic spline closest to the "ordinary" C^2 cubic spline (interpolating the same data points) from a topological viewpoint.

If G^2 smoothness is sufficient and a convex set of data points satisfying certain assumptions is specified, together with compatible initial/final unit tangent vectors, an alternative spline formulation based upon PH cubic (Tschirnhaus) segments was proposed in [112]. The construction requires solution of a non–linear tridiagonal system of equations. Since it is essentially based on piecing together segments of a unique curve, this formulation has a rather limited domain of applicability.

The construction of spatial C^2 PH quintic spline interpolants is more challenging, since it is characterized by a system of N quadratic vector equations in N quaternion unknowns. In [77], an additional N scalar conditions were imposed to formulate a Newton–like scheme that produces reasonable spline interpolants.

The study [113] considers the problem of constructing a single PH curve segment of degree n in d–dimensional Euclidean space by solving a Lagrange interpolation problem and its variant with specified boundary unit tangent vectors. The PH nature of the interpolant is imposed by constraint equations, rather than through the usual forms (3) and (5) appropriate to $\mathbb{R}^2$ and $\mathbb{R}^3$. This approach requires the solution of a rather complicated system of non–linear equations, and homotopy theory is used to establish the existence and (in some cases) the number of admissible solutions.

5.2 PH Curves with Prescribed Arc Lengths

Since polynomial PH curves possess polynomial cumulative arc length functions, it is possible to impose an *exact* specified total arc length in the construction of curves interpolating prescribed geometrical data. It is perhaps surprising, however, that this capability has only recently been exploited. The imposition of arc length constraints in the construction of PH curves was first considered in [106], motivated by the problem of measuring physical shapes using the *Morphosense*, a ribbon–like device equipped with electronic sensors at known positions along its length. In computing a spline fit to data obtained from the Morphosense laid against a smooth surface, the sensor spacings must be imposed as the arc lengths between spline nodes.

With the Hopf map representation, the construction of an N–segment C^2 spatial PH quintic spline interpolating points $\mathbf{p}_0, \ldots, \mathbf{p}_N$ with segment arc lengths $l_1, \ldots, l_N$ can be formulated as the solution of a system of N complex and $2N$ real equations in $2N + 4$ complex coefficients. Instead of using spline end conditions to address the excess degrees of freedom, the solution is obtained by optimizing a shape measure, under constraints specified by the N complex and $2N$ real interpolation conditions. The construction used in [106] is overtly numerical in nature, and the complexity of the equations makes questions of existence and uniqueness of solutions difficult to analyze. They are more amenable to analysis in the context of individual PH quintic segments interpolating planar or spatial G^1 Hermite data, as described below.

The construction of planar PH quintics $\mathbf{r}(\xi)$, $\xi \in [\,0, 1\,]$ with specified end points $\mathbf{p}_0, \mathbf{p}_1$ and tangents $\mathbf{t}_0, \mathbf{t}_1$ and a prescribed total arc length L was addressed in [47], under the assumption of equal–magnitude end derivatives, i.e., $|\mathbf{r}'(0)| = |\mathbf{r}'(1)| = w^2$ where $w \neq 0$ (so symmetric data yields symmetric interpolants). Transforming the data to canonical form with $\mathbf{p}_0 = (0, 0)$ and $\mathbf{p}_1 = (1, 0)$, setting $\mathbf{t}_0 = (\cos\theta_0, \sin\theta_0)$ and $\mathbf{t}_1 = (\cos\theta_1, \sin\theta_1)$, and expressing the Bernstein coefficients of the complex quadratic polynomial $\mathbf{w}(\xi)$ as

$$\mathbf{w}_0 = w \exp(\mathrm{i}\tfrac{1}{2}\theta_0)\,, \quad \mathbf{w}_1 = u + \mathrm{i}\,v\,, \quad \mathbf{w}_2 = w \exp\left(\mathrm{i}\tfrac{1}{2}\theta_1\right)\,,$$

the construction can be reduced to finding the solutions of the system of quadratic equations

$$\begin{aligned} 4u^2 + 6(c_0 + c_1)uw + (6c_0^2 + 6c_1^2 + 2c_0c_1)w^2 - 15(L+1) &= 0\,, \\ 4uv + 3(s_0 + s_1)uw + 3(c_0 + c_1)vw + (6c_0s_0 + 6c_1s_1 + c_0s_1 + c_1s_0)w^2 &= 0\,, \\ 4v^2 + 6(s_0 + s_1)vw + (6s_0^2 + 6s_1^2 + 2s_0s_1)w^2 - 15(L-1) &= 0\,, \end{aligned}$$

in the three real unknowns u, v, w where $(c_0, s_0) = (\cos\frac{1}{2}\theta_0, \sin\frac{1}{2}\theta_0)$ and $(c_1, s_1) = (\cos\frac{1}{2}\theta_1, \sin\frac{1}{2}\theta_1)$. These equations admit further reduction to just a single quadratic equation in w^2 that always has two positive roots, but only the smaller root identifies real solutions to the above system [47]. Hence, in general, there are two distinct PH quintic interpolants to the data $\mathbf{p}_0, \mathbf{p}_1, \mathbf{t}_0, \mathbf{t}_1, L$ of which one has attractive shape, and

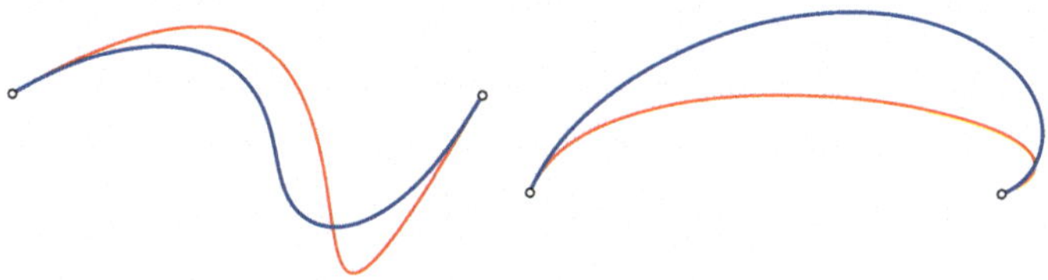

Fig. 5 Planar PH quintics (blue) with end points $\mathbf{p}_i = (0, 0)$, $\mathbf{p}_f = (1, 0)$ and tangent directions and arc lengths $\theta_0 = 30°$, $\theta_1 = 60°$, $L = 1.35$ (left); and $\theta_0 = 67.5°$, $\theta_1 = -157.5°$, $L = 1.50$ (right). The "ordinary" cubics with the same end points and tangents are also shown (red) for comparison

the other is discarded due to undesired loops. Figure 5 illustrates some examples of the constructed PH quintics—for comparison, the "ordinary" cubics with the same end points and tangents are also shown. This procedure has been employed in [48] to construct helical paths (or *curves of constant slope*) between given end points and motion directions, through a "lifting" process. Such curves can be used to specify a flight path at constant speed for an aerial vehicle with a limited climb rate.

The generalization of the analysis in [47] to the case of *spatial* PH quintics $\mathbf{r}(\xi)$, $\xi \in [\,0, 1\,]$ with end points $\mathbf{p}_0, \mathbf{p}_1$ and tangents $\mathbf{t}_0, \mathbf{t}_1$ and total arc length L, again with the assumption that $|\mathbf{r}'(0)| = |\mathbf{r}'(1)| = w^2$, is more challenging [49]. In the standard C^1 Hermite interpolation by spatial PH quintics [56], it is known that the solutions depend on two free parameters ψ_0, ψ_2 and their arc lengths are determined (within a certain range) by the *difference* $\psi_2 - \psi_0$. Compared to C^1 Hermite interpolation, the two degrees of freedom gained in the present context by relaxing from C^1 to G^1 data are used up by imposing the arc length L, which need only satisfy $L > |\,\mathbf{p}_1 - \mathbf{p}_0|$, and the assumption $|\mathbf{r}'(0)| = |\mathbf{r}'(1)| = w^2$. Thus, we may expect solutions to the present problem (if they exist for any given data) to also exhibit two free parameters.

Based on canonical–form data, the construction of G^1 spatial PH quintic Hermite interpolants of given arc length L was reduced in [49] to finding the positive roots of a quadratic equation in w^2 with data–dependent coefficients, and these roots must satisfy certain data–dependent bounds. It was shown [49] that this equation always has two positive roots, of which only the smaller root satisfies the stipulated bounds. Solutions are thus obtained for any given data G^1 Hermite $\mathbf{p}_0, \mathbf{p}_1, \mathbf{t}_0, \mathbf{t}_1$ and any arc length $L > |\,\mathbf{p}_1 - \mathbf{p}_0|$, incorporating two free parameters ψ_0, ψ_2. Since the interpolant shape can be quite sensitive to the values of these parameters, exploiting them to optimize a suitable measure of the intrinsic curve shape is desirable.

It is appropriate to also mention here the "rectifying control polygon" for planar PH curves, introduced in [123]. Since a PH curve of degree n has fewer degrees of freedom than a general polynomial curve of degree n, an arbitrary modification of its Bézier control polygon will destroy its PH nature. Also, as observed in Sect. 2, only the simplest non–trivial (cubic) case admits an intuitive characterization of PH curves in terms of the geometry of the Bézier control polygon.

The rectifying control polygon is defined so as to have the same arc length and degrees of freedom as the associated PH curve. The control points of the rectifying control polygon are convex combinations of the Bézier control points. It matches the curve end points, but not the tangents, and the familiar shape–preserving properties of the Bézier form (which arise from its linear nature) do not hold. The problem of determining a PH curve from a given rectifying control polygon is also addressed in [123]. As usual with PH curves, this problem admits a multiplicity of solutions, and a suitable shape measure must be invoked to identify the "good" solution.

5.3 Algorithms for Geometric Design Applications

The advantageous properties of PH curves have proven useful in a variety of basic geometric design problems. For example, they have been employed to define *fillet* or *transition* curves that smoothly join simple geometrical elements, such as linear, circular, or spiral path segments [95, 101, 151, 205, 206]. Such transition curves are of interest in applications such as the layout of highways and railways, or smoothing of sharp corners in piecewise–linear toolpaths for CNC machining (see Sect. 7.1), and may have to satisfy various constraints—a prescribed total arc length, the order of continuity with the segments they connect, monotonicity of curvature, etc.

Planar polynomial PH curves have also been applied [158] to solve the classical *isoperimetric problem*, i.e., to find the closed curve of a given length encompassing the largest possible area. It is well known that, among all possible planar curves, the optimal solution to this problem is a circle. The use of polynomial PH curves in this context is motivated by the availability of simple closed–form expressions for the total arc length. In [158] the problem is studied in the context of PH curves $\mathbf{r}(\xi)$, $\xi \in [\,-1,+1\,]$ with $\mathbf{r}(+1) = \mathbf{r}(-1)$ and various degrees of smoothness at this juncture point. It is interesting to note that the best PH curve solution of a given degree is just C^0 even though solutions with higher orders of continuity still exhibit convergence of the geodesic distance from a circle with increasing degree.

The dual form of rational planar PH curves and their offsets has also been applied [1] to construct ovals and rosettes of constant width. An oval of constant width w is a convex curve free of self–intersections that, for any orientation, touches parallel lines a distance w apart—such curves have applications in mechanism design. The constructions used in [1] employ G^2 piecewise rational PH curves of class 4, whose segments are portions of the involutes to Tschirnhaus cubics, and the constant width property is achieved by parallel translations of the control lines in the dual form.

5.4 Reverse Engineering of PH Curves

To fully utilize the advantageous properties of PH curves, it is necessary to know their *pre–image polynomials*—namely, the complex polynomial $\mathbf{w}(\xi)$ in (3) for a planar PH curve, and the quaternion polynomial $\mathcal{A}(\xi)$ in (5) for a spatial PH curve (or equivalently the two complex polynomials $\boldsymbol{\alpha}(\xi)$, $\boldsymbol{\beta}(\xi)$ in the Hopf map form (8)). However, this information is lost when a polynomial PH curve $\mathbf{r}(\xi)$ is communicated to an application using only its control points, and in fact there is no immediate way of knowing whether or not a given curve $\mathbf{r}(\xi)$ *is* a PH curve.

In order to address this problem, the idea of "reverse engineering" PH curves was introduced in [65]. The first step is to *identify* whether given control points define a polynomial PH curve, and to determine its degree (since the control points may define a degree–elevated form). With an affirmative outcome, the second step is to construct the pre–image polynomial(s) from the given control points.

The simplest identification is based on the polynomial arc length of PH curves, instead of analyzing the control polygon. The parametric speed $|\mathbf{r}'(\xi)| = \mathrm{d}s/\mathrm{d}\xi$ of a polynomial curve $\mathbf{r}(\xi)$ is the derivative of arc length s with respect to the parameter ξ, and it is a polynomial $\sigma(\xi)$ if and only if $\mathbf{r}(\xi)$ is a PH curve. A quadrature scheme with m nodes on $\xi \in [0, 1]$ is said to have *precision* n if it yields exact integral values for polynomials of degree $\leq n$ from a weighted sum of values sampled at the nodes. On applying such a scheme to $|\mathbf{r}'(\xi)|$, the arc length estimates will "saturate" as m increases if $\mathbf{r}(\xi)$ is a PH curve, since its parametric speed is a polynomial.

The *Gauss–Legendre* and *Newton–Cotes* rules are two commonly used numerical quadratures, which exhibit saturation of the arc length estimates for PH curves of degree n when $m \geq \lceil \frac{1}{2} n \rceil$ and $2 \lfloor \frac{1}{2} (m + 1) \rfloor > n$, respectively. Hence, for PH quintics, saturation occurs for $m \geq 3$ with the former, and $m \geq 5$ with the latter, although the former involves more complicated nodes and weights. Note that this identification scheme applies equally to both planar and spatial PH curves.

Once the PH nature and degree of a polynomial curve $\mathbf{r}(\xi)$ have been verified, the Bernstein coefficients of the pre–image polynomial can be determined from a system of quadratic equations. In the case of a planar PH quintic with the complex control points $\mathbf{p}_k = x_k + \mathrm{i}\, y_k$, for example, $\mathbf{r}'(\xi)$ has coefficients $\mathbf{d}_k = n\,(\mathbf{p}_{k+1} - \mathbf{p}_k)$, and the coefficients $\mathbf{w}_0, \mathbf{w}_1, \mathbf{w}_2$ of the quadratic polynomial $\mathbf{w}(\xi)$ must satisfy

$$\mathbf{w}_0^2 = \mathbf{d}_0\,, \quad \mathbf{w}_0\mathbf{w}_1 = \mathbf{d}_1\,, \quad 2\,\mathbf{w}_1^2 + \mathbf{w}_0\mathbf{w}_2 = 3\,\mathbf{d}_2\,, \quad \mathbf{w}_1\mathbf{w}_2 = \mathbf{d}_3\,, \quad \mathbf{w}_2^2 = \mathbf{d}_4\,.$$

Although this system appears overdetermined, the known PH nature of $\mathbf{r}(\xi)$ ensures its consistency, and one may use only the first (or last) 3 equations. The $\mathbf{w}_0, \mathbf{w}_1, \mathbf{w}_2$ values are then easily obtained (up to a sign indeterminacy).

As an example, the method was applied to the planar PH quintic defined [80] by the initial and final pairs of control points $\mathbf{p}_0 = (1.0, 1.0), \mathbf{p}_1 = (2.5, -0.5)$ and $\mathbf{p}_4 = (2.5, 4.5), \mathbf{p}_5 = (4.0, 3.0)$ with Newton–Cotes quadrature. The arc lengths computed in double–precision arithmetic saturate as expected:

$$\begin{aligned}
m &= 2, \quad S = 10.60660171779822,\\
m &= 3, \quad S = 6.88667502260488,\\
m &= 4, \quad S = 6.09350707561590,\\
m &= 5, \quad S = 5.45897271802472,\\
m &= 6, \quad S = 5.45897271802472,\\
m &= 7, \quad S = 5.45897271802472,\\
m &= 8, \quad S = 5.45897271802472,
\end{aligned}$$

and the Bernstein coefficients of $\mathbf{w}(\xi)$ are obtained as

$$\begin{aligned}\mathbf{w}_0 &= 3.00887036259443 - 1.24631491160906\,\mathrm{i}\,,\\ \mathbf{w}_1 &= 0.00383089626255 + 4.56753122870050\,\mathrm{i}\,,\\ \mathbf{w}_2 &= 3.00887036259443 - 1.24631491160906\,\mathrm{i}\,.\end{aligned}$$

An analogous method holds for determining the coefficients of the quaternion pre–image polynomial $\mathcal{A}(\xi)$ in the case of spatial PH curves [65].

6 Surface Constructions Based on PH Curves

The distinguishing property of a PH curve $\mathbf{r}(\xi)$ is the particularly simple form of its *parametric speed*, which defines the fundamental relation between the parameter ξ and curve arc length s. This property has no immediate, intuitive generalization to a parametric surface $\mathbf{r}(u, v)$, although there are definitions—such as the PN surfaces discussed in Sect. 3.5—that inherit certain aspects of PH curves. Nevertheless, PH curves can still play an important role in numerous surface construction contexts, as illustrated by the representative examples described below.

6.1 Rational Patches Bounded by Lines of Curvature

A curve $\mathbf{c}(\xi)$ on a parametric surface $\mathbf{s}(u, v)$ is defined by specifying the two surface parameters as functions $u(\xi)$, $v(\xi)$ of another parameter ξ identifying position along the curve. The *Darboux frame* $(\mathbf{t}(\xi), \mathbf{h}(\xi), \mathbf{n}(\xi))$ along the curve $\mathbf{c}(\xi) = \mathbf{s}(u(\xi), v(\xi))$ is defined [144] in terms of the surface partial derivatives $\mathbf{s}_u, \mathbf{s}_v$ as

$$\mathbf{t} = \frac{u'\,\mathbf{s}_u + v'\,\mathbf{s}_v}{|\,u'\,\mathbf{s}_u + v'\,\mathbf{s}_v\,|}\,, \quad \mathbf{h} = \frac{\mathbf{s}_u \times \mathbf{s}_v}{|\,\mathbf{s}_u \times \mathbf{s}_v\,|} \times \mathbf{t}\,, \quad \mathbf{n} = \frac{\mathbf{s}_u \times \mathbf{s}_v}{|\,\mathbf{s}_u \times \mathbf{s}_v\,|}\,. \tag{26}$$

It consists of the *tangent* $\mathbf{t}(\xi)$ to the curve $\mathbf{c}(\xi)$, the *surface normal* $\mathbf{n}(\xi)$ along it, and the *tangent normal* $\mathbf{h}(\xi) = \mathbf{n}(\xi) \times \mathbf{t}(\xi)$—a vector lying in the surface tangent plane that is orthogonal to the curve tangent $\mathbf{t}(\xi)$ at each point. The variation of the Darboux frame along $\mathbf{c}(\xi)$ is described [144, 199] by the equations

$$\begin{bmatrix}\mathbf{t}'\\ \mathbf{h}'\\ \mathbf{n}'\end{bmatrix} = \sigma \begin{bmatrix} 0 & \kappa_{\mathrm{g}} & \kappa_{\mathrm{n}} \\ -\kappa_{\mathrm{g}} & 0 & -\tau_{\mathrm{g}} \\ -\kappa_{\mathrm{n}} & \tau_{\mathrm{g}} & 0 \end{bmatrix} \begin{bmatrix}\mathbf{t}\\ \mathbf{h}\\ \mathbf{n}\end{bmatrix}, \tag{27}$$

where the parametric speed, normal curvature, geodesic curvature, and geodesic torsion of $\mathbf{c}(\xi)$ are defined by

$$\sigma = |u'\mathbf{s}_u + v'\mathbf{s}_v|, \quad \kappa_n = \frac{\mathbf{n} \cdot \mathbf{t}'}{\sigma}, \quad \kappa_g = \frac{\mathbf{h} \cdot \mathbf{t}'}{\sigma}, \quad \tau_g = \frac{\mathbf{h} \cdot \mathbf{n}'}{\sigma}.$$

The variation of the Darboux frame may alternatively be expressed in terms of its angular velocity

$$\boldsymbol{\omega} = -\tau_g \mathbf{t} - \kappa_n \mathbf{h} + \kappa_g \mathbf{n}, \tag{28}$$

as

$$\frac{d\mathbf{t}}{ds} = \boldsymbol{\omega} \times \mathbf{t}, \quad \frac{d\mathbf{h}}{ds} = \boldsymbol{\omega} \times \mathbf{h}, \quad \frac{d\mathbf{n}}{ds} = \boldsymbol{\omega} \times \mathbf{n}.$$

A *line of curvature* on a smooth surface is a locus whose tangent coincides with a principal curvature direction of the surface at every point [144]. There are two families of lines of curvature, tangent to each of the principal directions, that form an "orthogonal net" over the surface.[7] An alternative characterization [144] is that a surface curve $\mathbf{c}(\xi)$ is a line of curvature if and only if its geodesic torsion satisfies $\tau_g \equiv 0$. From (28) this implies that $\boldsymbol{\omega} \cdot \mathbf{t} \equiv 0$, and hence $\mathbf{c}(\xi)$ is a line of curvature if and only if its Darboux frame is rotation–minimizing with respect to **t**, i.e., **h** and **n** exhibit no instantaneous rotation about **t** along the curve $\mathbf{c}(\xi)$.

This property facilitates construction of rational tensor–product surface patches with boundary curves that are lines of curvature [13], specified by the RRMF curves described in Sect. 4.1. The construction commences with the four corner points and associated Darboux frames for the desired surface patch. Four quintic RRMF curves are then constructed to define the patch boundary curves, and their RMFs determine the variation of the surface normal and the tangent normal on them. The patch interior is then "filled in" using a Coons interpolation scheme [38], and the residual free parameters may be used to optimize the interior shape.

6.2 *Rational Swept Surface Constructions*

The construction of surfaces through *sweep operations* offers an intuitive approach to specifying a diverse family of surface types that are commonly used in geometric design. The basic idea is to use a *sweep curve* $\mathbf{s}(v)$ to specify a continuous family of geometrical transformations that act on a *profile curve* $\mathbf{p}(u)$. These transformations may be combinations of translations, rotations, scalings, and more general actions. The profile and sweep curves are typically defined as polynomial or rational forms, and in order to ensure compatibility with prevailing CAD representations, the sweep operation should generate a rational swept surface $\mathbf{r}(u, v)$.

[7]The system of lines of curvature is singular at *umbilic points*, where the principal directions are undefined [144].

A systematic approach to rational swept surface constructions was first proposed in [105]. For rational profile and sweep curves $\mathbf{p}(u)$ and $\mathbf{s}(v)$, this approach defines a rational swept surface $\mathbf{r}(u, v)$ whose homogeneous coordinates are bilinear in those of $\mathbf{p}(u)$ and $\mathbf{s}(v)$. When the profile and sweep curves are rational cubics, they may be represented by 4×4 coefficient matrices, and by contracting these matrices with a $4 \times 4 \times 4$ *selector matrix*, which defines a particular type of sweep operation, one obtains the $4 \times 4 \times 4$ matrix that defines a rational bicubic swept surface $\mathbf{r}(u, v)$.

This formalism allows familiar sweep forms, such as translational, rotational, and conical surfaces, to be automatically generated, eliminating the possibility of errors in deriving them "by hand" [32, 176]. However, for general polynomial or rational curves, restricting $\mathbf{r}(u, v)$ to a bilinear dependence on the homogeneous coordinates of $\mathbf{p}(u)$ and $\mathbf{s}(v)$ excludes many desirable sweep types. For example, to translate $\mathbf{p}(u)$ along $\mathbf{s}(v)$ such that the former always lies in the normal plane of the latter, the sweep curve must have a rational unit tangent vector, i.e., it must be a PH curve.

Since polynomial PH curves possess polynomial parametric speed and arc length functions, and rational unit tangents and curvatures, using them as sweep curves vastly enlarges the morphology of sweep types that generate rational swept surfaces [81]. The examples described below illustrate the diversity of rational swept surface types that are possible with PH sweep curves (see [81] for more complete details). In these examples, the parameters u and v indicate whether a function refers to the profile curve or sweep curve.

- *Oriented–translation sweep*. Given a profile curve $\mathbf{p}(u) = x(u)\,\mathbf{i} + z(u)\,\mathbf{k}$ in the (x, z) plane, and a PH sweep curve $\mathbf{s}(v) = x(v)\,\mathbf{i} + y(v)\,\mathbf{j}$ in the (x, y) plane with parametric speed $\sigma(v)$ and normal $\mathbf{n}(v)$, the surface

$$\mathbf{r}(u, v) = \mathbf{s}(v) + x(u)\,\mathbf{n}(v) + z(u)\,\mathbf{k}$$

  is generated by translating $\mathbf{p}(u)$ along $\mathbf{s}(v)$, while orienting $\mathbf{p}(u)$ so as to always lie in the normal plane of $\mathbf{s}(v)$. To guarantee that $\mathbf{r}(u, 0) = \mathbf{p}(u)$, the sweep curve must satisfy $\mathbf{s}(0) = \mathbf{0}$ and $\mathbf{s}'(0)/\sigma(0) = \mathbf{j}$.

- *Offset–translation sweep*. In place of a sweep curve $\mathbf{s}(v)$, this surface type uses a distance function $d(v)$ to specify a continuous family of offset curves to a planar PH profile curve $\mathbf{p}(u) = x(u)\,\mathbf{i} + y(u)\,\mathbf{j}$ with unit normal $\mathbf{n}(u)$, which are uniformly translated orthogonal to the plane of $\mathbf{p}(u)$. The resulting *offset–translation* swept surface is defined by

$$\mathbf{r}(u, v) = \mathbf{p}(u) + d(v)\,\mathbf{n}(u) + v\,\mathbf{k}\,.$$

- *Oriented–involute sweep*. An *involute* of a given plane curve $\mathbf{s}(v)$ is a curve $\mathbf{c}(v)$ within the same plane, such that $\mathbf{s}(v)$ is its locus of centers of curvature. Involutes are fundamental in the design of gear tooth shapes that achieve *conjugate action* (i.e., an exactly constant angular velocity ratio) of meshing gears. If $s(v)$ and $\mathbf{t}(v)$ are the arc–length function and tangent of $\mathbf{s}(v)$, the involute is defined by

$$\mathbf{c}(v) \;=\; \mathbf{s}(v) - s(v)\,\mathbf{t}(v)\,.$$

For a profile curve $\mathbf{p}(u) = x(u)\,\mathbf{i} + z(u)\,\mathbf{k}$ in the (x, z) plane and a PH sweep curve $\mathbf{s}(v) = x(v)\,\mathbf{i} + y(v)\,\mathbf{j}$ in the (x, y) plane, we define the oriented–involute sweep by

$$\mathbf{r}(u, v) \;=\; \mathbf{c}(v) - x(u)\,\mathbf{t}(v) + z(u)\,\mathbf{k}\,.$$

This surface is obtained by translating $\mathbf{p}(u)$ along the involute $\mathbf{c}(v)$, and orienting it in the normal plane to $\mathbf{c}(v)$ at each point. To ensure that $\mathbf{r}(u, 0) = \mathbf{p}(u)$, we must have $\mathbf{s}(0) = \mathbf{0}$ and $\mathbf{s}'(0)/\sigma(0) = -\mathbf{i}$.

- *Generalized conical sweep.* The oriented–translation sweep may be generalized by also subjecting the profile curve to a uniform scaling, linearly dependent on the sweep curve arc length $s(v)$. If S is the total arc length of $\mathbf{s}(v)$ and c_0, c_1 are positive initial and final scale factors, we introduce the scaling function

$$c(v) \;=\; \frac{c_0(S - s(v)) + c_1 s(v)}{S}\,, \qquad v \in [\,0, 1\,]\,,$$

and the generalized conical swept surface is defined by

$$\mathbf{r}(u, v) \;=\; \mathbf{s}(v) + c(v)\,[\,x(u)\,\mathbf{n}(v) + z(u)\,\mathbf{k}\,]\,.$$

For each v, the translated/oriented copy of the profile curve $\mathbf{p}(u)$ is scaled by the factor $c(v)$. To ensure that $\mathbf{r}(u, 0) = \mathbf{p}(u)$, we must have $\mathbf{s}(0) = \mathbf{0}$, $\mathbf{s}'(0)/\sigma(0) = \mathbf{j}$, and $c_0 = 1$.

When $\mathbf{p}(u)$ is a polynomial curve, $\mathbf{s}(v)$ is a polynomial PH curve, and $d(v)$ is a polynomial function, the homogeneous coordinates of the swept surfaces $\mathbf{r}(u, v)$ in each of the above examples are polynomials in u and v. In the case of the generalized conical sweep, for example, they are specified by

$$\begin{aligned}
W(u, v) &= S\sigma(v)\,,\\
X(u, v) &= S\sigma(v)x(v) + [\,c_0 S + (c_1 - c_0)s(v)\,]\,y'(v)x(u)\,,\\
Y(u, v) &= S\sigma(v)y(v) - [\,c_0 S + (c_1 - c_0)s(v)\,]\,x'(v)x(u)\,,\\
Z(u, v) &= [\,c_0 S + (c_1 - c_0)\,s(v)\,]\,\sigma(v)z(u)\,.
\end{aligned}$$

The above examples are by no means exhaustive, and are only intended to illustrate the rich variety of rational swept surface types that become possible with the use of PH sweep curves. In fact, unlike the matrix formalism [105] used to generate swept surfaces specified by bilinear forms in the profile and sweep curve homogeneous coordinates, this richness makes it difficult to categorize the complete set of rational swept surfaces based on the use of PH sweep curves.

A real–time interpolator algorithm that machines the above surface types directly from their high–level procedural definitions (i.e., the sweep type and the profile and

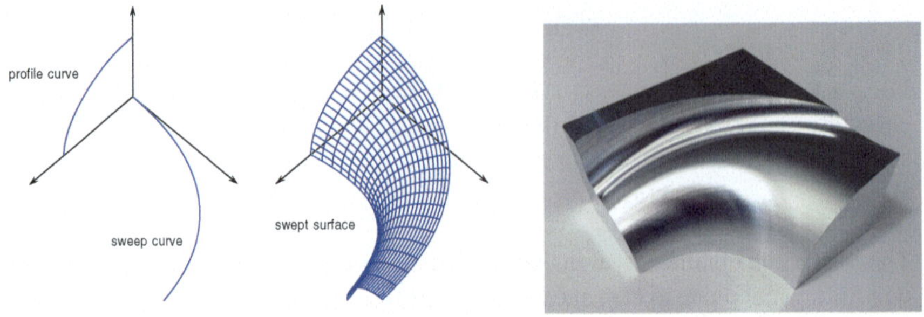

Fig. 6 Left: construction of a generalized conical sweep surface from a profile curve and a PH sweep curve. Right: the surface machined in aluminum using its high–level procedural definition

sweep curve definitions) was described in [166]. This bypasses the need to generate voluminous, approximate G code part programs, or to re–generate them when the cutting tool is changed. Figure 6 illustrates an example generalized conical sweep surface, and its machining in aluminum based on this methodology.

6.3 Surface Patches with PH Isoparametric Curves

A method for constructing tensor–product surface patches $\mathbf{s}(u, v)$ for which the $v =$ constant isoparametric loci are spatial PH curves was developed in [83], based on integrating the form

$$\mathbf{s}_u(u, v) = \mathcal{A}(u, v)\,\mathbf{i}\,\mathcal{A}^*(u, v)\,, \tag{29}$$

with respect to u, where $\mathcal{A}(u, v)$ is a bivariate quaternion polynomial expressed in terms of the Bernstein bases of degree m and n in u and v as

$$\mathcal{A}(u, v) = \sum_{i=0}^{m}\sum_{j=0}^{n} \mathcal{A}_{ij}\, b_i^m(u) b_j^n(v)\,. \tag{30}$$

The product (29) yields the pure vector expression

$$\mathcal{A}(u, v)\,\mathbf{i}\,\mathcal{A}^*(u, v) = \sum_{i=0}^{2m}\sum_{j=0}^{2n} \mathbf{a}_{ij}\, b_i^{2m}(u)\, b_j^{2n}(v)\,, \tag{31}$$

where

$$\mathbf{a}_{ij} = \sum_{k=\max(0,i-m)}^{\min(m,i)} \; \sum_{l=\max(0,j-n)}^{\min(n,j)} \frac{\binom{m}{k}\binom{m}{i-k}}{\binom{2m}{i}} \frac{\binom{n}{l}\binom{n}{j-l}}{\binom{2n}{j}} \mathcal{A}_{kl}\,\mathbf{i}\,\mathcal{A}^*_{i-k,j-l} \tag{32}$$

for $i = 0, \ldots, 2m$ and $j = 0, \ldots, 2n$. Integration with respect to u then gives

$$\mathbf{s}(u, v) = \sum_{i=0}^{2m+1} \sum_{j=0}^{2n} \mathbf{p}_{ij} \, b_i^{2m+1}(u) \, b_j^{2n}(v) \,, \tag{33}$$

where $\mathbf{p}_{0j}$ for $j = 0, \ldots, 2n$ are freely chosen to define the boundary curve $\mathbf{s}(0, v)$, and the remaining control points are given by

$$\mathbf{p}_{ij} = \mathbf{p}_{i-1,j} + \frac{1}{2m+1} \, \mathbf{a}_{i-1,j} \tag{34}$$

for $i = 1, \ldots, 2m + 1$ and $j = 0, \ldots, 2n$. The focus in [83] is on the particular case $m = n = 2$, yielding tensor–product patches of degree (5, 4) in (u, v).

Upon fixing $\mathcal{A}_{00}, \mathcal{A}_{10}, \mathcal{A}_{20}$ and $\mathcal{A}_{02}, \mathcal{A}_{12}, \mathcal{A}_{22}$ by constructing the two boundary curves $\mathbf{s}(u, 0)$ and $\mathbf{s}(u, 1)$ as PH quintic Hermite interpolants, $\mathcal{A}_{01}, \mathcal{A}_{11}, \mathcal{A}_{21}$ remain to be determined. As demonstrated in [83], these coefficients can be used to fix the interior control points $\mathbf{p}_{51}, \mathbf{p}_{52}, \mathbf{p}_{53}$ defining the boundary curve $\mathbf{s}(1, v)$ with $\mathbf{p}_{50}, \mathbf{p}_{54}$ having already been determined in fixing $\mathbf{s}(u, 0)$ and $\mathbf{s}(u, 1)$. In particular, $\mathcal{A}_{01}, \mathcal{A}_{21}$ can be eliminated to obtain a single quadratic equation in $\mathcal{A}_{11}$ of the form

$$\mathcal{A}_{11} \mathcal{P} \mathcal{A}_{11}^* + \mathcal{A}_{11} \mathcal{Q} + \mathcal{R} \mathcal{A}_{11}^* = \mathcal{S} \,, \tag{35}$$

with quaternion coefficients $\mathcal{P}, \mathcal{Q}, \mathcal{R}, \mathcal{S}$. This equation is rather unusual, since the quaternion unknown and its conjugate exhibit left and right coefficients, while the quadratic term has a coefficient interposed between them. A comprehensive analysis of its solution space has been presented in [53].

The above construction yields a degree (5, 4) tensor–product patch $\mathbf{s}(u, v)$ with quintic PH curves as its $v =$ constant isoparametric loci, and pre–defined boundary curves. The coefficients $\mathcal{P}, \mathcal{Q}, \mathcal{R}, \mathcal{S}$ in (35) include three free scalar parameters, that can be exploited to optimize the interior shape of the surface patch.

7 Applications of Pythagorean–Hodograph Curves

Applications play an important role in computer aided geometric design, motivating development of new theory and algorithms to serve practical needs. In minimizing the need to invoke numerical approximations, PH curves can enhance the accuracy, efficiency, and robustness of applications. The present section reviews two important application domains for PH curves: *real–time motion control* and *path planning*.

7.1 Real–Time Motion Control

The distinctive features of PH curves make them well–suited to problems of real–time motion control in robotics, manufacturing, inspection, and related applications. Given a desired feedrate (speed) along a prescribed path, the *real–time interpolator* algorithm generates a reference point (i.e., commanded machine position) in each sampling interval of the controller. The difference between the commanded position and actual machine position (as measured by encoders on the machine axes) defines the instantaneous *position error*, which is the basic control variable.

PH curves admit essentially exact real–time interpolator algorithms for various feedrate dependencies on elapsed time, arc length, path curvature, etc. [74, 90, 202]. However, they also possess many other advantages in motion control, for example in *inverse dynamics* error compensation schemes [36, 186]. The physical limitations (axes inertia and friction) of real machines prevent them from exactly executing a desired path $\mathbf{r}(\xi)$, but if the machine physical properties and controller parameters are known, one may compute a *modified* path that, subject to the machine dynamics, will induce the machine to exactly execute the original desired path $\mathbf{r}(\xi)$.

For a basic P (proportional) controller, the modified path can be exactly specified as a rational curve when the given path $\mathbf{r}(\xi)$ is a polynomial PH curve. In fact, the modification amounts [36] to simply adding the "correction" term

$$\Delta\mathbf{r} = b\,(V/\sigma)^2\,\mathbf{r}'' + (V/\sigma)\,[\,b\,(V/\sigma)' + c\,]\,\mathbf{r}' \tag{36}$$

to $\mathbf{r}(\xi)$, where $\sigma(\xi) = |\mathbf{r}'(\xi)|$ is the parametric speed, V is the feedrate (which may be non–constant), and b, c are constants[8] determined by the machine and controller properties. Experiments on a CNC mill with an open–architecture controller [186] show that the inverse dynamics scheme substantially improves the tracking accuracy along strongly–curved paths, as shown by the plots of *contour error* (the geometrical deviation of the executed path from the commanded path) in Fig. 7.

A further important application of the PH curves is in the *high–speed cornering* problem. It is physically impossible for a machine to execute a sharp corner at finite speed, and the need to decelerate into and accelerate out of corners can substantially increase overall execution times for piecewise–linear paths, and impose undesirable stresses on the axis drive systems. The typical solution to this problem is to smooth out the path with *corner rounding curves* that satisfy a prescribed tolerance, and to impose a suitable feedrate modulation along these corner curves.

PH curves are especially well–suited to the corner rounding problem [82, 107, 108, 167, 191, 198, 208] since they can define G^2 corner rounding curves between linear segments that have smooth curvature distributions, and are amenable to real–time feedrate modulation dependent on position or curvature along the corner curve. A "canonical form" G^2 PH quintic corner curve $\mathbf{r}(\xi)$, $\xi \in [\,0, 1\,]$ with side length L and corner angle θ between the initial and final control points $\mathbf{p}_0 = (0, 0)$ and

[8]The expression (36) assumes that these constants are the same for each machine axis, but this assumption is not essential.

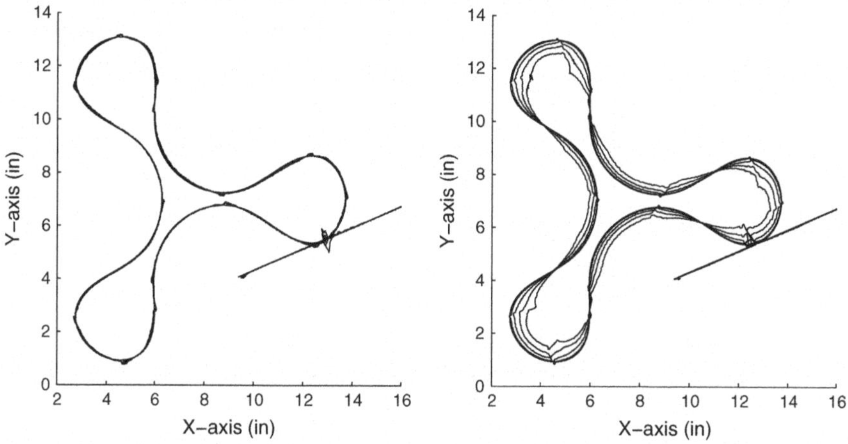

Fig. 7 Measured path contour error magnified 200× along a C^2 PH spline curve, for feedrates of 200, 400, 600, 800 in/min, with the inverse dynamics scheme switched *on* (left) and *off* (right). Linear "lead–in" and "lead–out" segments are also included, so as to minimize transient effects

$\mathbf{p}_5 = (L + L\cos\theta, L\sin\theta)$ is defined [82] by the interior control points

$$\mathbf{p}_1 = \mathbf{p}_2 = \left(\frac{6L\cos\frac{1}{2}\theta}{6\cos\frac{1}{2}\theta + 1}, 0 \right), \quad \mathbf{p}_3 = \mathbf{p}_4 = \left(L + \frac{L\cos\theta}{6\cos\frac{1}{2}\theta + 1}, \frac{L\sin\theta}{6\cos\frac{1}{2}\theta + 1} \right)$$

and expressions for its maximum curvature, total arc length, and deviation from the exact corner point $\mathbf{p}_c = (L, 0)$ can be easily derived [82]. Figure 8 shows examples of these corner curves, and the curvature–dependent feedrate

$$V(\xi) = \frac{f\kappa_{\max} V_0}{(1-f)\,\kappa(\xi) + f\kappa_{\max}},$$

which for $0 < f < 1$ achieves a reduction from V_0 to $f\,V_0$ at the mid–point extremum of the curvature, $\kappa_{\max} = \kappa(\frac{1}{2})$. Experiments on a variety of corner–rounded paths [167] indicate reductions in overall execution times of ∼24–32% compared to the "full stop" execution of piecewise–linear paths without corner rounding.

A number of other real–time motion control strategies and applications based on the use of PH curves have been described in [21, 92, 100, 109, 110, 121].

7.2 *Path Planning Applications*

Path planning is a key component of diverse applications, such as CNC machining, automated assembly or inspection, the coordinated motion of autonomous vehicles,

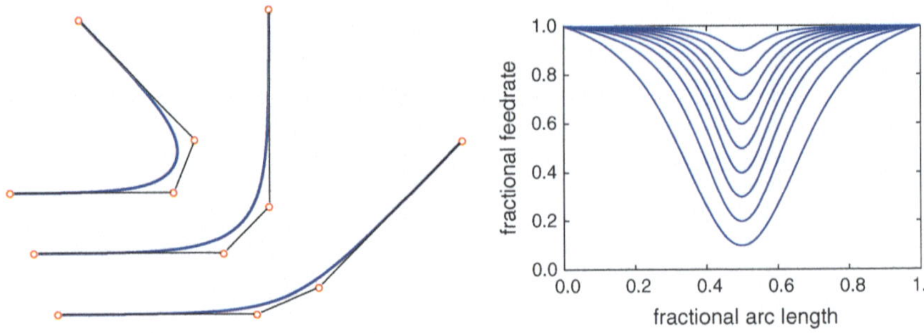

Fig. 8 Left: canonical form G^2 PH quintic corner curves for the turning angles 45°, 90°, 135°. Right: curvature–dependent feedrate functions for the corner curves, corresponding to feedrate reduction factors f of 10%, ..., 90% relative to the nominal value V_0

robotic surgery, and computer animation. Such problems may be concerned only with issues of position, or with position *and* orientation, and may entail satisfaction of various constraints—correlation of position and orientation, obstacle avoidance, maintenance of minimum safe separations, bounds on linear or angular velocity and acceleration, etc. We briefly review here the growth of interest over the past decade in exploiting the properties of PH curves for path planning applications.

7.2.1 Unmanned or Autonomous Vehicles

The use of PH curves in planning paths for unmanned or autonomous aerial, land, or submarine vehicles has been discussed by several authors [6, 19, 23, 24, 31, 59, 93, 154, 163–165, 189, 190, 201, 211, 212]. The availability of cheap "drones" equipped with surveillance devices and wireless communications has spawned an interest in diverse applications—e.g., remote sensing or surveying for agricultural, geophysical, and meteorological assessment; search–and–rescue missions; marine biology and oceanography studies; assessment and management of disaster relief plans; law enforcement; and automated consumer products delivery.

The path generation problem may be subject to a variety of constraints, such as avoidance of environmental obstacles; satisfaction of bounds on the path curvature or climb angle; and maintaining safe separations within, or ensuring simultaneous arrivals of, vehicle swarms. However, as noted in [59], some early studies indicate misconceptions concerning the basic properties and advantages of PH curves in the path planning context. These attributes of PH curves may be summarized as follows.

- The ability to construct paths with precisely–specified arc lengths between given end states (see Sect. 5.2) allows predictable arrival times for vehicles that travel at a constant speed. This can ensure simultaneous arrival of multiple vehicles that follow independent paths with common departure and destination points.
- The polynomial arc length property facilitates assessment of nearest approaches of vehicles following different paths at constant speed—paths that intersect may

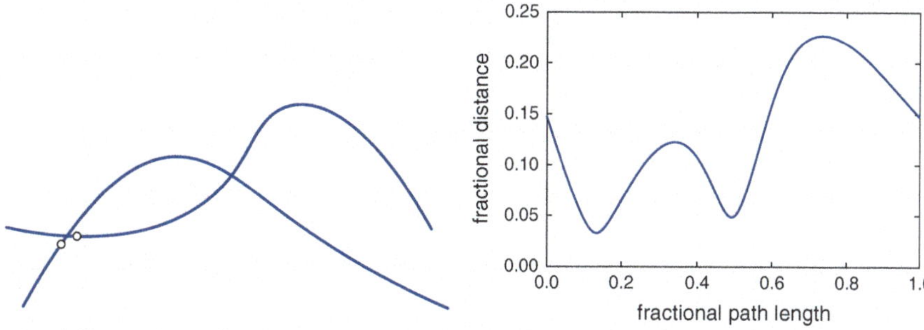

Fig. 9 Left: intersecting planar PH quintic paths of equal arc length, traversed at constant speed (the dots indicate the closest approach). Right: path separation as a function of the path arc length

nevertheless satisfy safe minimum separations, since the intersections correspond to different timings along them (see Fig. 9). The rational offsets to planar PH paths define "exclusion zones" that can be used to discount a priori the possibility of unsafe close approach if they do not overlap.

- Algorithms for planning routes through obstacle fields typically yield piecewise–linear paths with sharp corners, which cannot be physically realized at non–zero speed by vehicles with finite inertia. The PH curve corner–rounding scheme (see Sect. 7.1) can be used to address this problem, and ensure that physical bounds on acceleration or steering rates are not exceeded at a given desired speed.
- The determination of the curvature extrema of PH paths can be formulated as a polynomial root solving problem, whose solutions can be used to assign feasible constant speeds under known acceleration bounds. For aerial vehicles, helical PH paths can be used to set safe constant speeds under known bounds on climb rates.

Further studies of path planning with PH curves, subject to obstacle avoidance constraints, have been presented in [34, 96]. A key preliminary step is to construct a piecewise–linear path in the free space that connects prescribed initial and final positions. A PH quintic G^1 [96] or G^2 [34] spline interpolation scheme with tension is then employed to define a smoother admissible path passing through the vertices of the piecewise–linear path. An off–line feedrate scheduler incorporating kinematic constraints was proposed in [97], that produces a C^2 spline time–dependent feedrate function for planar paths specified as PH splines, that is compatible with standard real–time PH curve interpolator algorithms.

7.2.2 Constrained Spatial Rigid–Body Motions

A spatial motion of a rigid body involves specifying the variation of *position* and *orientation* with time t. The path of a specific point (e.g., the center of mass) may be used to describe the variation of position as a curve $\mathbf{r}(t)$, and an orthonormal frame $(\mathbf{f}_1(t), \mathbf{f}_2(t), \mathbf{f}_3(t))$ embedded within the body can be used to specify the variation of orientation. In general, $\mathbf{r}(t)$ and $(\mathbf{f}_1(t), \mathbf{f}_2(t), \mathbf{f}_3(t))$ are independent, and a number of studies [140, 142, 177] address rational motions of this type. However, we will focus here on "constrained" spatial motions, involving a correlation of position and orientation, which are important in many application contexts.

The first two derivatives of the path $\mathbf{r}(t)$ define the velocity $\mathbf{v}$ and acceleration $\mathbf{a}$ of the body, while the first two derivatives of the frame $(\mathbf{f}_1(t), \mathbf{f}_2(t), \mathbf{f}_3(t))$ specify the angular velocity $\boldsymbol{\omega}$ and angular acceleration $\boldsymbol{\alpha}$. As observed in Sect. 4.1, an adapted *rotation–minimizing motion* satisfies the constraint $\mathbf{v} \cdot \boldsymbol{\omega} \equiv 0$ (i.e., there is no instantaneous rotation about the direction of motion), and the simplest *rational* rotation–minimizing motions correspond to the quintic RRMF curves. Preliminary results on the construction of RRMF quintics that interpolate given G^1 Hermite data (end points and unit tangent vectors) were presented in [63].

The construction of rational rotation–minimizing motions, with given initial and final positions and orientations, was studied in [58] using the characterization (18) of RRMF quintics in terms of their quaternion pre–image polynomial coefficients, and it was shown that the problem can be reduced to finding the positive real roots of a degree 6 polynomial with coefficients dependent on the given data. However, it was observed that positive real roots do not exist for all possible data sets.

An alternative approach based on the degree 7 PH curves that have rotation–minimizing ERFs, as identified by the quaternion coefficient constraints (14), was developed in [71]. In this case, it was possible to reduce the construction to solving a system of four quadratic equations in four real variables. These equations incorporate two free parameters specifying the magnitudes of the end derivatives, which can be used to optimize the curve shape. Although the curves are of higher degree than the RRMF quintics, their rational RMFs are actually of lower degree, since the rational normal–plane rotation (15) is not required. Two distinct solutions are typically observed for specified end points, frames, and derivative magnitudes. Because of the non–linear nature of the problem, a formal proof for existence of interpolants to arbitrary given data—for *some* values of the free parameters—is non–trivial. However, empirical evidence supporting this was presented in [71]. Figure 10 illustrates some examples of the constructed rational rotation–minimizing motions.

As noted in Sect. 4.2, the construction of an RMF on a given curve is an initial value problem, and consequently an RMF along a pre–defined curve cannot match both initial *and* final orientations—in the methods described above, the curve was not pre–defined, but was rather an outcome of the construction. To address this issue, a method

Fig. 10 Two examples of rotation–minimizing rigid body motions along degree 7 PH curves that possess rational RMFs, matching prescribed initial and final positions and orientations

has been developed in [162] to match desired initial/final orientations for rigid body motion along a *prescribed* curve, using the *minimal twist frame* (MTF) introduced in Sect. 4.2, and was also extended to minimal twist *spline motions*, interpolating orientations at a sequence of discrete locations along a specified curve, with the least possible twist between consecutive locations.

The study [64] considers the construction of rational spatial motions specified by rotation–minimizing *directed*, rather than adapted, frames (see Sect. 4.4). These motions are based upon a family of quartic P curves, characterized by the condition (24), that possess rational rotation–minimizing directed frames. The quartic P curves can interpolate initial/final positions and orientations of a camera, together with an initial motion direction, and by piecing such P quartics together one can define a G^1 camera motion that yields the least possible image rotation of a stationary object. Further constructions of this type of motions are described in [115, 141, 185].

7.3 `PHquintic` *Software Library*

To encourage greater adoption of PH curves in applications, a C/C++ open–source software library for constructing, modifying, and analyzing planar PH quintics has been published [35]. Using the complex representation, a planar PH quintic segment is defined by a structure of the form

```
struct PHquintic{
complex p[6] ; /* Bezier control points */
complex w[3] ; /* pre-image coefficients */
double sigma[5] ; /* parametric speed coefficients */
double s[6] ; /* arc length coefficients */
};
```

This definition is redundant, since the initial control point $\mathbf{p}_0$ and the Bernstein coefficients $\mathbf{w}_0, \mathbf{w}_1, \mathbf{w}_2$ of $\mathbf{w}(\xi)$ suffice to reconstruct all the other information, but it serves to improve the efficiency of subsequent algorithms. The functions provided in this library include:

- construction of first–order PH quintic Hermite interpolants, as specified by initial and final control–point pairs $\mathbf{p}_0, \mathbf{p}_1$ and $\mathbf{p}_4, \mathbf{p}_5$ with the *absolute rotation index* [80] used to identify the "good" solution;
- construction of open and closed C^2 PH quintic spline curves that interpolate a sequence of points $\mathbf{p}_0, \ldots, \mathbf{p}_n$ (with $\mathbf{p}_n = \mathbf{p}_0$ in the case of a closed curve) using the efficient method described in [73];
- construction of the control points and weights defining the exact rational offsets to planar PH quintics [85];
- computation of the exact elastic bending energy of PH quintic segments [40].

In addition to these high–level C functions, which can be easily integrated with applications software, an interactive C++ interface was developed for the Windows environment, based on the Microsoft Foundation Class Library and Open Graphics Library. This allows the user to input data points by mouse, and interactively modify PH curves using the mouse to move them (exact data point coordinates can also be manually typed in). Key properties of the constructed PH curves (arc length, bending energy, etc.) are reported, and their offset curves can also be constructed.

7.4 Other Applications

For brevity, we mention only briefly a selection of other recent applications for the PH curves, which include: aesthetic design of artistic patterns [125]; vision–based command generation and disturbance compensation in planar contour tracking [21]; smooth tracking of head motions for virtual reality applications [5]; improving the efficiency of robotic pick–and–place operations [200]; and modeling of continuum manipulators [192].

8 Closure

As documented in this survey, the past decade—since the appearance of [43]—has been very fruitful in broadening and deepening the basic theory of Pythagorean–hodograph curves; in the formulation of algorithms for constructing, modifying, and analyzing them; and in demonstrating their advantages in practical applications.

The increasing adoption of PH curves in practical applications in recent years is a propitious development, since these applications frequently suggest and motivate new lines of investigation for the development of basic theory and novel algorithms. As the 30th anniversary of the original Pythagorean hodograph concept approaches [85], it is evident that it has served as a very fruitful paradigm, for both theoretical and practical research, and may continue as such for many years to come.

Acknowledgements C. Giannelli and A. Sestini are members of the INdAM Research group GNCS. The INdAM support through GNCS and Finanziamenti Premiali SUNRISE is gratefully acknowledged.

References

1. Ait-Haddou, R., Herzog, W., Biard, L.: Pythagorean-hodograph ovals of constant width. Comput. Aided Geom. Des. **25**, 258–273 (2008)
2. Ait-Haddou, R., Mazure, M.-L.: Sparse Pythagorean-hodograph curves. Comput. Aided Geom. Des. **55**, 84–103 (2017)

3. Albrecht, G., Beccari, C.V., Canonne, J.-C., Romani, L.: Planar Pythagorean-hodograph B-spline curves. Comput. Aided Geom. Des. **57**, 57–77 (2017)
4. Albrecht, G., Farouki, R.T.: Construction of C^2 Pythagorean-hodograph interpolating splines by the homotopy method. Adv. Comput. Math. **5**, 417–442 (1996)
5. Amamra, A.: Smooth head tracking for virtual reality applications. Signal Image Video Proc. **11**, 479–486 (2017)
6. Askari, A., Mortazavi, M., Talebi, H.A., Motamedi, A.: A new approach in UAV path planning using Bézier-Dubins continuous curvature path. J. Aerosp. Eng. **230**, 1103–1113 (2016)
7. Barton, M., Jüttler, B., Wang, W.: Construction of rational curves with rational rotation-minimizing frames via Möbius transformations. Mathematical Methods for Curves and Surfaces 2008. Lecture Notes in Computer Science, vol. 5862, pp. 15–25. Springer, Berlin (2010)
8. Bastl, B., Bizzarri, M., Krajnc, M., Lavicka, M., Siaba, K., Sir, Z., Vitrih, V., Zagar, E.: C^1 Hermite interpolation with spatial Pythagorean-hodograph cubic biarcs. J. Comput. Appl. Math. **257**, 65–78 (2014)
9. Bastl, B., Bizzarri, M., Ferjancic, K., Kovac, B., Krajnc, M., Lavicka, M., Michalkova, K., Sir, Z., Zagar, E.: C^2 Hermite interpolation by Pythagorean-hodograph quintic triarcs. Comput. Aided Geom. Des. **31**, 412–426 (2014)
10. Bastl, B., Jüttler, B., Lavicka, M., Kosinka, J.: Computing exact rational offsets of quadratic triangular Bézier surface patches. Comput. Aided Des. **40**, 197–209 (2008)
11. Bastl, B., Slaba, K., Byrtus, M.: Planar C^1 Hermite interpolation with uniform and non-uniform TC biarcs. Comput. Aided Geom. Des. **30**, 58–77 (2013)
12. Beltran, J.V., Monterde, J.: A characterization of quintic helices. J. Comput. Appl. Math. **206**, 116–121 (2007)
13. Biard, L., Farouki, R.T., Szafran, N.: Construction of rational surface patches bounded by lines of curvature. Comput. Aided Geom. Des. **27**, 359–371 (2010)
14. Birkhoff, G., de Boor, C.: Piecewise polynomial interpolation and approximation. In: Garabedian, H.L. (ed.) Approximation of Functions, pp. 164–190. Elsevier, Amsterdam (1965)
15. Bishop, R.L.: There is more than one way to frame a curve. Am. Math. Mon. **82**, 246–251 (1975)
16. Bizzarri, M., Lavicka, M., Kosinka, J.: Medial axis transforms yielding rational envelopes. Comput. Aided Geom. Des. **46**, 92–102 (2016)
17. Bizzarri, M., Lavicka, M., Vrsek, J., Kosinka, J.: A direct and local method for computing polynomial Pythagorean-normal patches with global G^1 continuity. Comput. Aided Des. **102**, 44–51 (2018)
18. Blum, H.: New algorithm for medial axis transform of plane domain. Graph. Models Image Proc. **59**, 463–483 (1967)
19. Bruyninckx, H., Reynaerts, D.: Path planning for mobile and hyper–redundant robots using Pythagorean–hodograph curves. In: Proceedings, International Conference on Advanced Robotics (ICAR 97), Monterey, CA, pp. 595–600 (1997)
20. Byrtus, M., Bastl, B.: G^1 Hermite interpolation by PH cubics revisited. Comput. Aided Geom. Des. **27**, 622–630 (2010)
21. Chen, C.C., Shieh, S.S., Cheng, M.Y., Su, K.H.: Vision-based Pythagorean hodograph spline command generation and adaptive disturbance compensation for planar contour tracking. Int. J. Adv. Manuf. Tech. **65**, 1185–1199 (2013)
22. Cheng, C.C.-A., Sakkalis, T.: On new types of rational rotation–minimizing frame space curves. J. Symb. Comput. **74**, 400–407 (2016)
23. Choe, R., Puig-Navarro, J., Cichella, V., Xargay, E., Hovakimyan, N.: Trajectory generation using spatial Pythagorean hodograph Bézier curves. In: Proceedings, AIAA Guidance, Navigation, and Control Conference, Kissimmee, FL (2015)
24. Choe, R., Puig-Navarro, J., Cichella, V., Xargay, E., Hovakimyan, N.: Cooperative trajectory generation using Pythagorean hodograph Bézier curves. J. Guid. Control Dyn. **39**, 1744–1763 (2015)
25. Choi, H.I., Farouki, R.T., Kwon, S.H., Moon, H.P.: Topological criterion for selection of quintic Pythagorean-hodograph Hermite interpolants. Comput. Aided Geom. Des. **25**, 411–433 (2008)

26. Choi, H.I., Han, C.Y.: Euler-Rodrigues frames on spatial Pythagorean-hodograph curves. Comput. Aided Geom. Des. **19**, 603–620 (2002)
27. Choi, H.I., Han, C.Y., Moon, H.P., Roh, K.H., Wee, N.-S.: Medial axis transform and offset curves by Minkowski Pythagorean hodograph curves. Comput. Aided Des. **31**, 59–72 (1999)
28. Choi, H.I., Kwon, S.H.: Absolute hodograph winding number and planar PH quintic splines. Comput. Aided Geom. Des. **25**, 230–246 (2008)
29. Choi, H.I., Moon, H.P.: Weierstrass-type approximation theorems with Pythagorean hodograph curves. Comput. Aided Geom. Des. **25**, 305–319 (2008)
30. Choi, H.I., Lee, D.S., Moon, H.P.: Clifford algebra, spin representation, and rational parameterization of curves and surfaces. Adv. Comput. Math. **17**, 5–48 (2002)
31. Chu, K., Kim, J., Jo, K., Sunwoo, M.: Real-time path planning of autonomous vehicles for unstructured road navigation. Int. J. Autom. Tech. **16**, 653–668 (2015)
32. Cobb, J.E.: Letter to the editor. Comput. Aided Geom. Des. **6**, 85–86 (1989)
33. Dietz, R., Hoschek, J., Jüttler, B.: An algebraic approach to curves and surfaces on the sphere and on other quadrics. Comput. Aided Geom. Des. **10**, 211–229 (1993)
34. Donatelli, M., Giannelli, C., Mugnaini, D., Sestini, A.: Curvature continuous path planning and path finding based on PH splines with tension. Comput. Aided Des. **88**, 14–30 (2017)
35. Dong, B., Farouki, R.T.: Algorithm 952: PHquintic: a library of basic functions for the construction and analysis of planar quintic Pythagorean–hodograph curves. ACM Trans. Math. Softw. **41**, Article 28 (2015)
36. Ernesto, C.A., Farouki, R.T.: Solution of inverse dynamics problems for contour error minimization in CNC machines. Int. J. Adv. Manuf. Tech. **49**, 589–604 (2010)
37. Fang, L., Wang, G.: Geometric characteristics of planar quintic Pythagorean-hodograph curves. J. Comput. Appl. Math. **330**, 117–127 (2018)
38. Farin, G.: Curves and Surfaces for Computer Aided Geometric Design, 4th edn. Academic Press, San Diego (1997)
39. Farouki, R.T.: The conformal map $z \to z^2$ of the hodograph plane. Comput. Aided Geom. Des. **11**, 363–390 (1994)
40. Farouki, R.T.: The elastic bending energy of Pythagorean-hodograph curves. Comput. Aided Geom. Des. **13**, 227–241 (1996)
41. Farouki, R.T.: Pythagorean–hodograph curves. In: Farin, G., Hoschek, J., Kim, M.-S. (eds.) Handbook of Computer Aided Geometric Design, pp. 405–427, Elsevier, Amsterdam (2002)
42. Farouki, R.T.: Exact rotation-minimizing frames for spatial Pythagorean-hodograph curves. Graph. Models **64**, 382–395 (2002)
43. Farouki, R.T.: Pythagorean-Hodograph Curves: Algebra and Geometry Inseparable. Springer, Berlin (2008)
44. Farouki, R.T.: Quaternion and Hopf map characterizations for the existence of rational rotation-minimizing frames on quintic space curves. Adv. Comput. Math. **33**, 331–348 (2010)
45. Farouki, R.T.: Arc lengths of rational Pythagorean-hodograph curves. Comput. Aided Geom. Des. **34**, 1–4 (2015)
46. Farouki, R.T.: Rational rotation-minimizing frames – recent advances and open problems. Appl. Math. Comput. **272**, 80–91 (2016)
47. Farouki, R.T.: Construction of G^1 planar Hermite interpolants with prescribed arc lengths. Comput. Aided Geom. Des. **46**, 64–75 (2016)
48. Farouki, R.T.: Helical polynomial curves interpolating G^1 data with prescribed axes and pitch angles. Comput. Aided Geom. Des. **56**, 4–15 (2017)
49. Farouki, R.T.: Existence of Pythagorean–hodograph quintic interpolants to spatial G^1 Hermite data with prescribed arc lengths. J. Symb. Comput. **95**, 202–216 (2019)
50. Farouki, R.T., al-Kandari, M., Sakkalis, T.: Structural invariance of spatial Pythagorean hodographs. Comput. Aided Geom. Des. **19**, 395–407 (2002)
51. Farouki, R.T., al-Kandari, M., Sakkalis, T.: Hermite interpolation by rotation-invariant spatial Pythagorean-hodograph curves. Adv. Comput. Math. **17**, 369–383 (2002)
52. Farouki, R.T., Dospra, P., Sakkalis, T.: Scalar-vector algorithm for the roots of quadratic quaternion polynomials, and the characterization of quintic rational rotation-minimizing frame curves. J. Symb. Comput. **58**, 1–17 (2013)

53. Farouki, R.T., Gentili, G., Giannelli, C., Sestini, A., Stoppato, C.: Solution of a quadratic quaternion equation with mixed coefficients. J. Symb. Comput. **74**, 140–151 (2016)
54. Farouki, R.T., Gentili, G., Giannelli, C., Sestini, A., Stoppato, C.: A comprehensive characterization of the set of polynomial curves with rational rotation-minimizing frames. Adv. Comput. Math. **43**, 1–24 (2017)
55. Farouki, R.T., Giannelli, C.: Spatial camera orientation control by rotation-minimizing directed frames. Comput. Anim. Virtual Worlds **20**, 457–472 (2009)
56. Farouki, R.T., Giannelli, C., Manni, C., Sestini, A.: Identification of spatial PH quintic Hermite interpolants with near-optimal shape measures. Comput. Aided Geom. Des. **25**, 274–297 (2008)
57. Farouki, R.T., Giannelli, C., Manni, C., Sestini, A.: Quintic space curves with rational rotation-minimizing frames. Comput. Aided Geom. Des. **26**, 580–592 (2009)
58. Farouki, R.T., Giannelli, C., Manni, C., Sestini, A.: Design of rational rotation-minimizing rigid body motions by Hermite interpolation. Math. Comput. **81**, 879–903 (2012)
59. Farouki, R.T., Giannelli, C., Mugnaini, D., Sestini, A.: Path planning with Pythagorean-hodograph curves for unmanned or autonomous vehicles. J. Aerosp. Eng. **232**, 1361–1372 (2018)
60. Farouki, R.T., Giannelli, C., Sampoli, M.L., Sestini, A.: Rotation-minimizing osculating frames. Comput. Aided Geom. Des. **31**, 27–42 (2014)
61. Farouki, R.T., Giannelli, C., Sestini, A.: Helical polynomial curves and double Pythagorean hodographs I. Quaternion and Hopf map representations. J. Symb. Comput. **44**, 161–179 (2009)
62. Farouki, R.T., Giannelli, C., Sestini, A.: Helical polynomial curves and double Pythagorean hodographs II. Enumeration of low–degree curves. J. Symb. Comput. **44**, 307–322 (2009)
63. Farouki, R.T., Giannelli, C., Sestini, A.: Geometric design using space curves with rational rotation–minimizing frames. In: Daehlen, M., et al. (eds.) Lecture Notes in Computer Science, vol. 5862, pp. 194–208 (2010)
64. Farouki, R.T., Giannelli, C., Sestini, A.: An interpolation scheme for designing rational rotation-minimizing camera motions. Adv. Comput. Math. **38**, 63–82 (2013)
65. Farouki, R.T., Giannelli, C., Sestini, A.: Identification and "reverse engineering" of Pythagorean-hodograph curves. Comput. Aided Geom. Des. **34**, 21–36 (2015)
66. Farouki, R.T., Giannelli, C., Sestini, A.: Local modification of Pythagorean-hodograph quintic spline curves using the B-spline form. Adv. Comput. Math. **42**, 199–225 (2016)
67. Farouki, R.T., Giannelli, C., Sestini, A.: Rational minimal–twist rigid body motions along space curves with rotation–minimizing Euler–Rodrigues frames (Preprint)
68. Farouki, R.T., Gutierrez, R.: Geometry of the ringed surfaces in R^4 that generate spatial Pythagorean hodographs. J. Symb. Comput. **37**, 87–103 (2016)
69. Farouki, R.T., Han, C.Y.: Rational approximation schemes for rotation-minimizing frames on Pythagorean-hodograph curves. Comput. Aided Geom. Des. **20**, 435–454 (2003)
70. Farouki, R.T., Han, C.Y., Manni, C., Sestini, A.: Characterization and construction of helical polynomial space curves. J. Comput. Appl. Math. **162**, 365–392 (2004)
71. Farouki, R.T., Han, C.Y., Dospra, P., Sakkalis, T.: Rotation-minimizing Euler-Rodrigues rigid-body motion interpolants. Comput. Aided Geom. Des. **30**, 653–671 (2013)
72. Farouki, R.T., Jüttler, B., Manni, C.: Preface: Pythagorean-hodograph curves and related topics. Comput. Aided Geom. Des. **25**, 203–204 (2008)
73. Farouki, R.T., Kuspa, B.K., Manni, C., Sestini, A.: Efficient solution of the complex quadratic tridiagonal system for C^2 PH quintic splines. Numer. Algorithms **27**, 35–60 (2001)
74. Farouki, R.T., Manjunathaiah, J., Nicholas, D., Yuan, G.-F., Jee, S.: Variable-feedrate CNC interpolators for constant material removal rates along Pythagorean-hodograph curves. Comput. Aided Des. **30**, 631–640 (1998)
75. Farouki, R.T., Manni, C., Pelosi, F., Sampoli, M.L.: Design of C^2 Spatial Pythagorean-hodograph quintic spline curves by control polygons. Lecture Notes in Computer Science, vol. 6920, pp. 253–269 (2011)

76. Farouki, R.T., Manni, C., Sampoli, M.L., Sestini, A.: Shape-preserving interpolation of spatial data by Pythagorean-hodograph quintic spline curves. IMA J. Numer. Anal. **35**, 478–498 (2015)
77. Farouki, R.T., Manni, C., Sestini, A.: Spatial C^2 PH quintic splines. In: Lyche, T., Mazure, M.-L., Schumaker, L.L. (eds.) Curve and Surface Design: Saint-Malo 2002, pp. 147–156. Nashboro Press, Brentwood (2003)
78. Farouki, R.T., Manni, C., Sestini, A.: Shape-preserving interpolation by G^1 and G^2 PH quintic splines. IMA J. Numer. Anal. **23**, 175–195 (2003)
79. Farouki, R.T., Moon, H.P.: Rational frames of minimal twist along space curves under specified boundary conditions. Adv. Comput. Math. (In press)
80. Farouki, R.T., Neff, C.A.: Hermite interpolation by Pythagorean-hodograph quintics. Math. Comput. **64**, 1589–1609 (1995)
81. Farouki, R.T., Nittler, K.M.: Rational swept surface constructions based on differential and integral sweep curve properties. Comput. Aided Geom. Des. **33**, 1–16 (2015)
82. Farouki, R.T., Nittler, K.M.: Efficient high–speed cornering motions based on continuously–variable feedrates I. Real-time interpolator algorithm. Int. J. Adv. Manuf. Tech. **87**, 3557–3568 (2016)
83. Farouki, R.T., Pelosi, F., Sampoli, M.L., Sestini, A.: Tensor-product surface patches with Pythagorean-hodograph isoparametric curves. IMA J. Numer. Anal. **36**, 1389–1409 (2016)
84. Farouki, R.T., Peters, J.: Smooth curve design with double-Tschirnhausen cubics. Ann. Numer. Math. **3**, 63–82 (1996)
85. Farouki, R.T., Sakkalis, T.: Pythagorean hodographs. IBM J. Res. Develop. **34**, 736–752 (1990)
86. Farouki, R.T., Rampersad, J.: Cycles upon cycles: an anecdotal history of higher curves in science and engineering. In: Daehlen, M., Lyche, T., Schumaker, L.L. (eds.) Mathematical Methods for Curves and Surfaces II, pp. 95–116. Vanderbilt University Press, Nashville (1998)
87. Farouki, R.T., Sakkalis, T.: Pythagorean-hodograph space curves. Adv. Comput. Math. **2**, 41–66 (1994)
88. Farouki, R.T., Sakkalis, T.: Rational rotation-minimizing frames on polynomial space curves of arbitrary degree. J. Symb. Comput. **45**, 844–856 (2010)
89. Farouki, R.T., Sakkalis, T.: A complete classification of quintic space curves with rational rotation-minimizing frames. J. Symb. Comput. **47**, 214–226 (2012)
90. Farouki, R.T., Shah, S.: Real-time CNC interpolators for Pythagorean-hodograph curves. Comput. Aided Geom. Des. **13**, 583–600 (1996)
91. Farouki, R.T., Sir, Z.: Rational Pythagorean-hodograph space curves. Comput. Aided Geom. Des. **28**, 75–88 (2011)
92. Farouki, R.T., Srinathu, J.: A real-time CNC interpolator algorithm for trimming and filling planar offset curves. Comput. Aided Des. **86**, 1–11 (2017)
93. Fernandes, D.de.A., Soensen, A.J., Donha, D.C.: Path generation for high–performance motion of ROVs based on a reference model. Model. Identif. Control **36**, 81–101 (2013)
94. Fiorot, J.C., Gensane, T.: Characterizations of the set of rational parametric curves with rational offsets. In: Laurent, P.J., Le Méhauté, A., Schumaker, L.L. (eds.) Curves and Surfaces in Geometric Design, pp. 153–160. AK Peters, Wellesley (1994)
95. Goodman, T.N.T., Meek, D.S., Walton, D.J.: An involute spiral that matches G^2 Hermite data in the plane. Comput. Aided Geom. Des. **26**, 733–756 (2009)
96. Giannelli, C., Mugnaini, D., Sestini, A.: Path planning with obstacle avoidance by G^1 PH quintic splines. Comput. Aided Des. **75–76**, 47–60 (2016)
97. Giannelli, C., Mugnaini, D., Sestini, A.: C^2 continuous time-dependent feedrate scheduling with configurable kinematic constraints. Comput. Aided Geom. Des. **63**, 78–95 (2018)
98. Gonzalez, C., Albrecht, G., Paluszny, M., Lentini, M.: Design of C^2 algebraic-trigonometric Pythagorean hodograph splines with shape parameters. Comput. Appl. Math. **37**, 1472–1495 (2018)
99. Gravesen, J., Jüttler, B., Sir, Z.: On rationally supported surfaces. Comput. Aided Geom. Design **25**, 320–331 (2008)

100. Guo, J.: Design of the real-time motion controller for Pythagorean hodograph curves based on FPGA. Comput. Meas. Control **20**, 1562–1568 (2012)
101. Habib, Z., Sakai, M.: Transition between concentric or tangent circles with a single segment of G^2 PH quintic curve. Comput. Aided Geom. Des. **25**, 247–257 (2008)
102. Han, C.Y.: Nonexistence of rational rotation-minimizing frames on cubic curves. Comput. Aided Geom. Des. **25**, 298–304 (2008)
103. Han, C.Y.: Geometric Hermite interpolation by monotone helical quintics. Comput. Aided Geom. Des. **27**, 713–719 (2010)
104. Han, C.Y., Kwon, S.-H.: Cubic helical splines with Frenet-frame continuity. Comput. Aided Geom. Des. **28**, 395–406 (2011)
105. Hinds, J.K., Kuan, L.P.: Surfaces defined by curve transformations. In: Proceedings of the 15th Numerical Control Society Annual Meeting & Technical Conference, pp. 325–340 (1978)
106. Huard, M., Farouki, R.T., Sprynski, N., Biard, L.: C^2 interpolation of spatial data subject to arc-length constraints using Pythagorean-hodograph quintic splines. Graph. Models **76**, 30–42 (2014)
107. Imani, B.M., Ghandehariun, A.: Real-time PH-based interpolation algorithm for high speed CNC machining. Int. J. Adv. Manuf. Tech. **56**, 619–629 (2011)
108. Jahanpour, J., Imani, B.M.: Real-time PH curve CNC interpolators for high speed cornering. Int. J. Adv. Manuf. Tech. **39**, 302–316 (2008)
109. Jahanpour, J., Tsai, M.C., Cheng, M.Y.: High-speed contouring control with NURBS-based C^2 PH spline curves. Int. J. Adv. Manuf. Tech. **49**, 663–674 (2010)
110. Jahanpour, J., Ghadirifar, A.: The improved NURBS-based C^2 PH spline curve contour following task with PDFF controller. Int. J. Adv. Manuf. Tech. **70**, 995–1007 (2014)
111. Jaklic, G., Kozak, J., Krajnc, M., Vitrih, V., Zagar, E.: Geometric Lagrange interpolation by planar cubic Pythagorean-hodograph curves. Comput. Aided Geom. Des. **25**, 720–728 (2008)
112. Jaklic, G., Kozak, J., Krajnc, M., Vitrih, V., Zagar, E.: On interpolation by planar cubic G^2 Pythagorean-hodograph spline curves. Math. Comput. **79**, 305–326 (2010)
113. Jaklic, G., Kozak, J., Krajnc, M., Vitrih, V., Zagar, E.: An approach to geometric interpolation by Pythagorean-hodograph curves. Adv. Comput. Math. **37**, 123–150 (2012)
114. Jaklic, G., Kozak, J., Krajnc, M., Vitrih, V., Zagar, E.: Interpolation by G^2 quintic Pythagorean-hodograph curves. Numer. Math. Theor. Methods Appl. **7**, 374–398 (2014)
115. Jaklic, G., Sampoli, M.L., Sestini, A., Zagar, E.: C^1 rational interpolation of spherical motions with rational rotation-minimizing directed frames. Comput. Aided Geom. Des. **30**, 159–173 (2013)
116. Jüttler, B.: Generating rational frames of space curves via Hermite interpolation with Pythagorean hodograph cubic splines. In: Choi, D.P., Choi, H.I., Kim, M.S., Martin, R.R. (eds.) Differential Topological Techniques in Geometric Modeling and Processing '98, pp. 83–106. Bookplus Press, Seoul (1998)
117. Jüttler, B.: Triangular Bézier surface patches with a linear normal vector field. In: Cripps, R. (ed.) The Mathematics of Surfaces VIII, pp. 431–446. Information Geometers, Winchester (1998)
118. Jüttler, B.: Hermite interpolation by Pythagorean-hodograph curves of degree seven. Math. Comput. **70**, 1089–1111 (2001)
119. Jüttler, B., Mäurer, C.: Cubic Pythagorean hodograph spline curves and applications to sweep surface modelling. Comput. Aided Des. **31**, 73–83 (1999)
120. Jüttler, B., Sampoli, M.L.: Hermite interpolation by piecewise polynomial surfaces with rational offsets. Comput. Aided Geom. Des. **17**, 361–385 (2000)
121. Karimi, M., Jahanpour, J., Ilbeigi, S.: A novel scheme for flexible NURBS-based C^2 PH spline curve contour following task using neural network. Int. J. Precis. Eng. Manuf. **15**, 2659–2672 (2014)
122. Kim, G.I., Lee, S.: Pythagorean-hodograph preserving mappings. J. Comput. Appl. Math. **216**, 217–226 (2008)
123. Kim, S.H., Moon, H.P.: Rectifying control polygon for planar Pythagorean hodograph curves. Comput. Aided Geom. Des. **54**, 1–14 (2017)

124. Klok, F.: Two moving coordinate frames for sweeping along a 3D trajectory. Comput. Aided Geom. Des. **3**, 217–229 (1986)
125. Klar, G., Valasek, G.: Employing Pythagorean hodograph curves for artistic patterns. Acta Cybern. **20**, 101–110 (2011)
126. Kong, J.H., Jeong, S.P., Lee, S., Kim, G.I.: C^1 Hermite interpolation with simple planar PH curves by speed reparametrization. Comput. Aided Geom. Des. **25**, 214–229 (2008)
127. Kong, J.H., Lee, S., Kim, G.: Minkowski Pythagorean-hodograph preserving mappings. J. Comput. Appl. Math. **308**, 166–176 (2016)
128. Kosinka, J., Jüttler, B.: G^1 Hermite interpolation by Minkowski Pythagorean hodograph cubics. Comput. Aided Geom. Des. **23**, 401–418 (2006)
129. Kosinka, J., Jüttler, B.: C^1 Hermite interpolation by Pythagorean hodograph quintics in Minkowski space. Adv. Comput. Math. **30**, 123–140 (2009)
130. Kosinka, J., Lavicka, M.: On rational Minkowski Pythagorean hodograph curves. Comput. Aided Geom. Des. **27**, 514–524 (2010)
131. Kosinka, J., Lavicka, M.: A unified Pythagorean hodograph approach to the medial axis transform and offset approximation. J. Comput. Appl. Math. **235**, 3413–3424 (2011)
132. Kosinka, J., Lavicka, M.: Pythagorean hodograph curves: a survey of recent advances. J. Geom. Graph. **18**, 23–34 (2014)
133. Kosinka, J., Sir, Z.: C^2 Hermite interpolation by Minkowski Pythagorean hodograph curves and medial axis transform approximations. Comput. Aided Geom. Des. **27**, 631–643 (2010)
134. Kozak, J., Krajnc, M., Vitrih, V.: Dual representation of spatial rational Pythagorean-hodograph curves. Comput. Aided Geom. Des. **31**, 43–56 (2014)
135. Kozak, J., Krajnc, M., Rogina, M., Vitrih, V.: Pythagorean-hodograph cycloidal curves. J. Numer. Math. **23**, 345–360 (2015)
136. Kozak, J., Krajnc, M., Vitrih, V.: Parametric curves with Pythagorean binormal. Adv. Comput. Math. **41**, 813–832 (2015)
137. Kozak, J., Krajnc, M., Vitrih, V.: A quaternion approach to polynomial PN surfaces. Comput. Aided Geom. Des. **47**, 172–188 (2015)
138. Kozak, J., Krajnc, M., Vitrih, V.: G^1 interpolation by rational cubic PH curves in $\mathbb{R}^3$. Comput. Aided Geom. Des. **42**, 7–22 (2016)
139. Krajnc, M.: Interpolation with spatial rational Pythagorean-hodograph curves of class 4. Comput. Aided Geom. Des. **56**, 16–34 (2017)
140. Krajnc, M., Pockaj, K., Vitrih, V.: Construction of low degree rational motions. J. Comput. Appl. Math. **256**, 92–103 (2014)
141. Krajnc, M., Sampoli, M.L., Sestini, A., Zagar, E.: C^1 interpolation by rational biarcs with rational rotation minimizing directed frames. Comput. Aided Geom. Design **31**, 427–440 (2014)
142. Krajnc, M., Vitrih, V.: Motion design with Euler-Rodrigues frames of quintic Pythagorean-hodograph curves. Math. Comput. Simul. **82**, 1696–1711 (2012)
143. Krasauskas, R.: Unifying theory of Pythagorean-normal surfaces based on geometric algebra. Adv. Appl. Clifford Algebr. **27**, 491–502 (2017)
144. Kreyszig, E.: Differential Geometry. University of Toronto Press (1959)
145. Kwon, S.H.: Solvability of G^1 Hermite interpolation by spatial Pythagorean-hodograph cubics and its selection scheme. Comput. Aided Geom. Des. **27**, 138–149 (2010)
146. Lavicka, M., Bastl, B.: Rational hypersurfaces with rational convolutions. Comput. Aided Geom. Des. **24**, 410–426 (2007)
147. Lavicka, M., Bastl, B.: PN surfaces and their convolutions with rational surfaces. Comput. Aided Geom. Des. **25**, 763–774 (2009)
148. Lavicka, M., Sir, Z., Vrsek, J.: Smooth surface interpolation using patches with rational offsets. Comput. Aided Geom. Des. **48**, 75–85 (2016)
149. Lee, H.C., Jung, E.K., Kim, G.: Planar C^1 Hermite interpolation with PH cuts of degree (1, 3) of Laurent series. Comput. Aided Geom. Des. **31**, 689–700 (2014)
150. Lee, S., Lee, H.C., Lee, M.R., Jeong, S., Kim, G.: Hermite interpolation using Möbius transformations of planar Pythagorean–hodograph cubics. Abstr. Appl. Anal. Article No. 560246 (2012)

151. Li, Z., Ait-Haddou, R., Biard, L.: Pythagorean hodograph spline spirals that match G^3 Hermite data from circles. J. Comput. Appl. Math. **278**, 162–180 (2015)
152. Lü, W.: Offset-rational parametric plane curves. Comput. Aided Geom. Des. **12**, 601–616 (1995)
153. Lu, X.J., Zheng, J.M., Cai, Y.Y., Zhao, G.: Geometric characteristics of a class of cubic curves with rational offsets. Comput. Aided Des. **70**, 36–45 (2016)
154. Macharet, D.G., Neto, A.A., Campos, M.F.M.: On the generation of feasible paths for aerial robots in environments with obstacles. In: Proceedings, 2009 IEEE/RSJ International Conference on Intelligent Robots and Systems, Saint Louis, MO, pp. 3380–3385 (2009)
155. Mäurer, C., Jüttler, B.: Rational approximation of rotation minimizing frames using Pythagorean-hodograph cubics. J. Geom. Graph. **3**, 141–159 (1999)
156. Monterde, J.: A characterization of helical polynomial curves of any degree. Adv. Comput. Math. **30**, 61–78 (2009)
157. Monterde, J.: Salkowski curves revisited: a family of curves with constant curvature and non-constant torsion. Comput. Aided Geom. Des. **26**, 271–278 (2009)
158. Monterde, J., Ongay, F.: An isoperimetric type problem for primitive Pythagorean hodograph curves. Comput. Aided Geom. Des. **29**, 626–647 (2012)
159. Moon, H.P.: Computing rational offsets via medial axis transform using polynomial speed curves in $\mathbb{R}^{2,1}$. In: Geometric Modeling and Processing '98, pp. 187–203. Bookplus Press (1998)
160. Moon, H.P.: Medial axis transform and Minkowski geometry. Ph.D. Dissertation, Seoul National University (1998)
161. Moon, H.P.: Minkowski Pythagorean hodographs. Comput. Aided Geom. Des. **16**, 739–753 (1999)
162. Moon, H.P., Farouki, R.T.: C^1 and C^2 interpolation of orientation data along spatial Pythagorean-hodograph curves using rational adapted spline frames. Comput. Aided Geom. Des. **66**, 1–15 (2018)
163. Neto, A.A., Macharet, D.G., Campos, M.F.M.: Feasible RRT–based path planning using seventh order Bézier curves. In: Proceedings, 2010 IEEE/RSJ International Conference on Intelligent Robots and Systems, Taipei, Taiwan, pp. 1445–1450 (2010)
164. Neto, A.A., Macharet, D.G., Campos, M.F.M.: On the generation of trajectories for multiple UAVs in environments with obstacles. J. Intell. Robot. Syst. **57**, 123–141 (2010)
165. Neto, A.A., Macharet, D.G., Campos, M.F.M.: Feasible path planning for fixed-wing UAVs using seventh order Bézier curves. J. Braz. Comput. Soc. **19**, 193–203 (2013)
166. Nittler, K.M., Farouki, R.T.: A real-time surface interpolator methodology for precision CNC machining of swept surfaces. Int. J. Adv. Manuf. Tech. **83**, 561–574 (2016)
167. Nittler, K.M., Farouki, R.T.: Efficient high–speed cornering motions based on continuously–variable feedrates. II. Implementation and performance analysis. Int. J. Adv. Manuf. Tech. **88**, 159–174 (2017)
168. Pelosi, F., Farouki, R.T., Manni, C., Sestini, A.: Geometric Hermite interpolation by spatial Pythagorean-hodograph cubics. Adv. Comput. Math. **22**, 325–352 (2005)
169. Pelosi, F., Sampoli, M.L., Farouki, R.T., Manni, C.: A control polygon scheme for design of planar C^2 PH quintic spline curves. Comput. Aided Geom. Des. **24**, 28–52 (2007)
170. Perwass, C.B.U., Farouki, R.T., Noakes, L.: A geometric product formulation for spatial Pythagorean-hodograph curves with applications to Hermite interpolation. Comput. Aided Geom. Des. **24**, 220–237 (2007)
171. Perwass, C.: Geometric Algebra with Applications in Engineering. Springer, Berlin (2009)
172. Peternell, M.: Rational two-parameter families of spheres and rational offset surfaces. J. Symb. Comput. **45**, 1–18 (2010)
173. Peternell, M., Odehnal, B.: Convolution surfaces of quadratic triangular Bézier surfaces. Comput. Aided Geom. Des. **25**, 116–129 (2008)
174. Peternell, M., Odehnal, B., Sampoli, M.L.: On quadratic two-parameter families of spheres and their envelopes. Comput. Aided Geom. Des. **25**, 342–355 (2008)

175. Peternell, M., Pottmann, H.: A Laguerre geometric approach to rational offsets. Comput. Aided Geom. Des. **15**, 223–249 (1998)
176. Piegl, L.: The sphere as a rational Bézier surface. Comput. Aided Geom. Des. **3**, 45–52 (1986)
177. Pockaj, K.: Hermite G^1 rational spline motions of degree six. Numer. Algorithms **66**, 721–739 (2014)
178. Pottmann, H.: Applications of the dual Bézier representation of rational curves and surfaces. In: Laurent, P.-J., Le Méhauté, A., Schumaker, L.L. (eds.) Curves and Surfaces in Geometric Design, AK Peters, Wellesley, MA, pp. 153–160 (1994)
179. Pottmann, H.: Rational curves and surfaces with rational offsets. Comput. Aided Geom. Des. **12**, 175–192 (1995)
180. Pottmann, H.: Curve design with rational Pythagorean-hodograph curves. Adv. Comput. Math. **3**, 147–170 (1995)
181. Romani, L., Montagner, F.: Algebraic–trigonometric Pythagorean–hodograph space curves. Adv. Comput. Math. (In press)
182. Romani, L., Saini, L., Albrecht, G.: Algebraic-trigonometric Pythagorean-hodograph curves and their use for Hermite interpolation. Adv. Comput. Math. **40**, 977–1010 (2014)
183. Sakkalis, T., Farouki, R.T.: Pythagorean-hodograph curves in Euclidean spaces of dimension greater than 3. J. Comput. Appl. Math. **236**, 4375–4382 (2012)
184. Sampoli, M.L., Peternell, M., Jüttler, B.: Rational surfaces with linear normals and their convolutions with rational surfaces. Comput. Aided Geom. Des. **23**, 179–192 (2006)
185. Sampoli, M.L., Sestini, A., Jaklic, G., Zagar, E.: A theoretical analysis of an improved rational spline scheme for spherical camera motions. Lecture Notes in Computer Science, vol. 8177, pp. 442–455 (2014)
186. Schraeder, T.F., Farouki, R.T.: Experimental performance analysis of an inverse dynamics CNC compensation scheme for high-speed execution of curved toolpaths. Int. J. Adv. Manuf. Tech. **73**, 195–208 (2014)
187. Sestini, A., Ferjancic, K., Manni, C., Sampoli, M.L.: A fully data-dependent criterion for free angles selection in spatial PH cubic biarc Hermite interpolation. Comput. Aided Geom. Des. **31**, 398–411 (2014)
188. Sestini, A., Landolfi, L., Manni, C.: On the approximation order of a space data-dependent PH quintic Hermite interpolation scheme. Comput. Aided Geom. Des. **30**, 148–158 (2013)
189. Shanmugavel, M., Tsourdos, A., White, B.A., Zbikowski, R.: Differential geometric path plannings of multiple UAVs. ASME J. Dyn. Syst. Meas. Control **129**, 620–632 (2007)
190. Shanmugavel, M., Tsourdos, A., Zbikowski, R., White, B.A., Rabbath, C.A., Léchevin, N.: A solution to simultaneous arrival of multiple UAVs using Pythagorean hodograph curves. In: Proceedings, 2006 American Control Conference, pp. 2813–2818 (2006)
191. Shi, J., Bi, Q.Z., Zhu, L.M., Wang, Y.H.: Corner rounding of linear five-axis tool path by dual PH curves blending. Int. J. Mach. Tools Manuf. **88**, 223–236 (2015)
192. Singh, I., Amara, Y., Melingui, A., Pathak, P.M., Merzouki, R.: Modeling of continuum manipulators using Pythagorean hodograph curves. Soft Robotics (In press)
193. Sir, Z., Bastl, B., Lavicka, M.: Hermite interpolation by hypocycloids and epicycloids with rational offsets. Comput. Aided Geom. Des. **27**, 405–417 (2010)
194. Sir, Z., Feichtinger, R., Jüttler, B.: Approximating curves and their offsets using biarcs and Pythagorean hodograph quintics. Comput. Aided Des. **38**, 608–618 (2006)
195. Sir, Z., Gravesen, J., Jüttler, B.: Curves and surfaces represented by polynomial support functions. Theor. Comput. Sci. **392**, 141–157 (2008)
196. Sir, Z., Jüttler, B.: Spatial Pythagorean hodograph quintics and the approximation of pipe surfaces. In: Martin, R., Bez, H., Sabin, M. (eds.) Mathematics of Surfaces XI. Lecture Notes in Computer Science, vol. 3604, pp. 364–380 (2005)
197. Sir, Z., Jüttler, B.: C^2 Hermite interpolation by spatial Pythagorean hodograph curves. Math. Comput. **76**, 1373–1391 (2007)
198. Sir, Z., Wings, E., Jüttler, B.: Rounding spatial G-code tool paths using Pythagorean hodograph curves. ASME J. Comput. Info. Sci. Eng. **7**, 186–191 (2007)

199. Struik, D.J.: Lectures on Classical Differential Geometry. Dover Publications (reprint), New York (1961)
200. Su, T.T., Cheng, L., Wang, Y.K., Liang, X., Zheng, J., Zhang, H.J.: Time-optimal trajectory planning for Delta robot based on quintic Pythagorean-hodograph curves. IEEE Access **6**, 28530–28539 (2018)
201. Subchan, S., White, B.A., Tsourdos, A., Shanmugavel, M., Zbikowski, R.: Pythagorean hodograph (PH) planning for tracking airborne contaminant using sensor swarm. In: Proceedings, IEEE International Instrumentation and Measurement Control Technology Conference (2008)
202. Tsai, Y.-F., Farouki, R.T., Feldman, B.: Performance analysis of CNC interpolators for time-dependent feedrates along PH curves. Comput. Aided Geom. Des. **18**, 245–265 (2001)
203. Ueda, K.: Spherical Pythagorean-hodograph curves. In: Daehlen, M., Lyche, T., Schumaker, L.L. (eds.) Mathematical Methods for Curves and Surfaces II, pp. 485–492. Vanderbilt University Press, Nashville (1998)
204. Wagner, M.G., Ravani, B.: Curves with rational Frenet-Serret motion. Comput. Aided Geom. Des. **15**, 79–101 (1997)
205. Walton, D.J., Meek, D.S.: G^2 blends of linear segments with cubics and Pythagorean-hodograph quintics. Int. J. Comput. Math. **86**, 1498–1511 (2009)
206. Walton, D.J., Meek, D.S.: Curve design with more general planar Pythagorean-hodograph quintic spiral segments. Comput. Aided Geom. Des. **30**, 707–721 (2013)
207. Wang, G., Fang, L.: On control polygons of quartic Pythagorean-hodograph curves. Comput. Aided Geom. Des. **26**, 1006–1015 (2009)
208. Wang, Q., Chen, Y., Li, W., Wei, H.: Corner smoothing using PH curve for CNC system. Acta Aero. Astro. Sinica **31**, 1481–1487 (2010)
209. Wang, W., Joe, B.: Robust computation of the rotation minimizing frame for sweep surface modelling. Comput. Aided Des. **29**, 379–391 (1997)
210. Wang, W., Jüttler, B., Zheng, D., Liu, Y.: Computation of rotation minimizing frames. ACM Trans. Graph. **27**, Article 2 (2008)
211. Zeng, W.H., Yi, J., Rao, X., Zheng, Y.: A two-stage path planning approach for multiple car-like robots based on PH curves and a modified harmony search algorithm. Eng. Optim. **49**, 1995–2012 (2017)
212. Zhang, Y., Yang, X.X., Zhou, W.W., Zhao, H.W.: Study of three-dimensional on-line path planning for UAV based on Pythagorean-hodograph curve. Int. J. Smart Sens. Intell. Syst. **8**, 1641–1666 (2015)
213. Zheng, Z., Wang, G., Yang, P.: On control polygons of Pythagorean-hodograph septic curves. J. Comput. Appl. Math. **296**, 212–227 (2016)

Tchebycheffian B-Splines Revisited: An Introductory Exposition

Tom Lyche, Carla Manni and Hendrik Speleers

Abstract Tchebycheffian splines are smooth piecewise functions where the different pieces are drawn from extended Tchebycheff spaces. They are a natural generalization of polynomial splines and can be represented in terms of an interesting set of basis functions, the so-called *Tchebycheffian B-splines*, which generalize the standard polynomial B-splines. We provide an accessible and self-contained exposition of Tchebycheffian B-splines and their main properties. Our construction is based on an integral recurrence relation and allows for the use of different extended Tchebycheff spaces on different intervals. The special class of *generalized B-splines* is also discussed in detail.

1 Introduction

Extended Tchebycheff (ET-) spaces are natural generalizations of algebraic polynomial spaces [14, 30]. Any nontrivial element of an ET-space of dimension $p + 1$ has at most p zeros counting multiplicity. Extended complete Tchebycheff (ECT-) spaces are an important subclass that can be generated through a set of positive weight functions [25, 30] and allow for defining generalized power functions [18]. Relevant examples are nullspaces of linear differential operators on suitable intervals [10, 30].

Similarly to the polynomial spline case, Tchebycheffian splines are smooth piecewise functions whose pieces are drawn from ET-spaces [26, 30]. They share many properties with the classical polynomial splines but also offer a more flexible framework, due to the wide variety of ET-spaces. As it is difficult to trace all the works

T. Lyche
Department of Mathematics, University of Oslo, Oslo, Norway
e-mail: tom@math.uio.no

C. Manni · H. Speleers (✉)
Department of Mathematics, University of Rome Tor Vergata, Rome, Italy
e-mail: speleers@mat.uniroma2.it

C. Manni
e-mail: manni@mat.uniroma2.it

C. Giannelli and H. Speleers (eds.), *Advanced Methods for Geometric Modeling and Numerical Simulation*, Springer INdAM Series 35,
https://doi.org/10.1007/978-3-030-27331-6_8

on Tchebycheffian splines, we refer the reader to [30] for an extended bibliography on the topic. Multivariate extensions of Tchebycheffian splines can be easily obtained via (local) tensor-product structures [5–7]. Besides their theoretical interest, Tchebycheffian splines have application in several branches of the sciences, including geometric modeling and numerical simulations; see, e.g., [20–23, 31].

Most of the results known for polynomial splines extend in a natural way to the Tchebycheffian setting. There are basically two main categories of extensions: the various pieces are drawn either from the same ET-space or from different ET-spaces. In the former case, Tchebycheffian splines always admit a representation in terms of basis functions, called Tchebycheffian B-splines, with similar properties to polynomial B-splines. In this context, Tchebycheffian B-splines were introduced in 1968 by Karlin [13] using generalized divided differences. We refer the reader to the historical notes in [30, Chaps. 9 and 11] for further details. The existence of a basis with similar properties for splines with pieces taken from different ET-spaces requires constraints on the various ET-spaces. Necessary and sufficient conditions for the existence of Tchebycheffian B-splines in this more general piecewise structure—sometimes referred to in the literature as piecewise Tchebycheffian B-splines [27]—has been obtained by means of blossoms; see [26, 27, 29] and references therein. Due to the intrinsic "piecewise" structure of splines, we prefer to use the term Tchebycheffian B-splines also in this more general piecewise setting. Various approaches can be used to construct Tchebycheffian B-splines whenever they exist, such as generalized divided differences [30], Hermite interpolation [8, 28], integral recurrence relations [4], de Boor-like recurrence relations [11, 18], and blossoming [26]. Each of these definitions has advantages according to the problem one has to face or to the properties to be proved. All these constructions lead to the same functions, up to a proper scaling.

This paper aims to provide a self-contained exposition of Tchebycheffian B-splines and their main properties, which are often scattered, and sometimes hidden, in the literature. Our construction of Tchebycheffian B-splines is based on an integral recurrence relation and allows for the use of different ET-spaces on different intervals in order to be able to completely exploit the rich variety of ET-spaces, as often required in applications. Although the construction and the properties we present are already known, the corresponding proofs—just based on elementary calculus—are largely new, resulting in an accessible, homogeneous, and original presentation. On the other hand, due to space limitation, some properties of Tchebycheffian B-splines are not treated in this short overview. In particular, we do not discuss the Marsden identity, dual functionals and construction of quasi-interpolants; see, e.g., [1, 30].

Our presentation of the Tchebycheffian B-spline setting strongly relies on properties of ECT-spaces related to the generating weight functions. On the other hand, any ET-space on a bounded and closed interval is an ECT-space (see [24, Corollary 2.12] and [26]), and therefore it is also equipped with weight functions. In view of this important result, since only closed and bounded intervals are of interest to define spline spaces, we could avoid mentioning the concept of ECT-spaces for constructing Tchebycheffian B-splines, as it is sometimes the case in the literature (see, e.g., [26]). However, for the sake of completeness and clarity, we prefer to present the material in terms of ECT-spaces (similar to, e.g., [3, 30]). Our approach requires that the E(C)T-spaces we are dealing with are identified by a sequence of weights which connect smoothly across the different segments. How to construct these weights is well known for a single ET-space [13, 26] but it can be an issue whenever different ET-spaces are considered [27]. However, admissible weights can be easily obtained for an interesting class of Tchebycheffian splines which allows for the use of different ET-spaces, the so-called generalized splines. For generalized splines, the presented construction readily applies without requiring any additional work.

The remainder of the paper is divided in three sections. Section 2 introduces ET-spaces and ECT-spaces. It also summarizes some of their properties to be used in the construction of Tchebycheffian B-splines. Section 3 contains the core of the paper: it defines Tchebycheffian B-splines through an integral recurrence relation and proves some of their main properties including non-negativity, smoothness, and knot insertion. Section 4 concludes the paper by discussing the interesting special class of generalized B-splines.

2 Extended Tchebycheff Spaces

In this section we introduce spaces that are a natural generalization of algebraic polynomial spaces, the so-called extended Tchebycheff spaces. In particular, we mainly focus on the subclass of extended complete Tchebycheff spaces. Such spaces can be spanned by a set of basis functions that are a natural generalization of the polynomial power basis.

2.1 Definition and Basic Properties

Suppose we have a $(p+1)$-dimensional subspace $\mathbb{U}_p(I)$ of $C^p(I)$ where I is a real interval. A **Hermite interpolation problem** in $\mathbb{U}_p(I)$ consists of finding an element $g \in \mathbb{U}_p(I)$ satisfying the following conditions:

$$D^j g(z_i) = f_{i,j}, \quad j = 0, \ldots, m_i - 1, \quad i = 0, \ldots, \ell, \tag{1}$$

where $z_0, \ldots, z_\ell$ are distinct points in I, and m_i are positive integers such that $\sum_{i=0}^{\ell} m_i = p + 1$, and $f_{i,j} \in \mathbb{R}$. We now define extended Tchebycheff spaces[1] on a real interval I.

Definition 1 Let I be an interval of the real line. Given an integer $p \geq 0$, a space $\mathbb{T}_p(I) \subset C^p(I)$ of dimension $p+1$ is an **extended Tchebycheff (ET-) space** on I

[1]The space $\mathbb{T}_p(I)$ is called a **Tchebycheff (T-) space** if any solution of (1) with $m_0 = \cdots = m_p = 1$ is unique in $\mathbb{T}_p(I)$. In such a case, (1) is a **Lagrange interpolation problem**.

if any Hermite interpolation problem with $p+1$ data on I has a unique solution in $\mathbb{T}_p(I)$.

The definition immediately implies that a $(p+1)$-dimensional subspace of $C^p(I)$ is an ET-space on I if and only if any nontrivial element of the space has at most p zeros in I counting multiplicity. Moreover, any ET-space on I is an ET-space of the same dimension on any nontrivial subinterval of I.

Example 1 The space $\mathbb{P}_p := \langle 1, x, \ldots, x^p \rangle$ of algebraic polynomials of degree less than or equal to p is an ET-space on the real line.

Example 2 The space $\langle \cos(x), \sin(x) \rangle$ is an ET-space on any interval $[a, a+\pi)$ with $a \in \mathbb{R}$. Indeed, the equation $c_1 \cos(x) + c_2 \sin(x) = 0$ has exactly one solution in the considered interval for any fixed c_1, c_2 not both equal to zero. On the other hand, on any interval $[a, a+\pi]$ or larger, this space is not an ET-space anymore.

We now focus on a special subclass of ET-spaces.

Definition 2 Let I be an interval of the real line. Given an integer $p \geq 0$, the space $\mathbb{T}_p(I) \subset C^p(I)$ of dimension $p+1$ is an **extended complete Tchebycheff (ECT-) space** if there exists a basis $\{u_0, \ldots, u_p\}$ of $\mathbb{T}_p(I)$ such that every subspace $\langle u_0, \ldots, u_k \rangle$ is an ET-space on I for $k = 0, \ldots, p$. The basis $\{u_0, \ldots, u_p\}$ is called an **ECT-system**.

Example 3 Taking $u_k(x) = x^k, k = 0, \ldots, p$, we see from Example 1 that the space $\mathbb{P}_p$ is an ECT-space on any interval of the real line.

Example 4 An ECT-space is clearly an ET-space, but the converse is not always true. It is sufficient to consider the space $\langle \cos(x), \sin(x) \rangle$. This is an ET-space on $[0, \pi)$, see Example 2, but not an ECT-space on $[0, \pi)$. However, the space is an ECT-space on $(0, \pi)$.

The next theorem shows that the classes of ECT-spaces and ET-spaces coincide in a very important case; see [24, Corollary 2.12] and [26] for details.

Theorem 1 *If I is a bounded closed interval, then any ET-space on I is an ECT-space on I.*

In the following we will provide a characterization of an ECT-space in terms of Wronskians. The **Wronskian** of $k+1$ functions $\{u_0, \ldots, u_k\}$ of class $C^k(I)$ is given by the determinant

$$W[u_0, \ldots, u_k](x) := \det\big(D^i u_j(x)\big)_{i,j=0}^{k}.$$

If there exists a point $\bar{x} \in I$ such that $W[u_0, \ldots, u_k](\bar{x}) \neq 0$, then the functions $\{u_0, \ldots, u_k\}$ are linearly independent. Wronskians can be used to characterize an ECT-space as follows; see [30, Theorem 9.1] for a proof.

Theorem 2 *A $(p+1)$-dimensional subspace of $C^p(I)$ is an ECT-space on I if and only if there exists a basis $\{u_0, \dots, u_p\}$ such that all the Wronskians are positive; more precisely,*

$$W[u_0, \dots, u_k](x) > 0, \quad k = 0, \dots, p, \quad x \in I.$$

Note that the basis $\{u_0, \dots, u_p\}$ in Theorem 2 is an ECT-system.

Theorem 2 gives a characterization of ECT-spaces in terms of Wronskians, but there is no similar characterization for ET-spaces. If $\langle u_0, \dots, u_p \rangle$ is a $(p+1)$-dimensional ET-space on I, then the Wronskian $W[u_0, \dots, u_p](x)$ is either positive for all $x \in I$ or negative for all $x \in I$. However, the converse does not hold. It is sufficient to consider the space $\langle \cos(x), \sin(x) \rangle$. This is an ET-space only on intervals of the form $[a, a+\pi)$ or subintervals (see Example 2), but $W[\cos, \sin](x) = 1$ for all $x \in \mathbb{R}$.

Example 5 In Example 3 we have shown that the space $\mathbb{P}_p$ is an ECT-space using the set $\{1, x, \dots, x^p\}$. More generally, $\mathbb{P}_p$ can be seen as the span of the **power basis**

$$\left\{1, x-y, \frac{(x-y)^2}{2}, \dots, \frac{(x-y)^p}{p!}\right\}, \tag{2}$$

for any fixed $y \in \mathbb{R}$. Indeed, the Wronskians of this set of functions are all equal to one.

2.2 Generalized Powers

In this section we introduce special functions that can be regarded as a generalization of the power basis in (2).

Definition 3 Let $(v_1, \dots, v_p)$ be a vector of continuous functions on an interval I. For the points x, y in I, the **repeated integral** G_p is defined recursively by

$$G_p[v_1, \dots, v_p](x, y) := \int_y^x v_1(t) G_{p-1}[v_2, \dots, v_p](t, y) \, dt, \quad p \geq 1, \tag{3}$$

starting with $G_0(x, y) := 1$.

From the definition we obtain

$$G_1[v_1](x, y) = \int_y^x v_1(t) \, dt,$$

and for $p > 1$,

$$G_p[v_1, \ldots, v_p](x, y) = \int_y^x v_1(t_1) \int_y^{t_1} v_2(t_2) \cdots \int_y^{t_{p-1}} v_p(t_p)\, dt_p \cdots dt_1.$$

We only list two basic properties of repeated integrals; for proofs and further properties we refer the reader to [18].

- **Diagonal Property**. Let $v_j \in C^{\max(p-1-j,0)}(I)$, $j = 1, \ldots, p$. For any $y \in I$ we have

$$\frac{\partial^r}{\partial x^r} G_p[v_1, \ldots, v_p](x, y)_{|x=y} = 0, \quad r = 0, \ldots, p-1. \tag{4}$$

- **Generalized Binomial Formula**. For any $x, y, c \in I$ we have

$$G_p[v_1, \ldots, v_p](x, y) = \sum_{j=0}^{p} (-1)^{p-j} G_j[v_1, \ldots, v_j](x, c) G_{p-j}[v_p, \ldots, v_{j+1}](y, c). \tag{5}$$

We are now ready to define a generalization of the classical power basis in (2).

Definition 4 Let $\boldsymbol{w} := (w_0, \ldots, w_q)$ be a vector of continuous functions on an interval I. For a nonnegative integer $p \le q$ and a fixed point y in I, we define the **generalized powers** by

$$\begin{aligned} u^{\boldsymbol{w}}_{0,p}(x, y) &:= w_p(x), \\ u^{\boldsymbol{w}}_{1,p}(x, y) &:= w_p(x) G_1[w_{p-1}](x, y), \\ u^{\boldsymbol{w}}_{2,p}(x, y) &:= w_p(x) G_2[w_{p-1}, w_{p-2}](x, y), \\ &\vdots \\ u^{\boldsymbol{w}}_{p,p}(x, y) &:= w_p(x) G_p[w_{p-1}, \ldots, w_0](x, y), \end{aligned} \tag{6}$$

with G_j the repeated integrals in (3).

We immediately obtain the following properties.

- **Recurrence Formula**. From the definition (3) of repeated integrals we deduce

$$u^{\boldsymbol{w}}_{j,p}(x, y) = w_p(x) \int_y^x u^{\boldsymbol{w}}_{j-1,p-1}(t, y)\, dt, \quad j = 1, \ldots, p, \quad x, y \in I. \tag{7}$$

- **Smoothness**. If $w_j \in C^j(I)$, $j = 0, \ldots, p$, then

$$u^{\boldsymbol{w}}_{j,p} \in C^p(I), \quad j = 0, \ldots, p. \tag{8}$$

  This follows from (7).

Example 6 If $w_{p-j} = \cdots = w_p = 1$ then

$$u^{\mathbf{w}}_{j,p}(x,y) = w_p(x) G_j[w_{p-1},\ldots,w_{p-j}](x,y) = \frac{(x-y)^j}{j!}. \tag{9}$$

In this case, (5) takes the form

$$\frac{(x-y)^p}{p!} = \sum_{j=0}^{p} (-1)^{p-j} \frac{(x-c)^j}{j!} \frac{(y-c)^{p-j}}{(p-j)!} = \frac{\big((x-c)-(y-c)\big)^p}{p!}.$$

The next theorem shows that the Wronskians of generalized powers can be expressed in a simple form; see [13, p. 278].

Theorem 3 *Let* $\mathbf{w} := (w_0,\ldots,w_p)$ *be a vector of weight functions on an interval* I *such that* $w_j \in C^j(I)$, $j = 0,\ldots,p$. *For any* $x, y \in I$ *and* $0 \le k \le p$ *we have*

$$W[u^{\mathbf{w}}_{0,p}(\cdot,y),\ldots,u^{\mathbf{w}}_{k,p}(\cdot,y)](x) = w_p^{k+1}(x) w_{p-1}^k(x) \cdots w_{p-k}(x).$$

Theorem 3 leads to the following properties.

- **Linear Independence**. Suppose there exists a point $\bar{x} \in I$ such that

$$w_p(\bar{x}) \cdots w_{p-k}(\bar{x}) \neq 0.$$

  Then, $W[u^{\mathbf{w}}_{0,p}(\cdot,y),\ldots,u^{\mathbf{w}}_{k,p}(\cdot,y)](\bar{x}) \neq 0$ and, as a consequence, the generalized powers $\{u^{\mathbf{w}}_{0,p}(\cdot,y),\ldots,u^{\mathbf{w}}_{k,p}(\cdot,y)\}$ are linearly independent for any $y \in I$.

- **Weight Functions**. Let $\mathbf{w} := (w_0,\ldots,w_p)$ be a vector of positive weight functions on an interval I such that $w_j \in C^j(I)$, $j = 0,\ldots,p$. Then, for any $x, y \in I$ the Wronskians of the generalized powers are positive, and we have

$$w_p(x) = u^{\mathbf{w}}_{0,p}(x,y), \quad w_{p-1}(x) = \frac{W[u^{\mathbf{w}}_{0,p}(\cdot,y), u^{\mathbf{w}}_{1,p}(\cdot,y)](x)}{\big(u^{\mathbf{w}}_{0,p}(x,y)\big)^2},$$

  and for $2 \le k \le p$,

$$w_{p-k}(x) = \frac{W[u^{\mathbf{w}}_{0,p}(\cdot,y),\ldots,u^{\mathbf{w}}_{k,p}(\cdot,y)](x)\, W[u^{\mathbf{w}}_{0,p}(\cdot,y),\ldots,u^{\mathbf{w}}_{k-2,p}(\cdot,y)](x)}{\big(W[u^{\mathbf{w}}_{0,p}(\cdot,y),\ldots,u^{\mathbf{w}}_{k-1,p}(\cdot,y)](x)\big)^2}.$$

In the following we discuss spaces spanned by generalized powers. The generalized binomial formula (5) and definition (6) imply

$$\langle u^{\mathbf{w}}_{0,p}(\cdot,y_1),\ldots,u^{\mathbf{w}}_{k,p}(\cdot,y_1)\rangle = \langle u^{\mathbf{w}}_{0,p}(\cdot,y_2),\ldots,u^{\mathbf{w}}_{k,p}(\cdot,y_2)\rangle, \quad y_1, y_2 \in I,$$

for $k = 0,\ldots,p$. This observation leads to the following well-posed definition.

Definition 5 Let $\boldsymbol{w} := (w_0, \ldots, w_q)$ be a vector of positive weight functions on an interval I such that $w_j \in C^j(I)$, $j = 0, \ldots, q$. For a nonnegative integer $p \leq q$, we define the space $\mathbb{T}_p^{\boldsymbol{w}}(I)$ on the interval I generated by the weight vector $\boldsymbol{w}$ by

$$\mathbb{T}_p^{\boldsymbol{w}}(I) := \langle u_{0,p}^{\boldsymbol{w}}(\cdot, y), \ldots, u_{p,p}^{\boldsymbol{w}}(\cdot, y)\rangle, \tag{10}$$

where $u_j^{\boldsymbol{w}}(\cdot, y)$, $j = 0, \ldots, p$ are given in (6) and y is any fixed point in I.

It is clear that the space $\mathbb{T}_p^{\boldsymbol{w}}(I)$ in (10) only depends on $w_0, \ldots, w_p$.

Example 7 From Example 6 we see that if $w_0 = \cdots = w_p = 1$ then

$$\mathbb{T}_p^{\boldsymbol{w}}(\mathbb{R}) = \left\langle 1, x - y, \frac{(x-y)^2}{2}, \ldots, \frac{(x-y)^p}{p!} \right\rangle = \mathbb{P}_p,$$

for any fixed $y \in \mathbb{R}$.

We now show that generalized powers defined in terms of positive weight functions span an ECT-space.

Theorem 4 *Let $\boldsymbol{w} := (w_0, \ldots, w_p)$ be a vector of positive weight functions on an interval I such that $w_j \in C^j(I)$, $j = 0, \ldots, p$. The space $\mathbb{T}_p^{\boldsymbol{w}}(I)$ is an ECT-space of dimension $p + 1$ on the interval I. In particular, the generalized powers $\{u_{0,p}^{\boldsymbol{w}}(\cdot, y), \ldots, u_{p,p}^{\boldsymbol{w}}(\cdot, y)\}$ are linearly independent.*

Proof By recalling (8) we see that $\mathbb{T}_p^{\boldsymbol{w}}(I) \subset C^p(I)$. From Theorem 3 it follows that all the Wronskians of the generalized powers $\{u_{0,p}^{\boldsymbol{w}}(\cdot, y), \ldots, u_{p,p}^{\boldsymbol{w}}(\cdot, y)\}$ are positive on I, and in particular that $\{u_{0,p}^{\boldsymbol{w}}(\cdot, y), \ldots, u_{p,p}^{\boldsymbol{w}}(\cdot, y)\}$ forms a basis for $\mathbb{T}_p^{\boldsymbol{w}}(I)$. Theorem 2 completes the proof. □

In view of Theorem 2, we can assume that any ECT-space $\mathbb{T}_p(I)$ is spanned by an ECT-system with positive Wronskians. The next theorem shows that $\mathbb{T}_p(I)$ is spanned by generalized powers associated with certain positive weight functions. The proof can be deduced from [13, Proof of Theorem 1.1, p. 276].

Theorem 5 *Let $\mathbb{T}_p(I)$ be any ECT-space of dimension $p + 1$ on the interval I, and let $\{u_0, \ldots, u_p\}$ be an ECT-system with positive Wronskians spanning $\mathbb{T}_p(I)$. We define the positive weight functions*

$$w_p(x) := u_0(x), \quad w_{p-1}(x) := \frac{W[u_0, u_1](x)}{(u_0(x))^2},$$
$$w_{p-k}(x) := \frac{W[u_0, \ldots, u_k](x) W[u_0, \ldots, u_{k-2}](x)}{\big(W[u_0, \ldots, u_{k-1}](x)\big)^2}, \quad k = 2, \ldots, p.$$

Then, the corresponding generalized powers $\{u_{0,p}^{\boldsymbol{w}}(\cdot, y), \ldots, u_{p,p}^{\boldsymbol{w}}(\cdot, y)\}$ form a basis for $\mathbb{T}_p(I)$ for any $y \in I$.

There exist different weight vectors generating the same ECT-space. Obviously, the weight vector $(c_0 w_0, \ldots, c_p w_p)$ generates the same space as the weight vector $(w_0, \ldots, w_p)$ for any positive constants c_j, $j = 0, \ldots, p$. A less trivial case is illustrated in Example 8. A nice construction of all possible weight vectors generating the same ECT-space is given in [25].

Example 8 The space $\langle 1, \cos(x), \sin(x) \rangle$ is an ECT-space of dimension 3 on $(-\pi/2, \pi/2)$. This can be shown by considering the weight vector $\boldsymbol{w} = (w_0, w_1, w_2)$ with

$$w_0(x) = \frac{1}{\cos^2(x)}, \quad w_1(x) = \cos(x), \quad w_2(x) = 1.$$

Then, the corresponding generalized powers with $y = 0$ are

$$\begin{aligned}
u^{\boldsymbol{w}}_{0,2}(x, 0) &= w_2(x) = 1, \\
u^{\boldsymbol{w}}_{1,2}(x, 0) &= \int_0^x \cos(t)\, \mathrm{d}t = \sin(x), \\
u^{\boldsymbol{w}}_{2,2}(x, 0) &= \int_0^x \cos(t_1) \int_0^{t_1} \frac{1}{\cos^2(t_2)}\, \mathrm{d}t_2 \mathrm{d}t_1 = \int_0^x \sin(t_1)\, \mathrm{d}t_1 = 1 - \cos(x).
\end{aligned}$$

Actually, with some additional effort we can prove that $\langle 1, \cos(x), \sin(x) \rangle$ is an ECT-space of dimension 3 on $(-\pi, \pi)$. To this end, consider the weight vector $\boldsymbol{w} = (w_0, w_1, w_2)$ with

$$w_0(x) = w_1(x) = 1 \Big/ \cos^2\left(\frac{x}{2}\right), \quad w_2(x) = \cos^2\left(\frac{x}{2}\right).$$

Then, for $x \in (-\pi, \pi)$ and any fixed $y \in (-\pi, \pi)$ we find the generalized powers

$$\begin{aligned}
u^{\boldsymbol{w}}_{0,2}(x, y) &= w_2(x) = \cos^2\left(\frac{x}{2}\right) = \frac{1 + \cos(x)}{2}, \\
u^{\boldsymbol{w}}_{1,2}(x, y) &= \sin(x) - (1 + \cos(x)) \tan\left(\frac{y}{2}\right), \\
u^{\boldsymbol{w}}_{2,2}(x, y) &= 2\sin^2\left(\frac{x - y}{2}\right) \Big/ \cos^2\left(\frac{y}{2}\right) = \frac{1 - \cos(x)\cos(y) - \sin(x)\sin(y)}{\cos^2(y/2)}.
\end{aligned}$$

Example 9 Let $\langle u, v \rangle$ be an ET-space of dimension 2 on $[a, b]$. We can assume without loss of generality that $u(a) = u(b) = 1$ and $v(a) = 0$, $v(b) \neq 0$. It turns out that u is positive on $[a, b]$ because otherwise it would have two zeros (counting multiplicity) in (a, b). Set

$$\begin{aligned}
w_0(x) &:= D\left(\frac{v}{u}\right)(x) = \frac{u(x)Dv(x) - v(x)Du(x)}{(u(x))^2} = \frac{W[u, v](x)}{(u(x))^2}, \\
w_1(x) &:= u(x).
\end{aligned}$$

Since $\langle u, v\rangle$ is an ET-space, the Wronskian $W[u, v]$ is either positive or negative on $[a, b]$. Hence, we can assume it is positive on $[a, b]$; if this is not the case, we change the sign of v. This implies that w_0 is positive on $[a, b]$. Moreover,

$$w_1(x)\int_a^x w_0(t)\,\mathrm{d}t = w_1(x)\left(\frac{v(x)}{u(x)} - \frac{v(a)}{u(a)}\right) = v(x). \tag{11}$$

Therefore, $\langle u, v\rangle$ is the space $\mathbb{T}_2^{\boldsymbol{w}}([a, b])$ with $\boldsymbol{w} = (w_0, w_1)$, and so from Theorem 4 we know it is an ECT-space on $[a, b]$. We conclude that any ET-space of dimension 2 on a closed interval $[a, b]$ is an ECT-space of dimension 2 on $[a, b]$. This is in agreement with Theorem 1. Note that the statement does not hold anymore if the interval is not closed; see Example 4.

Example 10 Let $U, V \in C^p([a, b])$ be given such that $\langle D^{p-1}U, D^{p-1}V\rangle$ is an ET-space on $[a, b]$. Then, the space $\langle 1, x, \ldots, x^{p-2}, U(x), V(x)\rangle$ for $p \geq 2$ is an ECT-space of dimension $p + 1$ on $[a, b]$. Indeed, it is the space $\mathbb{T}_p^{\boldsymbol{w}}([a, b])$ generated by the weight functions

$$w_0(x) = \frac{W[u, v](x)}{(u(x))^2}, \quad w_1(x) = u(x), \quad w_2(x) = \cdots = w_p(x) = 1,$$

where

$$u(x) := c_{0,u} D^{p-1}U(x) + c_{1,u} D^{p-1}V(x),$$
$$v(x) := c_{0,v} D^{p-1}U(x) + c_{1,v} D^{p-1}V(x),$$

such that $u(a) = u(b) = 1$ and $v(a) = 0$, $v(b) \neq 0$ (see Example 9).

Example 11 Let $\mathcal{L}_p$ be the linear differential operator defined by

$$\mathcal{L}_p f := D^{p+1} f + \sum_{j=0}^{p} a_j D^j f, \quad f \in C^{p+1}(I), \tag{12}$$

where $a_j \in C(I)$ and I is a real interval. Any operator of the form (12) is uniquely identified by its nullspace, denoted by $\mathbb{L}_p$. More details on linear differential operators can be found in [12, Chap. 5]. The nullspace $\mathbb{L}_p$ is an ECT-space on I if and only if there exist positive weight functions $w_j \in C^{j+1}(I)$, $j = 0, \ldots, p$ such that

$$\mathcal{L}_p f = w_0 \cdots w_p D_0 \cdots D_p f, \tag{13}$$

where

$$D_j f := D\left(\frac{f}{w_j}\right), \quad j = 0, \ldots, p;$$

see [10, Theorem 2, p. 91].

- If the coefficients a_j are equal to zero, then $\mathbb{L}_p = \mathbb{P}_p$ is an ECT-space on the real line; see Example 3.
- If the coefficients a_j are constants and the characteristic polynomial $\lambda^{p+1} + \sum_{j=0}^{p} a_j \lambda^j$ has only real roots, then $\mathbb{L}_p$ is an ECT-space on the real line; see [10, Proposition 16, p. 124]. For example, given distinct real numbers $\alpha_0 < \alpha_1 < \cdots < \alpha_p$, the space $\langle \mathrm{e}^{\alpha_0 x}, \mathrm{e}^{\alpha_1 x}, \ldots, \mathrm{e}^{\alpha_p x} \rangle$ is the nullspace of

$$\mathcal{L}_p f = (D - \alpha_0) \cdots (D - \alpha_p) f,$$

implying it is an ECT-space on the real line.
- If the coefficients a_j are constants and the characteristic polynomial has complex roots, then $\mathbb{L}_p$ is an ECT-space on a suitable interval. For example, the space $\langle 1, x, \ldots, x^{p-2}, \cos(x), \sin(x) \rangle$ with $p \geq 2$ is the nullspace of

$$\mathcal{L}_p f = D^{p+1} f + D^{p-1} f;$$

it is also an ECT-space on the interval $(-\pi/2, \pi/2)$ that can be generated by the weight functions

$$w_0(x) = \frac{1}{\cos^2(x)}, \quad w_1(x) = \cos(x), \quad w_2(x) = \cdots = w_p(x) = 1;$$

see Example 8 for the special case $p = 2$. The factorization (13) becomes

$$\mathcal{L}_p f = \frac{1}{\cos} D\left(\cos^2\left(D\left(\frac{1}{\cos} D^{p-1} f\right)\right)\right).$$

Actually, the space $\langle 1, x, \ldots, x^{p-2}, \cos(x), \sin(x) \rangle$ is an ECT-space on larger intervals whose maximum lengths increase with p; see Example 8 and [9].
- In the general case of nonconstant coefficients $a_j(x)$, $\mathbb{L}_p$ is an ECT-space on a suitable interval. For example, if I is compact, then $\mathbb{L}_p$ is an ECT-space on any subinterval of I of length less than $\min\left(1, \frac{1}{(p+1)M}\right)$, where

$$M \geq \max_{0 \leq j \leq p} \max_{x \in I} |a_j(x)|;$$

see [10, Proposition 1, p. 81].

3 Tchebycheffian B-Splines

In Sect. 2 we showed that ECT-spaces extend in a very natural way the space of algebraic polynomials. Now, we focus on smooth functions that are defined piecewise in ECT-spaces, and we define the so-called Tchebycheffian B-splines, which are a natural generalization of polynomial B-splines. Since we are interested in bounded and closed intervals, in view of Theorem 1, ET-spaces are ECT-spaces and so they are equipped with weight functions. Therefore, Tchebycheffian B-splines are actually defined piecewise in ET-spaces.

We start by introducing some preliminary notations. A function is called piecewise continuous on a finite interval I if it is bounded and continuous except at a finite number of points, where the value is obtained by taking the limit either from the left or the right. We denote the space of these functions by $C^{-1}(I)$. The right and left limits of a real number x are denoted by

$$x^+ := \lim_{\substack{t\to x\\ t>x}} t, \quad x^- := \lim_{\substack{t\to x\\ t<x}} t, \quad x \in \mathbb{R}.$$

Similarly, we denote right and left derivatives of a function f by

$$D_+ f(x) := \lim_{\substack{h\to 0\\ h>0}} \frac{f(x+h)-f(x)}{h}, \quad D_- f(x) := \lim_{\substack{h\to 0\\ h<0}} \frac{f(x+h)-f(x)}{h},$$

provided that the limits exist at the point $x \in \mathbb{R}$. Furthermore, assume f to be defined on $[a,b] \supsetneq [c,d]$. The notation $f \in C^j([c^+,d^-])$ means that f is a function of class C^j on the interval $[c,d]$ when considering the right/left limit in the left/right endpoint. Note that, in general, $D_+^r f(c) \neq D_-^r f(c)$ and $D_+^r f(d) \neq D_-^r f(d)$, $r = 0, \ldots, j$.

3.1 Definition and Basic Properties

In order to define Tchebycheffian B-splines we use the concept of knot sequences. Suppose for integers $n > p \geq 0$ that a knot sequence

$$\boldsymbol{\xi} := \{\xi_i\}_{i=1}^{n+p+1} = \{\xi_1 \leq \xi_2 \leq \cdots \leq \xi_{n+p+1}\}, \quad n \in \mathbb{N}, \quad p \in \mathbb{N}_0,$$

is given. This allows us to define a set of n Tchebycheffian B-splines of degree p.

Definition 6 Given a knot sequence $\boldsymbol{\xi}$, the functions $w_0, \ldots, w_p$ are called **Tchebycheffian B-spline weights** with respect to $\boldsymbol{\xi}$ if they are positive on $[\xi_1, \xi_{n+p+1}]$ and for $j = 0, \ldots, p$,

$$w_j \in C^j([\xi_i^+, \xi_{i+1}^-]), \quad \xi_i < \xi_{i+1}, \quad i = 1, \ldots, n+p, \tag{14}$$

and

$$w_j \in C^{\max(j-\mu_i,-1)}(\xi_i), \quad i = 2, \ldots, n+p, \tag{15}$$

where μ_i is the multiplicity of ξ_i in $\boldsymbol{\xi}$.

The Tchebycheffian B-spline weights $\boldsymbol{w} := (w_0, \ldots, w_p)$ have a piecewise structure, and define a set of local weights on each interval $[\xi_i^+, \xi_{i+1}^-]$, denoted by

$$w_{j,i}(x) := w_j(x), \quad x \in [\xi_i^+, \xi_{i+1}^-],$$

for $j = 0, \ldots, p$. In view of (14), Theorem 4 implies that these local weights $\boldsymbol{w}_i := (w_{0,i}, \ldots, w_{p,i})$ specify an E(C)T-space $\mathbb{T}_p^{\boldsymbol{w}_i}([\xi_i^+, \xi_{i+1}^-])$ of dimension $p+1$ for each $\xi_i < \xi_{i+1}$. We will see that the smoothness at the knots in (15) suffices to define Tchebycheffian B-splines with smoothness properties similar to the polynomial B-spline case.

Example 12 For $i = 1, \ldots, n+1$, let $\langle u_i, v_i \rangle$ be ET-spaces of dimension 2 on $[\xi_i, \xi_{i+1}]$, where

$$u_i, v_i \in C^1([\xi_i, \xi_{i+1}]), \quad u_i(\xi_i) = u_i(\xi_{i+1}) = 1, \quad v_i(\xi_i) = 0, \; v_i(\xi_{i+1}) = 1,$$

and, according to Example 9, we can define the local weights

$$w_{0,i}(x) := \frac{W[u_i, v_i](x)}{(u_i(x))^2}, \quad w_{1,i}(x) := u_i(x), \quad x \in [\xi_i, \xi_{i+1}].$$

The global weights

$$w_0(x) := w_{0,i}(x), \quad w_1(x) := w_{1,i}(x), \quad x \in [\xi_i, \xi_{i+1}), \quad i = 1, \ldots, n+1$$

satisfy (14) and (15) for $p = 1$. Hence, w_0, w_1 are Tchebycheffian B-spline weights with respect to $\boldsymbol{\xi}$, and they generate the given ET-spaces on each (nontrivial) interval $[\xi_i^+, \xi_{i+1}^-]$.

Definition 7 Let $\boldsymbol{w} := (w_0, \ldots, w_q)$ be a vector of Tchebycheffian B-spline weights with respect to a knot sequence $\boldsymbol{\xi}$. Suppose for a nonnegative integer $p \leq q$ and some integer j that $\xi_j \leq \xi_{j+1} \leq \cdots \leq \xi_{j+p+1}$ are $p+2$ real numbers taken from $\boldsymbol{\xi}$. The jth **Tchebycheffian B-spline** $B^{\boldsymbol{w}}_{j,p,\boldsymbol{\xi}} : \mathbb{R} \to \mathbb{R}$ of degree p is identically zero if $\xi_{j+p+1} = \xi_j$ and otherwise defined recursively by

$$B^{\boldsymbol{w}}_{j,p,\boldsymbol{\xi}}(x) := w_p(x) \left(\int_{\xi_j}^{x} \frac{B^{\boldsymbol{w}}_{j,p-1,\boldsymbol{\xi}}(y)}{\gamma^{\boldsymbol{w}}_{j,p-1,\boldsymbol{\xi}}} \, dy - \int_{\xi_{j+1}}^{x} \frac{B^{\boldsymbol{w}}_{j+1,p-1,\boldsymbol{\xi}}(y)}{\gamma^{\boldsymbol{w}}_{j+1,p-1,\boldsymbol{\xi}}} \, dy \right), \tag{16}$$

starting with

$$B^{\boldsymbol{w}}_{i,0,\boldsymbol{\xi}}(x) := \begin{cases} w_0(x), & \text{if } x \in [\xi_i, \xi_{i+1}), \\ 0, & \text{otherwise.} \end{cases} \tag{17}$$

Here, $\gamma^{\boldsymbol{w}}_{i,k,\boldsymbol{\xi}}$ is defined as the integral of $B^{\boldsymbol{w}}_{i,k,\boldsymbol{\xi}}$,

$$\gamma^{\boldsymbol{w}}_{i,k,\boldsymbol{\xi}} := \int_{\xi_i}^{\xi_{i+k+1}} B^{\boldsymbol{w}}_{i,k,\boldsymbol{\xi}}(y)\,\mathrm{d}y, \tag{18}$$

and we used the convention that if $\gamma^{\boldsymbol{w}}_{i,k,\boldsymbol{\xi}} = 0$ then

$$\int_{\xi_i}^{x} \frac{B^{\boldsymbol{w}}_{i,k,\boldsymbol{\xi}}(y)}{\gamma^{\boldsymbol{w}}_{i,k,\boldsymbol{\xi}}}\,\mathrm{d}y := \begin{cases} 1, & \text{if } x \geq \xi_{i+k+1}, \\ 0, & \text{otherwise.} \end{cases} \tag{19}$$

The Tchebycheffian B-spline $B^{\boldsymbol{w}}_{j,p,\boldsymbol{\xi}}$ is called **normalized** when $w_p = 1$.

In order to stress the similarity with the polynomial B-spline case, the term **degree** p is used in Definition 7 to identify the dimension $p+1$ of the underlying E(C)T-space. We also use the terms **linear** in case of $p = 1$, **quadratic** in case of $p = 2$, and so on. Furthermore, we use the notation

$$B[\xi_j, \ldots, \xi_{j+p+1}; w_0, \ldots, w_p] := B^{\boldsymbol{w}}_{j,p,\boldsymbol{\xi}},$$

showing explicitly on which knots and weight functions the Tchebycheffian B-spline depends.[2]

Definition 7 allows for the construction of Tchebycheffian B-splines where the different pieces are drawn from different ET-spaces. This can be done provided that we are able to construct Tchebycheffian B-spline weights as in Definition 6 which identify on each interval the desired ET-space. Example 12 shows that this is always the case if we consider ET-spaces of dimension two. In view of Example 10, this paves the path for the construction of the so-called generalized B-splines which will be discussed in Sect. 4. A more general setting can be addressed by relying on the elegant constructive procedure for finding all weight vectors associated with a given ET-space in a bounded closed interval presented in [25]. In particular, this procedure, which is based on the properties of Tchebycheffian Bernstein functions, has been exploited in [27] to construct normalized Tchebycheffian B-splines with pieces belonging to ET-spaces of dimension 5 (see also [2]).

Example 13 The **linear Tchebycheffian B-spline** is given by

$$\begin{aligned} B^{\boldsymbol{w}}_{j,1,\boldsymbol{\xi}}(x) &= B[\xi_j, \xi_{j+1}, \xi_{j+2}; w_0, w_1](x) \\ &= \begin{cases} w_1(x) \dfrac{\int_{\xi_j}^{x} w_0(y)\,\mathrm{d}y}{\int_{\xi_j}^{\xi_{j+1}} w_0(y)\,\mathrm{d}y}, & \text{if } x \in [\xi_j, \xi_{j+1}), \\[2ex] w_1(x) \dfrac{\int_{x}^{\xi_{j+2}} w_0(y)\,\mathrm{d}y}{\int_{\xi_{j+1}}^{\xi_{j+2}} w_0(y)\,\mathrm{d}y}, & \text{if } x \in [\xi_{j+1}, \xi_{j+2}), \\[2ex] 0, & \text{otherwise.} \end{cases} \end{aligned}$$

[2]Our Tchebycheffian B-spline construction follows the approach of [3, 4], while it differs from [26] in two ways: the indexing of the weight functions and the positioning of the weight functions with respect to the integration. This provides a more intuitive notation.

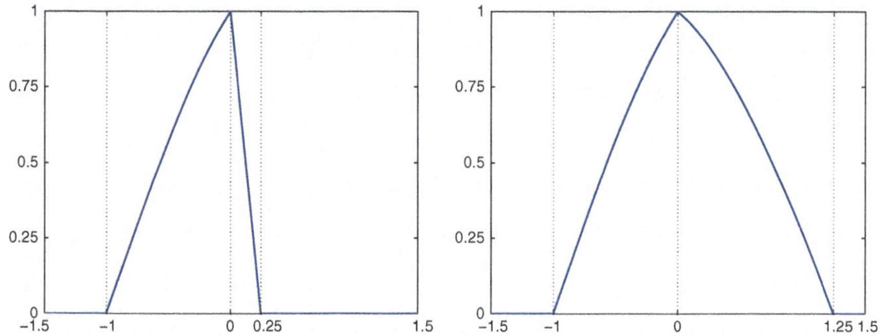

Fig. 1 Two linear trigonometric Tchebycheffian B-splines on a knot sequence of the form $\{-1, 0, \xi\}$. Left: $\xi = 1/4$, Right: $\xi = 5/4$. The knot positions are visualized by vertical dotted lines

In particular, when $w_0(x) = 1/\cos^2(x)$, $w_1(x) = \cos(x)$ and $-\pi/2 < \xi_j < \xi_{j+2} < \pi/2$, we have

$$B[\xi_j, \xi_{j+1}, \xi_{j+2}; w_0, w_1](x) = \begin{cases} \dfrac{\sin(x-\xi_j)\cos(\xi_{j+1})}{\sin(\xi_{j+1}-\xi_j)}, & \text{if } x \in [\xi_j, \xi_{j+1}), \\ \dfrac{\sin(\xi_{j+2}-x)\cos(\xi_{j+1})}{\sin(\xi_{j+2}-\xi_{j+1})}, & \text{if } x \in [\xi_{j+1}, \xi_{j+2}), \\ 0, & \text{otherwise.} \end{cases}$$

All spline pieces belong to the trigonometric space $\langle \cos(x), \sin(x) \rangle$. This is in agreement with Example 8. The spline function is discontinuous at a double knot and continuous at a simple knot. Two such linear trigonometric Tchebycheffian B-splines (with $\xi_1 = -1$, $\xi_2 = 0$, $\xi_3 = 1/4$ or $\xi_3 = 5/4$) are depicted in Fig. 1.

Example 14 The **quadratic (normalized) Tchebycheffian B-spline** defined on the uniform knot sequence $\{i\omega\}_{i=0}^{3}$ with $\omega < \pi$, and generated by the weight functions

$$w_0(x) = \frac{1}{(w_1(x))^2}, \quad w_1(x) = \frac{\cos(x-(i+1)\omega/2)}{\cos(\omega/2)}, \quad w_2(x) = 1, \qquad x \in [i\omega, (i+1)\omega), \tag{20}$$

is given by

$$B[0, \omega, 2\omega, 3\omega; w_0, w_1, 1](x) = \begin{cases} \dfrac{1-\cos(x)}{2(1-\cos(\omega))}, & \text{if } x \in [0, \omega), \\ \dfrac{\cos(2\omega - x) + \cos(x-\omega) - 2\cos(\omega)}{2(1-\cos(\omega))}, & \text{if } x \in [\omega, 2\omega), \\ \dfrac{1-\cos(3\omega - x)}{2(1-\cos(\omega))}, & \text{if } x \in [2\omega, 3\omega), \\ 0, & \text{otherwise.} \end{cases}$$

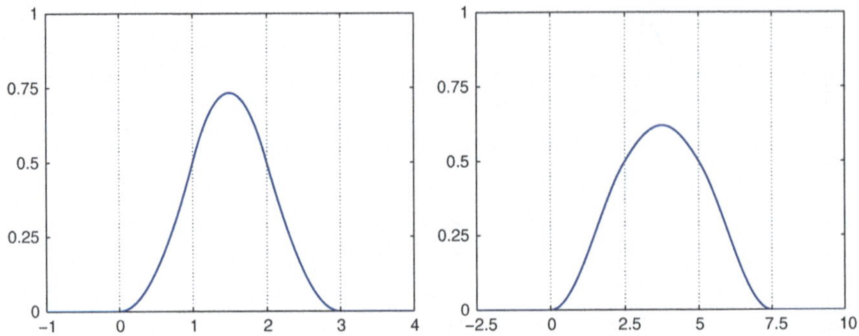

Fig. 2 Two quadratic trigonometric Tchebycheffian B-splines. Left: $\omega = 1$, Right: $\omega = 5/2$. The knot positions are visualized by vertical dotted lines

All spline pieces belong to the trigonometric space $\langle 1, \cos(x), \sin(x)\rangle$. The knots are simple and it can be verified that the function is continuous with a continuous first derivative for all $x \in \mathbb{R}$. Two such quadratic trigonometric Tchebycheffian B-splines (with $\omega = 1$ and $\omega = 5/2$) are depicted in Fig. 2. Observe that, to obtain quadratic Tchebycheffian B-splines with pieces belonging to $\langle 1, \cos(x), \sin(x)\rangle$, we could have used the simpler weight functions $w_0(x) = 1/\cos^2(x)$, $w_1(x) = \cos(x)$, $w_2(x) = 1$ instead of (20). However, this choice results in the restriction $\omega < \pi/6$ to ensure positivity of the weight functions on the interval $[0, 3\omega]$.

Let χ_i denote the characteristic function on the interval $[\xi_i, \xi_{i+1})$. The general explicit expression for a Tchebycheffian B-spline is quite complicated. Applying the recurrence relation in Definition 7 repeatedly we find

$$B^{\mathbf{w}}_{j,p,\boldsymbol{\xi}}(x) = \sum_{i=j}^{j+p} B^{\mathbf{w},\{i\}}_{j,p,\boldsymbol{\xi}}(x)\chi_i(x), \quad p \geq 0, \tag{21}$$

where $B^{\mathbf{w},\{i\}}_{j,p,\boldsymbol{\xi}}$ is defined on the interval $[\xi_i, \xi_{i+1})$ as the restriction of $B^{\mathbf{w}}_{j,p,\boldsymbol{\xi}}$ to that interval, and it is assumed to be zero if $\xi_i = \xi_{i+1}$. In particular, for the nontrivial cases we have

$$\begin{aligned}
B^{\mathbf{w},\{j\}}_{j,0,\boldsymbol{\xi}}(x) &= w_0(x), \\
B^{\mathbf{w},\{j\}}_{j,1,\boldsymbol{\xi}}(x) &= w_1(x)\frac{\int_{\xi_j}^{x} w_0(y)\,\mathrm{d}y}{\int_{\xi_j}^{\xi_{j+1}} w_0(y)\,\mathrm{d}y}, \\
B^{\mathbf{w},\{j+1\}}_{j,1,\boldsymbol{\xi}}(x) &= w_1(x)\frac{\int_{x}^{\xi_{j+2}} w_0(y)\,\mathrm{d}y}{\int_{\xi_{j+1}}^{\xi_{j+2}} w_0(y)\,\mathrm{d}y}.
\end{aligned}$$

For $p \geq 1$, in the nontrivial cases, it follows that the first and last piece are given by

$$\begin{aligned} B^{w,\{j\}}_{j,p,\xi}(x) &= w_p(x) G_p[w_{p-1}, \dots, w_0](x, \xi_j) \Big/ \prod_{i=1}^{p} \gamma^w_{j,i-1,\xi}, \\ B^{w,\{j+p\}}_{j,p,\xi}(x) &= w_p(x) G_p[w_{p-1}, \dots, w_0](\xi_{j+p+1}, x) \Big/ \prod_{i=1}^{p} \gamma^w_{j+p-i+1,i-1,\xi}, \end{aligned} \tag{22}$$

where G_p is defined in (3). If $\xi_j < \xi_{j+1} = \xi_{j+p+1}$ then (22) simplifies to

$$B^{w,\{j\}}_{j,p,\xi}(x) = w_p(x) \frac{G_p[w_{p-1}, \dots, w_0](x, \xi_j)}{G_p[w_{p-1}, \dots, w_0](\xi_{j+1}, \xi_j)}, \tag{23}$$

and if $\xi_j = \xi_{j+p} < \xi_{j+p+1}$ then

$$B^{w,\{j+p\}}_{j,p,\xi}(x) = w_p(x) \frac{G_p[w_{p-1}, \dots, w_0](\xi_{j+p+1}, x)}{G_p[w_{p-1}, \dots, w_0](\xi_{j+p+1}, \xi_{j+p})}. \tag{24}$$

In the following we list some basic properties of Tchebycheffian B-splines that can be directly derived from Definition 7.

- **Local Support**. A Tchebycheffian B-spline is locally supported on the interval given by the extreme knots used in its definition. More precisely,

$$B^w_{j,p,\xi}(x) = 0, \quad x \notin [\xi_j, \xi_{j+p+1}). \tag{25}$$

  This can be proved using induction on the recurrence relation (16).

- **Piecewise Structure**. A Tchebycheffian B-spline has a piecewise Tchebycheff structure, i.e.,

$$B^{w,\{m\}}_{j,p,\xi} \in \mathbb{T}^w_p([\xi_m, \xi_{m+1})), \quad m = j, \dots, j+p. \tag{26}$$

  Proof We proceed by induction on p. Clearly, the case $p = 0$ holds by the definition in (17). Suppose (26) holds for degree $p-1$. By (16) the function $B^{w,\{m\}}_{j,p,\xi}(x)$ for $x \in [\xi_m, \xi_{m+1})$ is a linear combination of

$$w_p(x) \int_{\xi_j}^{x} \frac{B^w_{j,p-1,\xi}(y)}{\gamma^w_{j,p-1,\xi}} \, \mathrm{d}y, \quad w_p(x) \int_{\xi_{j+1}}^{x} \frac{B^w_{j+1,p-1,\xi}(y)}{\gamma^w_{j+1,p-1,\xi}} \, \mathrm{d}y,$$

  where we recall

$$\int_{\xi_i}^{x} \frac{B^w_{i,p-1,\xi}(y)}{\gamma^w_{i,p-1,\xi}} \, \mathrm{d}y = 1, \quad x \geq \xi_{i+p}.$$

The result immediately follows from the induction hypothesis, the recurrence relation (7) and Definition 5. □

- **Local Partition of Unity**. The sum of the Tchebycheffian B-splines of degree p is given by

$$\sum_{j=m-p}^{m} B^{w}_{j,p,\xi}(x) = w_p(x), \quad x \in [\xi_m, \xi_{m+1}), \quad p+1 \leq m \leq n. \tag{27}$$

In particular, for normalized Tchebycheffian B-splines this relation simplifies to

$$\sum_{j=m-p}^{m} B^{w}_{j,p,\xi}(x) = 1, \quad x \in [\xi_m, \xi_{m+1}), \quad p+1 \leq m \leq n. \tag{28}$$

Proof For $p = 0$, the relation (27) follows from (17). For $p \geq 1$, we obtain from (16), (25) and (19) that

$$\begin{aligned}
&\sum_{j=m-p}^{m} B^{w}_{j,p,\xi}(x) \\
&= w_p(x) \sum_{j=m-p}^{m} \left(\int_{\xi_j}^{x} \frac{B^{w}_{j,p-1,\xi}(y)}{\gamma^{w}_{j,p-1,\xi}} \, \mathrm{d}y - \int_{\xi_{j+1}}^{x} \frac{B^{w}_{j+1,p-1,\xi}(y)}{\gamma^{w}_{j+1,p-1,\xi}} \, \mathrm{d}y \right) \\
&= w_p(x) \left(1 + \sum_{j=m-p+1}^{m} \int_{\xi_j}^{x} \frac{B^{w}_{j,p-1,\xi}(y)}{\gamma^{w}_{j,p-1,\xi}} \, \mathrm{d}y - \sum_{j=m-p}^{m-1} \int_{\xi_{j+1}}^{x} \frac{B^{w}_{j+1,p-1,\xi}(y)}{\gamma^{w}_{j+1,p-1,\xi}} \, \mathrm{d}y \right) \\
&= w_p(x).
\end{aligned}$$

In case of normalized Tchebycheffian B-splines we have $w_p = 1$. □

- **Differentiation**. The derivative of a Tchebycheffian B-spline can be simply expressed in terms of two consecutive Tchebycheffian B-splines of lower degree as

$$D_+ \left(\frac{B^{w}_{j,p,\xi}}{w_p} \right)(x) = \frac{B^{w}_{j,p-1,\xi}(x)}{\gamma^{w}_{j,p-1,\xi}} - \frac{B^{w}_{j+1,p-1,\xi}(x)}{\gamma^{w}_{j+1,p-1,\xi}}, \quad p \geq 1, \tag{29}$$

where fractions with zero denominator have value zero. In particular, for normalized Tchebycheffian B-splines the relation simplifies to

$$D_+ B^{w}_{j,p,\xi}(x) = \frac{B^{w}_{j,p-1,\xi}(x)}{\gamma^{w}_{j,p-1,\xi}} - \frac{B^{w}_{j+1,p-1,\xi}(x)}{\gamma^{w}_{j+1,p-1,\xi}}, \quad p \geq 1. \tag{30}$$

Example 15 The standard **polynomial B-splines** of degree p (see, e.g., [19, Definition 2]) are normalized Tchebycheffian B-splines of degree p generated by the weights $w_0 = \cdots = w_p = 1$ and defined on the same knot sequence $\boldsymbol{\xi}$. This is in agreement with Example 7 showing that $\mathbb{T}_p^{\boldsymbol{w}}([a,b]) = \mathbb{P}_p([a,b])$ when $w_0 = \cdots = w_p = 1$.

Example 16 Let $\boldsymbol{w} := (w_0, \ldots, w_p)$ be a vector of positive functions on the interval $[a,b]$ such that $w_j \in C^j([a,b])$, $j = 0, \ldots, p$, and consider the knot sequence

$$\boldsymbol{\xi} := \{a =: \xi_1 = \cdots = \xi_{p+1} < \xi_{p+2} = \cdots = \xi_{2p+2} := b\},$$

which consists of only two different knots (a and b) but both of multiplicity $p+1$. Then, for $p \geq 1$ the functions in Definition 7 are given by

$$\begin{aligned}
B^{\boldsymbol{w}}_{1,p,\boldsymbol{\xi}}(x) &= w_p(x)\left(1 - \int_a^x \frac{B^{\boldsymbol{w}}_{2,p-1,\boldsymbol{\xi}}(y)}{\gamma^{\boldsymbol{w}}_{2,p-1,\boldsymbol{\xi}}}\,\mathrm{d}y\right),\\
B^{\boldsymbol{w}}_{j,p,\boldsymbol{\xi}}(x) &= w_p(x)\left(\int_a^x \frac{B^{\boldsymbol{w}}_{j,p-1,\boldsymbol{\xi}}(y)}{\gamma^{\boldsymbol{w}}_{j,p-1,\boldsymbol{\xi}}}\,\mathrm{d}y - \int_a^x \frac{B^{\boldsymbol{w}}_{j+1,p-1,\boldsymbol{\xi}}(y)}{\gamma^{\boldsymbol{w}}_{j+1,p-1,\boldsymbol{\xi}}}\,\mathrm{d}y\right), \quad 2 \leq j \leq p,\\
B^{\boldsymbol{w}}_{p+1,p,\boldsymbol{\xi}}(x) &= w_p(x)\int_a^x \frac{B^{\boldsymbol{w}}_{p+1,p-1,\boldsymbol{\xi}}(y)}{\gamma^{\boldsymbol{w}}_{p+1,p-1,\boldsymbol{\xi}}}\,\mathrm{d}y.
\end{aligned}$$

These functions are called **Tchebycheffian Bernstein functions** of degree p and span the ECT-space $\mathbb{T}_p^{\boldsymbol{w}}([a,b])$ of dimension $p+1$. They reduce to the standard **Bernstein polynomials** of degree p when $w_0 = \cdots = w_p = 1$.

3.2 Further Properties of Tchebycheffian B-Splines

In this section we prove several properties of Tchebycheffian B-splines, in particular nonnegativity, smoothness and local linear independence. The most technical part is to prove that $\gamma^{\boldsymbol{w}}_{i,\ell,\boldsymbol{\xi}} > 0$ whenever the support of the corresponding Tchebycheffian B-spline is nontrivial, i.e., $\xi_i < \xi_{i+\ell+1}$. The construction of $B^{\boldsymbol{w}}_{j,p,\boldsymbol{\xi}}$ requires all the Tchebycheffian B-splines $B^{\boldsymbol{w}}_{i,\ell,\boldsymbol{\xi}}$ for $i = j, \ldots, j+p-\ell$ and $\ell = p-1, p-2, \ldots, 0$; this involves the corresponding $\gamma^{\boldsymbol{w}}_{i,\ell,\boldsymbol{\xi}}$.

We first note that from Definition 7 the function

$$\int_{\xi_j}^x \frac{B^{\boldsymbol{w}}_{j,p-1,\boldsymbol{\xi}}(y)}{\gamma^{\boldsymbol{w}}_{j,p-1,\boldsymbol{\xi}}}\,\mathrm{d}y, \quad p \geq 1$$

is of class $C^0([\xi_1, \xi_{n+p+1}])$ if $\gamma^{\boldsymbol{w}}_{j,p-1,\boldsymbol{\xi}} \neq 0$, and of class $C^{-1}([\xi_1, \xi_{n+p+1}])$ otherwise. The next lemma discusses the behavior of Tchebycheffian B-splines at the endpoints of their support.

Lemma 1 *Suppose $\gamma^{\boldsymbol{w}}_{i,\ell,\boldsymbol{\xi}} > 0$ whenever $\xi_i < \xi_{i+\ell+1}$ for $i = j, \ldots, j+p-\ell$ and $\ell = 0, \ldots, p-1$.*

(i) Let $1 \leq \mu_j \leq p+1$ *such that* $\xi_j = \cdots = \xi_{j+\mu_j-1} < \xi_{j+\mu_j}$. *We have*

$$D_+^r B_{j,p,\xi}^w(\xi_j) = 0, \quad r = 0, \ldots, p-\mu_j,$$

and

$$D_+^{p+1-\mu_j} B_{j,p,\xi}^w(\xi_j) = w_p(\xi_j) \prod_{k=\mu_j-1}^{p-1} \frac{w_k(\xi_j)}{\gamma_{j,k,\xi}^w}.$$

(ii) Let $1 \leq \mu_j \leq p+1$ *such that* $\xi_{j+p+1-\mu_j} < \xi_{j+p+2-\mu_j} = \cdots = \xi_{j+p+1}$. *We have*

$$D_-^r B_{j,p,\xi}^w(\xi_{j+p+1}) = 0, \quad r = 0, \ldots, p-\mu_j,$$

and

$$D_-^{p+1-\mu_j} B_{j,p,\xi}^w(\xi_{j+p+1}) = (-1)^{p+1-\mu_j} w_p(\xi_{j+p+1}) \prod_{k=\mu_j-1}^{p-1} \frac{w_k(\xi_{j+p+1})}{\gamma_{j+p-k,k,\xi}^w}.$$

Proof We focus on statement (i). For $\mu_j = p+1$ the result follows from the explicit expression in (24), and in particular the result holds for $p = 0$. Suppose now $1 \leq \mu_j \leq p$. It follows from the definition that $B_{j,p,\xi}^w(\xi_j) = 0$, and for $r \geq 1$ the differentiation formula (29) implies

$$D_+^r \left(\frac{B_{j,p,\xi}^w}{w_p} \right)(x) = \frac{D_+^{r-1} B_{j,p-1,\xi}^w(x)}{\gamma_{j,p-1,\xi}^w} - \frac{D_+^{r-1} B_{j+1,p-1,\xi}^w(x)}{\gamma_{j+1,p-1,\xi}^w}. \tag{31}$$

We proceed by induction on p. The case $p = 0$ was already shown before. Since $B_{j,p-1,\xi}^w$ and $B_{j+1,p-1,\xi}^w$ have a knot of multiplicity μ_j and $\mu_j - 1$ at ξ_j, respectively, we deduce from the induction hypothesis that

$$\begin{aligned} D_+^r B_{j,p-1,\xi}^w(\xi_j) &= 0, \quad r = 0, \ldots, p-\mu_j-1, \\ D_+^r B_{j+1,p-1,\xi}^w(\xi_j) &= 0, \quad r = 0, \ldots, p-\mu_j, \end{aligned}$$

and

$$D_+^{p-\mu_j} B_{j,p-1,\xi}^w(\xi_j) = w_{p-1}(\xi_j) \prod_{k=\mu_j-1}^{p-2} \frac{w_k(\xi_j)}{\gamma_{j,k,\xi}^w}.$$

Therefore, from (31) we obtain

$$D_+^r \left(\frac{B_{j,p,\xi}^w}{w_p} \right)(\xi_j) = 0, \quad r = 1, \ldots, p-\mu_j,$$

and

$$D_+^{p-\mu_j+1}\left(\frac{B^{\mathbf{w}}_{j,p,\boldsymbol{\xi}}}{w_p}\right)(\xi_j) = \prod_{k=\mu_j-1}^{p-1} \frac{w_k(\xi_j)}{\gamma^{\mathbf{w}}_{j,k,\boldsymbol{\xi}}}.$$

Recall that $B^{\mathbf{w}}_{j,p,\boldsymbol{\xi}}(\xi_j) = 0$, and so $\left(\frac{B^{\mathbf{w}}_{j,p,\boldsymbol{\xi}}}{w_p}\right)(\xi_j) = 0$. Finally, taking into account the smoothness of w_p in (14), the Leibniz rule gives

$$D_+^r B^{\mathbf{w}}_{j,p,\boldsymbol{\xi}}(\xi_j) = \sum_{k=0}^{r} \binom{r}{k} D_+^{r-k} w_p(\xi_j)\, D_+^k \left(\frac{B^{\mathbf{w}}_{j,p,\boldsymbol{\xi}}}{w_p}\right)(\xi_j), \quad r = 0, \ldots, p-\mu_j+1,$$

which completes the proof of statement (i). The proof of statement (ii) is similar. □

In the following we investigate the number of sign changes of linear combinations of Tchebycheffian B-splines. We first define what we mean by sign changes of a function.

Definition 8 The number of sign changes of a function $f : [a, b] \to \mathbb{R}$ is defined by

$$S^-(f) := \sup_{k\geq 2} \sup_{a\leq x_1<\cdots<x_k\leq b} S^-(f(x_1), \ldots, f(x_k)),$$

where $S^-(c_1, \ldots, c_k)$ denotes the number of (strict) sign changes in the sequence of real numbers $c_1, \ldots, c_k$.

Lemma 2 *For a given function $g \in C^{-1}([a, b])$, we set*

$$f(x) := f(a) + \int_a^x g(y)\, \mathrm{d}y, \quad x \in [a, b].$$

Suppose the function f has a finite number of sign changes on $[a, b]$, as defined in Definition 8. Then,

(i) $S^-(g) \geq S^-(f) + 1$ if $f(a) = f(b) = 0$ and $f \not\equiv 0$;
(ii) $S^-(g) \geq S^-(f)$ if $f(a)f(b) = 0$;
(iii) $S^-(g) \geq S^-(f) - 1$.

Proof From its definition it follows that f is continuous on $[a, b]$. We start by proving statement (i). Suppose $S^-(f) = k$. Since $f(a) = f(b) = 0$ and $f \not\equiv 0$, there exists a sequence of points

$$a = x_0 < x_1 < \cdots < x_k < x_{k+1} = b,$$

and $z_i \in (x_i, x_{i+1})$, $i = 0, \ldots, k$, such that

$$f(x_i) = 0, \quad i = 0, \ldots, k+1, \quad f(z_i) \neq 0, \quad i = 0, \ldots, k.$$

Since

$$\int_{x_i}^{z_i} g(y)\,\mathrm{d}y = f(z_i) - f(x_i) \neq 0, \quad \int_{x_i}^{x_{i+1}} g(y)\,\mathrm{d}y = f(x_{i+1}) - f(x_i) = 0,$$

the function g changes sign at least once in the interval (x_i, x_{i+1}). This implies that $S^-(g) \geq k+1 = S^-(f)+1$. With a similar line of arguments we can prove the statements (ii) and (iii). □

The proof of the next lemma is inspired by [3, Lemma 2.11].

Lemma 3 *Let $k \geq 0$. Suppose $\gamma^{w}_{i,\ell,\xi} > 0$ whenever $\xi_i < \xi_{i+\ell+1}$ for $i = j, \ldots, j+k+p-\ell$ and $\ell = 0, \ldots, p-1$. For $c_i \in \mathbb{R}$, $i = j, \ldots, j+k$, the function*

$$s(x) := \sum_{i=j}^{j+k} c_i B^{w}_{i,p,\xi}(x)$$

has at most k sign changes on the interval $[\xi_j, \xi_{j+k+p+1}]$.

Proof If $s \equiv 0$ there is nothing to prove. Otherwise we use induction on p. For $p = 0$ the result follows from the definition (17) of $B^{w}_{i,0,\xi}$ and the positivity of the weight function w_0. Assuming $p \geq 1$ and using (16), we can write

$$\begin{aligned} s(x) = c_j w_p(x) \int_{\xi_j}^{x} \frac{B^{w}_{j,p-1,\xi}(y)}{\gamma^{w}_{j,p-1,\xi}}\,\mathrm{d}y - c_{j+k} w_p(x) \int_{\xi_j}^{x} \frac{B^{w}_{j+k+1,p-1,\xi}(y)}{\gamma^{w}_{j+k+1,p-1,\xi}}\,\mathrm{d}y \\ + w_p(x) \sum_{i=j+1}^{j+k} (c_i - c_{i-1}) \int_{\xi_j}^{x} \frac{B^{w}_{i,p-1,\xi}(y)}{\gamma^{w}_{i,p-1,\xi}}\,\mathrm{d}y. \end{aligned} \tag{32}$$

Suppose now that for any k and any j it holds that $\sum_{i=j}^{j+k} c_i B^{w}_{i,p-1,\xi}$ has at most k sign changes. If one of the knots has multiplicity $p+1$ in the knot sequence $\{\xi_{j+1} \leq \cdots \leq \xi_{j+k+p}\}$, say $\xi_{j+\ell} = \xi_{j+\ell+p}$ for some $\ell \in \{1, \ldots, k\}$, then

$$s(x) = \sum_{i=j}^{j+\ell-1} c_i B^{w}_{i,p,\xi}(x) + \sum_{i=j+\ell}^{j+k} c_i B^{w}_{i,p,\xi}(x) =: s_1(x) + s_2(x).$$

From (25) we know that $s_1(x) = 0$ for $x \notin [\xi_j, \xi_{j+\ell})$ and $s_2(x) = 0$ for $x \notin [\xi_{j+\ell}, \xi_{j+k+p+1})$. Hence,

$$S^-(s) \leq S^-(s_1) + S^-(s_2) + 1. \tag{33}$$

Therefore, to show that $S^-(s) \leq k$ it suffices to show that $S^-(s_1) \leq \ell - 1$ and $S^-(s_2) \leq k - \ell$. These are two subproblems of the same structure. A repeated application of this argument allows us to eliminate all knots of multiplicity $p+1$ in the knot sequence $\{\xi_{j+1} \leq \cdots \leq \xi_{j+k+p}\}$. Therefore, it is enough to prove the result in the case

$$\xi_i < \xi_{i+p}, \quad i = j+1, \ldots, j+k. \tag{34}$$

Assuming that (34) holds, we consider four cases.

First case: $\xi_j < \xi_{j+p}$ and $\xi_{j+k+1} < \xi_{j+k+p+1}$. For $x \in [\xi_j, \xi_{j+k+p+1}]$, let

$$f(x) := \int_{\xi_j}^{x} g(y)\,\mathrm{d}y, \quad g(x) := \sum_{i=j}^{j+k+1} \left(\frac{c_i - c_{i-1}}{\gamma^{w}_{i,p-1,\xi}} \right) B^{w}_{i,p-1,\xi}(x),$$

with $c_{j-1} := 0$ and $c_{j+k+1} := 0$. Since $\xi_i < \xi_{i+p}$ for $i = j, \ldots, j+k+1$, the hypothesis ensures that $\gamma^{w}_{i,p-1,\xi} > 0$ for $i = j, \ldots, j+k+1$. Hence, g is well defined and belongs to $C^{-1}([\xi_j, \xi_{j+k+p+1}])$ because of (26). By the induction hypothesis, we also know that g has at most $k+1$ sign changes. As a consequence, f is a continuous function with a finite number of sign changes and $f(\xi_j) = 0$. From (32) it is clear that $s(x) = w_p(x) f(x)$, and the positivity of w_p implies $S^-(s) = S^-(f)$. Moreover, from the local support of the B-splines it follows that $f(\xi_{j+k+p+1}) = 0$. Thus, from statement (i) of Lemma 2 we get $S^-(s) = S^-(f) \leq S^-(g) - 1 \leq k$.

Second case: $\xi_j = \xi_{j+p}$ and $\xi_{j+k+1} = \xi_{j+k+p+1}$. For $x \in [\xi_j, \xi_{j+k+p+1}]$, let

$$f(x) := c_j + \int_{\xi_j}^{x} g(y)\,\mathrm{d}y, \quad g(x) := \sum_{i=j+1}^{j+k} \left(\frac{c_i - c_{i-1}}{\gamma^{w}_{i,p-1,\xi}} \right) B^{w}_{i,p-1,\xi}(x).$$

Since $\xi_i < \xi_{i+p}$ for $i = j+1, \ldots, j+k$, the hypothesis ensures that $\gamma^{w}_{i,p-1,\xi} > 0$ for $i = j+1, \ldots, j+k$. Hence, g is well defined and belongs to $C^{-1}([\xi_j, \xi_{j+k+p+1}])$. Since $\gamma^{w}_{j,p-1,\xi} = \gamma^{w}_{j+k+1,p-1,\xi} = 0$, and taking into account (19), we have $s(x) = w_p(x) f(x)$ for $x \in [\xi_j, \xi_{j+k+p+1})$. From (25) we see that $s(\xi_{j+k+p+1}) = 0$, and so the positivity of w_p implies $S^-(s) = S^-(f)$. With the same line of arguments as in the previous case, from statement (iii) of Lemma 2 and the induction hypothesis, we get $S^-(s) = S^-(f) \leq S^-(g) + 1 \leq k$ on $[\xi_j, \xi_{j+k+p+1}]$.

In the two remaining cases $\xi_j < \xi_{j+p}$ and $\xi_{j+k+1} = \xi_{j+k+p+1}$ or $\xi_j = \xi_{j+p}$ and $\xi_{j+k+1} < \xi_{j+k+p+1}$, the result follows in a similar way by using statement (ii) of Lemma 2. □

We are now ready to show nonnegativity of Tchebycheffian B-splines.

Lemma 4 *If* $\xi_j < \xi_{j+p+1}$ *then* $B^{w}_{j,p,\xi}(x) \geq 0$ *for* $x \in [\xi_j, \xi_{j+p+1}]$ *and moreover* $\gamma^{w}_{j,p,\xi} > 0$.

Proof We proceed by induction on p. For $p = 0$ the result follows from the definition (17) of $B^{w}_{i,0,\xi}$ and the positivity of the weight function w_0. Suppose now that $\gamma^{w}_{i,\ell,\xi}(x) > 0$ whenever $\xi_i < \xi_{i+\ell+1}$ for $i = j, \ldots, j+p-\ell$ and $\ell = 0, \ldots, p-1$. Let $1 \leq \mu_j \leq p+1$ such that $\xi_j = \cdots = \xi_{j+\mu_j-1} < \xi_{j+\mu_j}$. From Lemma 1 and the induction hypothesis we get

$$D_+^r B^w_{j,p,\xi}(\xi_j) = 0, \quad r = 0, \ldots, p - \mu_j, \quad D_+^{p+1-\mu_j} B^w_{j,p,\xi}(\xi_j) > 0.$$

Therefore, $B^w_{j,p,\xi}(x) > 0$ for $x \in (\xi_j, \xi_j + \epsilon)$ and some $\epsilon > 0$. Moreover, from Lemma 3 (with $k = 0$) it follows that $B^w_{j,p,\xi}$ has no sign changes on $[\xi_j, \xi_{j+p+1}]$. This means that $B^w_{j,p,\xi}$ is nonnegative on $[\xi_j, \xi_{j+p+1}]$ and it is a nontrivial function on an open subset of $[\xi_j, \xi_{j+p+1}]$. As a consequence, $\gamma^w_{j,p,\xi} > 0$. □

The positivity of $\gamma^w_{j,p,\xi}$ whenever $\xi_j < \xi_{j+p+1}$ shown in Lemma 4 implies that the assumptions of Lemmas 1 and 3 are always satisfied.

Theorem 6 (Nonnegativity) *A Tchebycheffian B-spline is nonnegative everywhere, and positive inside its support, i.e.,*

$$B^w_{j,p,\xi}(x) \geq 0, \quad x \in \mathbb{R}, \quad \textit{and} \quad B^w_{j,p,\xi}(x) > 0, \quad x \in (\xi_j, \xi_{j+p+1}). \tag{35}$$

Proof The nonnegativity of $B^w_{j,p,\xi}$ on $\mathbb{R}$ follows from the local support (25) and Lemma 4. Hence, it just remains to prove that $B^w_{j,p,\xi}(x) > 0$ for $\xi_j < x < \xi_{j+p+1}$.

If $\xi_j < \xi_{j+1} = \xi_{j+p+1}$ or $\xi_j = \xi_{j+p} < \xi_{j+p+1}$ the result follows immediately from the expression of the first and last piece in (23) and (24).

Now, suppose $\xi_j < \xi_{j+p}$ and $\xi_{j+1} < \xi_{j+p+1}$. From Lemma 4 we obtain $\gamma^w_{j,p-1,\xi} > 0$ and $\gamma^w_{j+1,p-1,\xi} > 0$, so that $B^w_{j,p,\xi}(\xi_j) = B^w_{j,p,\xi}(\xi_{j+p+1}) = 0$. Moreover, from the proof of the same lemma, we know that $B^w_{j,p,\xi}(x) > 0$ for $x \in (\xi_j, \xi_j + \epsilon)$ and some $\epsilon > 0$. In a similar way, we can also prove that $B^w_{j,p,\xi}(x) > 0$ for $x \in (\xi_{j+p+1} - \epsilon, \xi_{j+p+1})$ and some $\epsilon > 0$. Assume now that there exists a point $\bar{x} \in (\xi_j, \xi_{j+p+1})$ such that $B^w_{j,p,\xi}(\bar{x}) = 0$. We will show that this assumption leads to a contradiction. Let

$$B^w_{j,p,\xi}(x) = w_p(x) \int_{\xi_j}^x g(y)\, \mathrm{d}y, \quad g(x) := \frac{B^w_{j,p-1,\xi}(x)}{\gamma^w_{j,p-1,\xi}} - \frac{B^w_{j+1,p-1,\xi}(x)}{\gamma^w_{j+1,p-1,\xi}}.$$

Since $B^w_{j,p,\xi}(\xi_j) = B^w_{j,p,\xi}(\bar{x}) = B^w_{j,p,\xi}(\xi_{j+p+1}) = 0$, and $B^w_{j,p,\xi}$ is not identically zero on each of the two intervals $(\xi_j, \bar{x})$ and $(\bar{x}, \xi_{j+p+1})$, from statement (i) of Lemma 2 we deduce that g must have at least one sign change on $[\xi_j, \bar{x}]$ and at least another sign change on $[\bar{x}, \xi_{j+p+1}]$. On the other hand, from Lemma 3 it follows that g can have at most one sign change on $[\xi_j, \xi_{j+p+1}]$. This contradiction concludes the proof. □

We now describe the smoothness behavior of Tchebycheffian B-splines at the knots in their support.

Theorem 7 (Smoothness) *If ξ is a knot of $B^w_{j,p,\xi}$ of multiplicity $\mu \leq p+1$ then*

$$B^w_{j,p,\xi} \in C^{p-\mu}(\xi), \tag{36}$$

i.e., its derivatives of order $0, 1, \ldots, p - \mu$ are continuous at ξ. Moreover, if $\mu = p+1$ then $B^w_{j,p,\xi}(\xi)$ is bounded.

Proof By Lemma 1 (and Lemma 4) the result holds if $\xi = \xi_j$ or $\xi = \xi_{j+p+1}$. Suppose now $\xi_j < \xi < \xi_{j+p+1}$. Observe that $\mu \leq p$, $\xi_j < \xi_{j+p}$ and $\xi_{j+1} < \xi_{j+p+1}$, and therefore by Lemma 4 we have $\gamma^{w}_{j,p-1,\xi} > 0$ and $\gamma^{w}_{j+1,p-1,\xi} > 0$.

We first prove that $B^{w}_{j,p,\xi}$ is continuous at ξ whenever $\mu \leq p$. Indeed, by (15) we have $w_p \in C^0(\xi)$ and by (26) the integrands in (16) are bounded, and so $B^{w}_{j,p,\xi}$ and $\frac{B^{w}_{j,p,\xi}}{w_p}$ are continuous at ξ.

In order to show (36) for $\mu < p$, we proceed by induction on p. Both terms in the differentiation formula (29) have a knot of multiplicity at most μ at ξ, and from the induction hypothesis we obtain $D\Big(\frac{B^{w}_{j,p,\xi}}{w_p}\Big) \in C^{p-1-\mu}(\xi)$. Moreover, since $\frac{B^{w}_{j,p,\xi}}{w_p}$ is continuous at ξ, we can conclude that $\frac{B^{w}_{j,p,\xi}}{w_p} \in C^{p-\mu}(\xi)$ for $\mu < p$. Since $\xi_j < \xi < \xi_{j+p+1}$ then by (15) we have $w_p \in C^{p-\mu}(\xi)$, and so $B^{w}_{j,p,\xi} \in C^{p-\mu}(\xi)$. □

Finally, we show that Tchebycheffian B-splines are (locally) linearly independent on each knot interval and span the local E(C)T-space defined on such interval.

Theorem 8 (Local Linear Independence) *The set* $\{B^{w}_{j,p,\xi}\}_{j=m-p}^{m}$ *forms a basis for the E(C)T-space* $\mathbb{T}^{w}_{p}$ *on* $[\xi_m, \xi_{m+1})$ *for any* $p+1 \leq m \leq n$.

Proof By the piecewise Tchebycheff structure (26) of Tchebycheffian B-splines, it suffices to prove that the functions $\{B^{w}_{j,p,\xi}\}_{j=m-p}^{m}$ are linearly independent on $[\xi_m, \xi_{m+1})$ for any $p+1 \leq m \leq n$.

We use induction on p. The case $p = 0$ follows from (17). Now, let $p \geq 1$. Fix m such that $\xi_m < \xi_{m+1}$, and suppose that for all $x \in [\xi_m, \xi_{m+1})$,

$$\frac{1}{w_p(x)} \sum_{j=m-p}^{m} c_j B^{w}_{j,p,\xi}(x) = 0. \tag{37}$$

After differentiating (37), it follows from (29) and (25) that

$$\sum_{j=m-p+1}^{m} \left(\frac{c_j - c_{j-1}}{\gamma^{w}_{j,p-1,\xi}}\right) B^{w}_{j,p-1,\xi}(x) = 0.$$

Since $\xi_m < \xi_{m+1}$, Lemma 4 implies that $\gamma^{w}_{j,p-1,\xi} > 0$, $j = m-p+1, \ldots, m$. Then, the induction hypothesis gives us that $c_{m-p} = \cdots = c_m$, so

$$\frac{c_m}{w_p(x)} \sum_{j=m-p}^{m} B^{w}_{j,p,\xi}(x) = 0.$$

By relation (27) we get $c_m = 0$. As a consequence, all functions $B^{w}_{j,p,\xi}$ in (37) are linearly independent on $[\xi_m, \xi_{m+1})$. □

3.3 The Tchebycheffian Spline Space

In this section we focus on the span of the Tchebycheffian B-splines of degree p specified by the knot sequence $\boldsymbol{\xi} := \{\xi_i\}_{i=1}^{n+p+1}$ and the vector of Tchebycheffian B-spline weights $\boldsymbol{w}$, i.e.,

$$\mathbb{S}^{\boldsymbol{w}}_{p,\boldsymbol{\xi}} := \left\{ s : [\xi_{p+1}, \xi_{n+1}] \to \mathbb{R} : s = \sum_{j=1}^{n} c_j B^{\boldsymbol{w}}_{j,p,\boldsymbol{\xi}},\ c_j \in \mathbb{R} \right\}. \tag{38}$$

This is the space of **Tchebycheffian splines** spanned by the Tchebycheffian B-splines $\{B^{\boldsymbol{w}}_{1,p,\boldsymbol{\xi}}, \ldots, B^{\boldsymbol{w}}_{n,p,\boldsymbol{\xi}}\}$ over the interval $[\xi_{p+1}, \xi_{n+1}]$, which is called the **basic interval**. We define the Tchebycheffian B-splines to be left continuous at the right endpoint ξ_{n+1}, so as to avoid asymmetry in the construction of the space.

We now introduce some terminology to identify certain properties of knot sequences which are crucial in the study of the space (38).

- A knot sequence $\boldsymbol{\xi}$ is called $(p+1)$**-regular** if $\xi_j < \xi_{j+p+1}$ for $j = 1, \ldots, n$. By the local support (25) such a knot sequence ensures that all the Tchebycheffian B-splines in (38) are not identically zero.

- A knot sequence $\boldsymbol{\xi}$ is called $(p+1)$**-basic** if it is $(p+1)$-regular with $\xi_{p+1} < \xi_{p+2}$ and $\xi_n < \xi_{n+1}$. As we will show later, the Tchebycheffian B-splines in (38) defined on a $(p+1)$-basic knot sequence are linearly independent on the basic interval $[\xi_{p+1}, \xi_{n+1}]$.

From the results in the previous section we can immediately conclude the following list of properties of Tchebycheffian splines in the B-spline representation.

- **Smoothness**. If ξ is a knot of multiplicity μ then $s \in C^r(\xi)$ for any $s \in \mathbb{S}^{\boldsymbol{w}}_{p,\boldsymbol{\xi}}$, where $r + \mu = p$. This follows from the smoothness property of the Tchebycheffian B-splines (Theorem 7). Therefore, the relation between smoothness, multiplicity and degree is the same as in the polynomial B-spline case:

$$\text{“smoothness + multiplicity = degree”}. \tag{39}$$

- **Local Support**. The local support (25) of the Tchebycheffian B-splines implies

$$\sum_{j=1}^{n} c_j B^{\boldsymbol{w}}_{j,p,\boldsymbol{\xi}}(x) = \sum_{j=m-p}^{m} c_j B^{\boldsymbol{w}}_{j,p,\boldsymbol{\xi}}(x), \quad x \in [\xi_m, \xi_{m+1}), \quad p+1 \le m \le n, \tag{40}$$

and if $\xi_m < \xi_{m+p}$ then

$$\sum_{j=1}^{n} c_j B^{\boldsymbol{w}}_{j,p,\boldsymbol{\xi}}(\xi_m) = \sum_{j=m-p}^{m-1} c_j B^{\boldsymbol{w}}_{j,p,\boldsymbol{\xi}}(\xi_m), \quad p+1 \le m \le n+1. \tag{41}$$

- **Minimal Support**. From the smoothness properties it can be proved that if the support of $s \in \mathbb{S}^{w}_{p,\xi}$ is a proper subset of $[\xi_j, \xi_{j+p+1}]$ for some j then $s \equiv 0$. Therefore, the Tchebycheffian B-splines have minimal support.

- **Partition of Unity**. By (27) we have

$$\sum_{j=1}^{n} B^{w}_{j,p,\xi}(x) = w_p(x), \quad x \in [\xi_{p+1}, \xi_{n+1}]. \tag{42}$$

In particular, for normalized Tchebycheffian B-splines this relation simplifies to

$$\sum_{j=1}^{n} B^{w}_{j,p,\xi}(x) = 1, \quad x \in [\xi_{p+1}, \xi_{n+1}]. \tag{43}$$

Since these splines are nonnegative it follows that they form a **nonnegative partition of unity** on $[\xi_{p+1}, \xi_{n+1}]$.

- **Differentiation**. By (29) we have for $p \geq 1$,

$$D_+\left(\frac{1}{w_p}\sum_{j=1}^{n} c_j B^{w}_{j,p,\xi}\right)(x) = \sum_{j=2}^{n} c_j^{(1)} B^{w}_{j,p-1,\xi}(x), \quad x \in [\xi_{p+1}, \xi_{n+1}], \tag{44}$$

where

$$c_j^{(1)} := \frac{c_j - c_{j-1}}{\gamma^{w}_{j,p-1,\xi}}, \tag{45}$$

and fractions with zero denominator have value zero.

- **Linear Independence**. If $\boldsymbol{\xi}$ is $(p+1)$-basic, then the Tchebycheffian B-splines $\{B^{w}_{1,p,\xi}, \ldots, B^{w}_{n,p,\xi}\}$ are linearly independent on the basic interval. Thus, the spline space $\mathbb{S}^{w}_{p,\xi}$ is a vector space of dimension n.

Proof We must show that if

$$s(x) = \sum_{j=1}^{n} c_j B^{w}_{j,p,\xi}(x) = 0, \quad x \in [\xi_{p+1}, \xi_{n+1}],$$

then $c_j = 0$ for all j. Let us fix $1 \leq j \leq n$. Since $\boldsymbol{\xi}$ is $(p+1)$-regular, there is an integer m_j with $j \leq m_j \leq j+p$ such that $\xi_{m_j} < \xi_{m_j+1}$. Moreover, the assumptions $\xi_{p+1} < \xi_{p+2}$ and $\xi_n < \xi_{n+1}$ guarantee that $[\xi_{m_j}, \xi_{m_j+1})$ can be chosen in the basic interval. From the local support property (40) we know

$$s(x) = \sum_{i=m_j-p}^{m_j} c_i B^{w}_{i,p,\xi}(x) = 0, \quad x \in [\xi_{m_j}, \xi_{m_j+1}).$$

Theorem 8 implies $c_{m_j-p} = \cdots = c_{m_j} = 0$, and in particular $c_j = 0$. □

In the following we are looking for a characterization of the Tchebycheffian spline space $\mathbb{S}^{w}_{p,\xi}$ in terms of piecewise Tchebycheff functions with a certain smoothness.

Definition 9 Let Δ be a sequence of break points,

$$\Delta := \{\eta_0 < \eta_1 < \cdots < \eta_{\ell+1}\}, \quad a := \eta_0, \quad b := \eta_{\ell+1}, \tag{46}$$

and let $\boldsymbol{r} := (r_1, \ldots, r_\ell)$ be a vector of integers such that $-1 \le r_i \le p$ for $i = 1, \ldots, \ell$. Furthermore, let $\boldsymbol{w} := (w_0, \ldots, w_p)$ be a vector of positive weight functions on $[a, b]$ such that for $j = 0, \ldots, p$,

$$\begin{aligned} w_j &\in C^j([\eta_i^+, \eta_{i+1}^-]), \quad i = 0, \ldots, \ell, \\ w_j &\in C^{\max(j-p+r_i,-1)}(\eta_i), \quad i = 1, \ldots, \ell. \end{aligned} \tag{47}$$

The space $\mathbb{S}^{\boldsymbol{r},\boldsymbol{w}}_p(\Delta)$ of **piecewise Tchebycheff functions** of degree p with smoothness $\boldsymbol{r}$ over the partition Δ is defined by

$$\begin{aligned} \mathbb{S}^{\boldsymbol{r},\boldsymbol{w}}_p(\Delta) := \big\{s : [\eta_0, \eta_{\ell+1}] \to \mathbb{R} : \; & s \in \mathbb{T}^{\boldsymbol{w}}_p([\eta_i, \eta_{i+1})), \; i = 0, \ldots, \ell-1, \\ & s \in \mathbb{T}^{\boldsymbol{w}}_p([\eta_\ell, \eta_{\ell+1}]), \; s \in C^{r_i}(\eta_i), \; i = 1, \ldots, \ell\big\}. \end{aligned} \tag{48}$$

Any element $s \in \mathbb{S}^{\boldsymbol{r},\boldsymbol{w}}_p(\Delta)$ can be written in the form

$$s(x) = \sum_{j=0}^{p} c_{0,j} u^{\boldsymbol{w}}_{j,p}(x, \eta_0) + \sum_{i=1}^{\ell} \sum_{j=r_i+1}^{p} c_{i,j} u^{\boldsymbol{w}}_{j,p}(x, \eta_i)_+, \quad x \in [a, b], \tag{49}$$

where $u^{\boldsymbol{w}}_{j,p}(x, y)$ are generalized powers (see Definition 4) and

$$u^{\boldsymbol{w}}_{j,p}(x, y)_+ := \begin{cases} u^{\boldsymbol{w}}_{j,p}(x, y), & x > y, \\ 0, & x < y, \end{cases} \tag{50}$$

where the value at y is defined by taking the right limit. The functions in (50) are called **generalized truncated powers**. From the smoothness conditions in (47) we see that $w_{p-k} \in C^{\max(r_i-k,-1)}(\eta_i)$, $k = 0, \ldots, p$, and from Definition 4 we immediately get

$$u^{\boldsymbol{w}}_{j,p}(\eta_k, \eta_i) \in C^{r_i}(\eta_k), \quad k = i+1, \ldots, \ell, \quad j = 0, \ldots, p.$$

Moreover, since $w_{p-k} \in C^{p-k}([\eta_i^+, \eta_{i+1}^-])$, $k = 0, \ldots, p$, by combining (4) and (6) we have

$$D_+^l u_{j,p}^{\boldsymbol{w}}(\eta_i, \eta_i) = 0, \quad l = 0, \dots, r_i, \quad j = r_i + 1, \dots, p.$$

This shows that the function in (49) belongs to the space (48). The representation (49) implies

$$\dim(\mathbb{S}_p^{\boldsymbol{r},\boldsymbol{w}}(\Delta)) \le p + 1 + \sum_{i=1}^{\ell}(p - r_i). \tag{51}$$

The next theorem states that the Tchebycheffian spline space $\mathbb{S}_{p,\boldsymbol{\xi}}^{\boldsymbol{w}}$ is equal to the space $\mathbb{S}_p^{\boldsymbol{r},\boldsymbol{w}}(\Delta)$ with a prescribed partition Δ and smoothness $\boldsymbol{r}$.

Theorem 9 (Characterization of Spline Space) *Let $\boldsymbol{\xi} := \{\xi_i\}_{i=1}^{n+p+1}$ be a $(p+1)$-basic knot sequence. The space $\mathbb{S}_{p,\boldsymbol{\xi}}^{\boldsymbol{w}}$ spanned by Tchebycheffian B-splines of degree p defined over the knot sequence $\boldsymbol{\xi}$ is characterized by*

$$\mathbb{S}_{p,\boldsymbol{\xi}}^{\boldsymbol{w}} = \mathbb{S}_p^{\boldsymbol{r},\boldsymbol{w}}(\Delta),$$

where Δ is a partition as in (46) *defined from the knot sequence as follows,*

$$\xi_{p+1} =: \eta_0, \quad \xi_{p+2}, \dots, \xi_n =: \overbrace{\eta_1, \dots, \eta_1}^{\mu_1}, \dots, \overbrace{\eta_\ell, \dots, \eta_\ell}^{\mu_\ell}, \quad \xi_{n+1} =: \eta_{\ell+1},$$

and the smoothness $\boldsymbol{r}$ is defined by

$$r_i := p - \mu_i, \quad i = 1, \dots, \ell.$$

Proof Since we are dealing with a $(p+1)$-basic knot sequence $\boldsymbol{\xi}$, we have $\eta_0 < \eta_1$ and $\eta_\ell < \eta_{\ell+1}$. The Tchebycheffian B-spline weights $\boldsymbol{w}$ satisfy the smoothness conditions in (47); see Definition 6. From the piecewise structure (26) and the smoothness (36) of Tchebycheffian B-splines it follows that the space $\mathbb{S}_{p,\boldsymbol{\xi}}^{\boldsymbol{w}}$ is a subspace of $\mathbb{S}_p^{\boldsymbol{r},\boldsymbol{w}}(\Delta)$. Moreover, using (51) we arrive at

$$\dim(\mathbb{S}_{p,\boldsymbol{\xi}}^{\boldsymbol{w}}) = n = p + 1 + \sum_{i=1}^{\ell}(p - r_i) \ge \dim(\mathbb{S}_p^{\boldsymbol{r},\boldsymbol{w}}(\Delta)).$$

This concludes the proof. □

3.4 Knot Insertion

In this section we are addressing the problem of representing a given Tchebycheffian spline on a refined knot sequence. In particular, we focus on the special case where only a single knot is inserted. Since any refined knot sequence can be reached by repeatedly inserting one knot at a time, it suffices to deal with this case.

Without loss of generality, we assume that the spline $s = \sum_{j=1}^{n} c_j B^{w}_{j,p,\boldsymbol{\xi}}$ is given on a $(p+1)$-basic knot sequence $\boldsymbol{\xi} := \{\xi_i\}_{i=1}^{n+p+1}$. We want to insert a knot ξ in some subinterval $[\xi_m, \xi_{m+1})$ of $[\xi_{p+1}, \xi_{n+1})$, resulting in a new $(p+1)$-basic knot sequence $\tilde{\boldsymbol{\xi}} := \{\tilde{\xi}_i\}_{i=1}^{n+p+2}$ defined by

$$\tilde{\xi}_i := \begin{cases} \xi_i, & \text{if } 1 \le i \le m, \\ \xi, & \text{if } i = m+1, \\ \xi_{i-1}, & \text{if } m+2 \le i \le n+p+2. \end{cases} \tag{52}$$

We are interested in the Tchebycheffian B-spline form of s on the new knot sequence.

Lemma 5 *Let the $(p+1)$-basic knot sequence $\tilde{\boldsymbol{\xi}} := \{\tilde{\xi}_i\}_{i=1}^{n+p+2}$ be obtained from the $(p+1)$-basic knot sequence $\boldsymbol{\xi} := \{\xi_i\}_{i=1}^{n+p+1}$ by inserting just one knot ξ in $[\xi_{p+1}, \xi_{n+1})$. Then,*

$$B^{w}_{j,p,\boldsymbol{\xi}} = \alpha_{j,p,\boldsymbol{\xi}} B^{w}_{j,p,\tilde{\boldsymbol{\xi}}} + \beta_{j+1,p,\boldsymbol{\xi}} B^{w}_{j+1,p,\tilde{\boldsymbol{\xi}}}, \tag{53}$$

where

(i) $\alpha_{j,p,\boldsymbol{\xi}} = 1$ *and* $\beta_{j+1,p,\boldsymbol{\xi}} = 0$ *if* $\xi \ge \xi_{j+p+1}$*;*
(ii) $\alpha_{j,p,\boldsymbol{\xi}} > 0$ *and* $\beta_{j+1,p,\boldsymbol{\xi}} > 0$ *if* $\xi_j < \xi < \xi_{j+p+1}$*;*
(iii) $\alpha_{j,p,\boldsymbol{\xi}} = 0$ *and* $\beta_{j+1,p,\boldsymbol{\xi}} = 1$ *if* $\xi \le \xi_j$*.*

Proof From Theorem 9 it follows that $\mathbb{S}^{w}_{p,\boldsymbol{\xi}} \subseteq \mathbb{S}^{w}_{p,\tilde{\boldsymbol{\xi}}}$, so every $B^{w}_{j,p,\boldsymbol{\xi}}$ can be written as a linear combination of the Tchebycheffian B-splines defined over $\tilde{\boldsymbol{\xi}}$. If $\xi \ge \xi_{j+p+1}$ then $B^{w}_{j,p,\boldsymbol{\xi}} = B^{w}_{j,p,\tilde{\boldsymbol{\xi}}}$, which shows (53) in case (i). If $\xi \le \xi_j$ then $B^{w}_{j,p,\boldsymbol{\xi}} = B^{w}_{j+1,p,\tilde{\boldsymbol{\xi}}}$, which shows (53) in case (iii). In the remainder, we focus on the last case (ii) and assume $\xi_j < \xi < \xi_{j+p+1}$.

Fix j. We can write

$$B^{w}_{j,p,\boldsymbol{\xi}}(x) = \sum_{i=1}^{n+1} c_{j,i} B^{w}_{i,p,\tilde{\boldsymbol{\xi}}}(x).$$

If $x \in [\tilde{\xi}_k, \tilde{\xi}_{k+1})$ with $\tilde{\xi}_k \ge \xi_{j+p+1} = \tilde{\xi}_{j+p+2}$, then

$$0 = B^{w}_{j,p,\boldsymbol{\xi}}(x) = \sum_{i=k-p}^{k} c_{j,i} B^{w}_{i,p,\tilde{\boldsymbol{\xi}}}(x),$$

and by local linear independence we get $c_{j,i} = 0$ for any $i \ge j+2$ since $k - p \ge j+2$. Similarly, $c_{j,i} = 0$ for any $i \le j-1$. This implies (53) for some $\alpha_{j,p,\boldsymbol{\xi}}$ and $\beta_{j+1,p,\boldsymbol{\xi}}$.

Next, we show the positivity of $\alpha_{j,p,\boldsymbol{\xi}}$. Let μ_j be the multiplicity of ξ_j as a knot of $B^{w}_{j,p,\boldsymbol{\xi}}$. Then, ξ_j appears μ_j times as a knot of $B^{w}_{j,p,\tilde{\boldsymbol{\xi}}}$ and $\mu_j - 1$ times as a knot of $B^{w}_{j+1,p,\tilde{\xi}}$. We consider the $(p+1-\mu_j)$th derivative at ξ_j of the two sides in (53). By using the expression in statement (i) of Lemma 1 and recalling that the weight functions are positive, we get

$$\alpha_{j,p,\xi} = \begin{cases} 1, & \mu_j = p+1, \\ \dfrac{\gamma^{w}_{j,p-1,\tilde{\xi}} \cdots \gamma^{w}_{j,\mu_j-1,\tilde{\xi}}}{\gamma^{w}_{j,p-1,\xi} \cdots \gamma^{w}_{j,\mu_j-1,\xi}}, & \mu_j \leq p. \end{cases} \tag{54}$$

Note that by Lemma 4 all γ's involved in (54) are positive because $\xi_j < \xi_{j+\mu_j}$. The positivity of $\beta_{j+1,p,\xi}$ can be proved in a similar way. Let μ_{j+p+1} be the multiplicity of ξ_{j+p+1} as a knot of $B^{w}_{j,p,\xi}$. From statement (ii) of Lemma 1 we get

$$\beta_{j+1,p,\xi} = \begin{cases} 1, & \mu_{j+p+1} = p+1, \\ \dfrac{\gamma^{w}_{j+2,p-1,\tilde{\xi}} \cdots \gamma^{w}_{j+p+2-\mu_{j+p+1},\mu_{j+p+1}-1,\tilde{\xi}}}{\gamma^{w}_{j+1,p-1,\xi} \cdots \gamma^{w}_{j+p+1-\mu_{j+p+1},\mu_{j+p+1}-1,\xi}}, & \mu_{j+p+1} \leq p. \end{cases} \tag{55}$$

This completes the proof. □

Theorem 10 (Knot Insertion) *Let the $(p+1)$-basic knot sequence $\tilde{\boldsymbol{\xi}} := \{\tilde{\xi}_i\}_{i=1}^{n+p+2}$ be obtained from the $(p+1)$-basic knot sequence $\boldsymbol{\xi} := \{\xi_i\}_{i=1}^{n+p+1}$ by inserting just one knot ξ, such that $\xi_m \leq \xi < \xi_{m+1}$ with $p+1 \leq m \leq n$ as in* (52). *Then,*

$$s(x) = \sum_{j=1}^{n} c_j B^{w}_{j,p,\xi}(x) = \sum_{i=1}^{n+1} \tilde{c}_i B^{w}_{i,p,\tilde{\xi}}(x), \quad x \in [\xi_{p+1}, \xi_{n+1}], \tag{56}$$

where

$$\tilde{c}_i = \begin{cases} c_i, & \text{if } i \leq m-p, \\ \alpha_{i,p,\xi} c_i + \beta_{i,p,\xi} c_{i-1}, & \text{if } m-p < i \leq m, \\ c_{i-1}, & \text{if } i > m. \end{cases} \tag{57}$$

The values $\alpha_{i,p,\xi}$ and $\beta_{i,p,\xi}$ in (57) *are nonnegative, and*

$$\alpha_{i,p,\xi} + \beta_{i,p,\xi} = 1, \quad m-p < i \leq m. \tag{58}$$

Proof From Lemma 5 we immediately deduce that

$$\sum_{j=1}^{n} c_j B^{w}_{j,p,\xi} = \alpha_{1,p,\xi} c_1 B^{w}_{1,p,\tilde{\xi}} + \beta_{n+1,p,\xi} c_n B^{w}_{n+1,p,\tilde{\xi}} + \sum_{i=2}^{n} (\alpha_{i,p,\xi} c_i + \beta_{i,p,\xi} c_{i-1}) B^{w}_{i,p,\tilde{\xi}},$$

where the α's and β's are nonnegative. This gives (56) with

$$\tilde{c}_i = \alpha_{i,p,\xi} c_i + \beta_{i,p,\xi} c_{i-1}, \quad i = 2, \ldots, n. \tag{59}$$

First, recall from (42) that both sets of B-splines in (56) sum to w_p. Hence, in the case $s = w_p$, (59) implies

$$1 = \alpha_{i,p,\xi} + \beta_{i,p,\xi}, \quad i = 2, \dots, n.$$

Since $p+1 \le m \le n$, it follows that $\{m-p+1, \dots, m\} \subseteq \{2, \dots, n\}$ and we obtain (58). Furthermore, from case (i) in Lemma 5 we have $\alpha_{i-1,p,\xi} = 1$ and $\beta_{i,p,\xi} = 0$ for $2 \le i \le m-p$. We also observe from (54) that $\alpha_{m-p,p,\xi} = 1$. Indeed, if $\mu_{m-p} = p+1$ it is obvious, and otherwise we have $\gamma^w_{m-p,k-1,\tilde{\xi}} = \gamma^w_{m-p,k-1,\xi}$ for $k = \mu_{m-p}, \dots, p$. Similarly, from case (iii) in the same lemma we have $\alpha_{i,p,\xi} = 0$ and $\beta_{i+1,p,\xi} = 1$ for $m < i \le n$. If $\xi_m = \xi$ then this case also implies $\beta_{m+1,p,\xi} = 1$. If $\xi_m < \xi$ then we can conclude from (55) that $\beta_{m+1,p,\xi} = 1$. This completes the proof. □

From the proof we observe that the α's and β's in (57) are specified in (54) and (55), respectively.

4 Generalized B-Splines

In this section we consider a special subclass of normalized Tchebycheffian B-splines, the so-called generalized B-splines.[3] They can be seen as the minimal extension of classical polynomial splines still offering a wide variety of additional flexibility.

Definition 10 Given the partition $\Delta := \{\eta_0 < \eta_1 < \cdots < \eta_{\ell+1}\}$ and an integer $p \ge 2$, a **generalized polynomial space** of degree p is defined as a space of the form

$$\mathbb{P}_p^{U,V}(\Delta) := \langle 1, x, \dots, x^{p-2}, U(x), V(x) \rangle, \quad x \in [\eta_0, \eta_{\ell+1}], \tag{60}$$

where $U, V \in C^p([\eta_i^+, \eta_{i+1}^-])$ and $\langle D^{p-1}U, D^{p-1}V \rangle$ is an ET-space on $[\eta_i^+, \eta_{i+1}^-]$ for all $i = 0, \dots, \ell$.

From Example 10 we conclude that the restriction of the generalized polynomial space $\mathbb{P}_p^{U,V}(\Delta)$ on the interval $[\eta_i^+, \eta_{i+1}^-]$ is an ECT-space of dimension $p+1$ generated by weight functions of the form

$$w_{0,i}(x), \quad w_{1,i}(x), \quad w_{2,i}(x) = \cdots = w_{p,i}(x) = 1, \tag{61}$$

where

[3]The term "generalized splines" has several different meanings in the literature. For example, the splines considered here are much less general than those described in [30, Chap. 11]. We follow the definition given in [17]. This definition was already used before for special choices of U and V; see, for example, [15, 16].

$$w_{1,i}(\eta_i) = w_{1,i}(\eta_{i+1}) = 1. \tag{62}$$

Note that the space $\langle D^{p-1}U, D^{p-1}V\rangle$ is an ECT-space on $[\eta_i^+, \eta_{i+1}^-]$ generated by the weights $w_{0,i}$ and $w_{1,i}$ (see Example 9). The local weights in (61)–(62) allow us to define a global weight vector $\boldsymbol{w} := (w_0, \ldots, w_p)$ such that

$$w_j(x) := w_{j,i}(x), \quad x \in [\eta_i, \eta_{i+1}), \quad i = 0, \ldots, \ell, \quad j = 0, \ldots, p. \tag{63}$$

From the construction it is easy to check that each weight $w_j \in C^j([\eta_i^+, \eta_{i+1}^-])$, $i = 0, \ldots, \ell$ and that $w_j \in C^{j-1}(\eta_i)$, $i = 1, \ldots, \ell$.

We now define generalized B-splines of degree p associated with a knot sequence $\boldsymbol{\xi}$ and a generalized polynomial space $\mathbb{P}_p^{U,V}(\Delta)$. The knot sequence $\boldsymbol{\xi} := \{\xi_i\}_{i=1}^{n+p+1}$ is connected to the partition Δ as follows

$$\xi_1, \ldots, \xi_{n+p+1} = \overbrace{\eta_0, \ldots, \eta_0}^{\mu_0}, \ldots, \overbrace{\eta_{\ell+1}, \ldots, \eta_{\ell+1}}^{\mu_{\ell+1}}, \tag{64}$$

for some integers $\mu_0, \ldots, \mu_{\ell+1}$.

Definition 11 For a given partition Δ, let $\mathbb{P}_p^{U,V}(\Delta)$ be a generalized polynomial space of degree $p \geq 2$, and let $\boldsymbol{\xi}$ be a knot sequence connected to Δ as in (64). For any $\xi_i < \xi_{i+1}$, let u_i, v_i be the unique functions in $\langle D^{p-1}U, D^{p-1}V\rangle$ on $[\xi_i^+, \xi_{i+1}^-]$ satisfying

$$u_i(\xi_i) = 1, \quad u_i(\xi_{i+1}) = 0, \quad v_i(\xi_i) = 0, \quad v_i(\xi_{i+1}) = 1.$$

Suppose for some integer j that $\xi_j \leq \xi_{j+1} \leq \cdots \leq \xi_{j+p+1}$ are $p+2$ real numbers taken from $\boldsymbol{\xi}$. The jth **generalized B-spline** $B_{j,p,\boldsymbol{\xi}}^{U,V} : \mathbb{R} \to \mathbb{R}$ of degree p is identically zero if $\xi_{j+p+1} = \xi_j$ and otherwise defined recursively by

$$B_{j,p,\boldsymbol{\xi}}^{U,V}(x) := \int_{\xi_j}^x \frac{B_{j,p-1,\boldsymbol{\xi}}^{U,V}(y)}{\gamma_{j,p-1,\boldsymbol{\xi}}^{U,V}}\, dy - \int_{\xi_{j+1}}^x \frac{B_{j+1,p-1,\boldsymbol{\xi}}^{U,V}(y)}{\gamma_{j+1,p-1,\boldsymbol{\xi}}^{U,V}}\, dy, \tag{65}$$

starting with

$$B_{i,1,\boldsymbol{\xi}}^{U,V}(x) := \begin{cases} v_i(x), & \text{if } x \in [\xi_i, \xi_{i+1}), \\ u_{i+1}(x), & \text{if } x \in [\xi_{i+1}, \xi_{i+2}), \\ 0, & \text{otherwise.} \end{cases} \tag{66}$$

Here, $\gamma_{i,k,\boldsymbol{\xi}}^{U,V}$ is defined as the integral of $B_{i,k,\boldsymbol{\xi}}^{U,V}$,

$$\gamma_{i,k,\boldsymbol{\xi}}^{U,V} := \int_{\xi_i}^{\xi_{i+k+1}} B_{i,k,\boldsymbol{\xi}}^{U,V}(y)\, dy,$$

and we used the convention that if $\gamma_{i,k,\boldsymbol{\xi}}^{U,V} = 0$ then

$$\int_{\xi_i}^{x} \frac{B_{i,k,\xi}^{U,V}(y)}{\gamma_{i,k,\xi}^{U,V}} \,\mathrm{d}y := \begin{cases} 1, & \text{if } x \geq \xi_{i+k+1}, \\ 0, & \text{otherwise.} \end{cases}$$

We now show that generalized B-splines are a special instance of normalized Tchebycheffian B-splines, and therefore they enjoy all their properties.

Theorem 11 *Generalized B-splines are normalized Tchebycheffian B-splines generated by the Tchebycheffian B-spline weights $w_0, \dots, w_p$ given in* (63).

Proof We first note that the global weights $w_0, \dots, w_p$ in (63) satisfy the smoothness conditions in Definition 6, so they are actually Tchebycheffian B-spline weights with respect to $\boldsymbol{\xi}$. Let $\boldsymbol{w} := (w_0, \dots, w_p)$. A direct computation shows that $B_{i,1,\boldsymbol{\xi}}^{\boldsymbol{w}}(x) = B_{i,1,\boldsymbol{\xi}}^{U,V}(x)$ for all i. Indeed, if $\xi_i < \xi_{i+1}$ then from Definition 7 and the weight property (62) we know that

$$B_{i,1,\boldsymbol{\xi}}^{\boldsymbol{w}}(\xi_i) = 0, \qquad \lim_{\substack{x \to \xi_{i+1} \\ x < \xi_{i+1}}} B_{i,1,\boldsymbol{\xi}}^{\boldsymbol{w}}(x) = \lim_{\substack{x \to \xi_{i+1} \\ x < \xi_{i+1}}} w_1(x) = 1,$$

and from the piecewise Tchebycheff structure that $B_{i,1,\boldsymbol{\xi}}^{\boldsymbol{w}}$ belongs to the ET-space $\langle D^{p-1}U, D^{p-1}V \rangle$ on $[\xi_i^+, \xi_{i+1}^-]$. Since the same properties also hold for $B_{i,1,\boldsymbol{\xi}}^{U,V}(x)$, they must be identical on $[\xi_i, \xi_{i+1})$. A similar argument holds for the interval $[\xi_{i+1}, \xi_{i+2})$. As a consequence, taking into account that $w_2 = \dots = w_p = 1$, it follows clearly from their definitions that $B_{j,p,\boldsymbol{\xi}}^{\boldsymbol{w}}(x) = B_{j,p,\boldsymbol{\xi}}^{U,V}(x)$ for $p \geq 2$ and they are normalized. In other words, generalized B-splines are a special instance of normalized Tchebycheffian B-splines. □

Example 17 Any linear Tchebycheffian B-spline $B_{j,1,\boldsymbol{\xi}}^{\boldsymbol{w}}(x)$ can be written as $B_{j,1,\boldsymbol{\xi}}^{U,V}(x)$ in the form (66), up to the positive scaling factor $w_1(\xi_{j+1})$. In particular, when $U(x) = \cos(x)$, $V(x) = \sin(x)$, we have

$$B_{j,1,\boldsymbol{\xi}}^{U,V}(x) = \begin{cases} \dfrac{\sin(x - \xi_j)}{\sin(\xi_{j+1} - \xi_j)}, & \text{if } x \in [\xi_j, \xi_{j+1}), \\ \dfrac{\sin(\xi_{j+2} - x)}{\sin(\xi_{j+2} - \xi_{j+1})}, & \text{if } x \in [\xi_{j+1}, \xi_{j+2}), \\ 0, & \text{otherwise.} \end{cases}$$

This is the scaled version of the spline in Example 13, with scaling factor $w_1(\xi_{j+1}) = \cos(\xi_{j+1})$.

Example 18 The generalized B-spline of degree $p = 2$ on a knot sequence $\boldsymbol{\xi}$ consisting of simple knots is given by

$$B^{U,V}_{j,2,\boldsymbol{\xi}}(x) = \begin{cases} \delta_j \displaystyle\int_{\xi_j}^{x} v_j(y)\,\mathrm{d}y, & \text{if } x \in [\xi_j, \xi_{j+1}), \\ 1 - \delta_{j+1} \displaystyle\int_{\xi_{j+1}}^{x} v_{j+1}(y) - \delta_i \int_{x}^{\xi_{j+2}} u_{j+1}(y), & \text{if } x \in [\xi_{j+1}, \xi_{j+2}), \\ \delta_{j+1} \displaystyle\int_{x}^{\xi_{j+3}} u_{j+2}(y)\,\mathrm{d}y, & \text{if } x \in [\xi_{j+2}, \xi_{j+3}), \\ 0, & \text{otherwise,} \end{cases}$$

where

$$\delta_i := \left(\gamma^{U,V}_{i,1,\boldsymbol{\xi}}\right)^{-1} = \left(\int_{\xi_i}^{\xi_{i+1}} v_i(y)\,\mathrm{d}y + \int_{\xi_{i+1}}^{\xi_{i+2}} u_{i+1}(y)\,\mathrm{d}y\right)^{-1}.$$

The normalized Tchebycheffian B-spline defined in Example 14 is a special case, considering the functions $U(x) = \cos(x)$, $V(x) = \sin(x)$, and the uniform knot sequence $\{i\omega\}_{i=0}^{3}$.

Example 19 Consider the partition $\Delta = \{0, 1, 2, 3\}$, and

$$U(x) = \begin{cases} x, & \text{if } x \in [0,1), \\ \mathrm{e}^{\alpha x}, & \text{if } x \in [1,2), \\ x, & \text{if } x \in [2,3), \end{cases} \qquad V(x) = \begin{cases} x^2, & \text{if } x \in [0,1), \\ \mathrm{e}^{-\alpha x}, & \text{if } x \in [1,2), \\ x^2, & \text{if } x \in [2,3). \end{cases}$$

When taking $\boldsymbol{\xi} = \Delta$, we get for $p = 1$,

$$B^{U,V}_{1,1,\boldsymbol{\xi}}(x) = \begin{cases} x, & \text{if } x \in [0,1), \\ \dfrac{\sinh((2-x)\alpha)}{\sinh(\alpha)}, & \text{if } x \in [1,2), \\ 0, & \text{otherwise,} \end{cases}$$

$$B^{U,V}_{2,1,\boldsymbol{\xi}}(x) = \begin{cases} \dfrac{\sinh((x-1)\alpha)}{\sinh(\alpha)}, & \text{if } x \in [1,2), \\ 3 - x, & \text{if } x \in [2,3), \\ 0, & \text{otherwise,} \end{cases}$$

and for $p = 2$,

$$B^{U,V}_{1,2,\boldsymbol{\xi}}(x) = \frac{1}{1 + \dfrac{\sinh(\beta)}{\beta\cosh(\beta)}} \begin{cases} x^2, & \text{if } x \in [0,1), \\ 1 + \dfrac{\cosh(\beta) - \cosh((3-2x)\beta)}{\beta\sinh(\beta)}, & \text{if } x \in [1,2), \\ (x-3)^2, & \text{if } x \in [2,3), \\ 0, & \text{otherwise,} \end{cases} \tag{67}$$

where $\beta := \alpha/2$. The three nontrivial pieces of $B^{U,V}_{1,2,\boldsymbol{\xi}}$ belong to $\mathbb{P}_2$, $\langle 1, \mathrm{e}^{\alpha x}, \mathrm{e}^{-\alpha x}\rangle$, and $\mathbb{P}_2$. When α tends to zero, the quadratic generalized B-spline in (67) tends to the

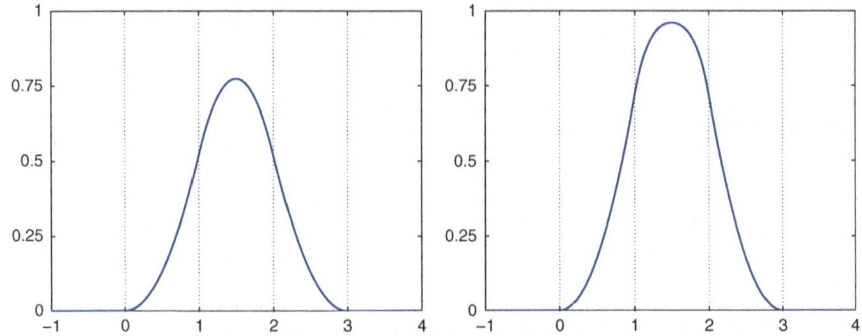

Fig. 3 Two quadratic generalized B-splines with pieces belonging to the spaces $\mathbb{P}_2$, $\langle 1, \mathrm{e}^{\alpha x}, \mathrm{e}^{-\alpha x}\rangle$, and $\mathbb{P}_2$. Left: $\alpha = 1$, Right: $\alpha = 5$. The knot positions are visualized by vertical dotted lines

quadratic polynomial cardinal B-spline. Two such quadratic generalized B-splines (with $\alpha = 1$ and $\alpha = 5$) are depicted in Fig. 3.

Example 20 If $U(x) = x^{p-1}$ and $V(x) = x^p$, then the space in (60) is nothing else than the polynomial space $\mathbb{P}_p$. In this case,

$$u_i(x) = \frac{\xi_{i+1} - x}{\xi_{i+1} - \xi_i}, \quad v_i(x) = \frac{x - \xi_i}{\xi_{i+1} - \xi_i}, \quad \xi_i < \xi_{i+1},$$

and Definition 11 results in the standard **polynomial B-splines** of degree p. This is in agreement with Example 15.

Acknowledgements C. Manni and H. Speleers are members of the Gruppo Nazionale Calcolo Scientifico—Istituto Nazionale di Alta Matematica (GNCS-INdAM), and acknowledge the MIUR Excellence Department Project awarded to the Department of Mathematics, University of Rome Tor Vergata (CUP E83C18000100006). The authors are grateful to the Centre International de Rencontres Mathématiques (CIRM, Luminy) and the Centro Internazionale per la Ricerca Matematica (CIRM, Trento) for the Research-in-Pairs support.

References

1. Barry, P.J.: de Boor-Fix dual functionals for Tchebycheffian B-spline curves. Constr. Approx. **12**, 385–408 (1996)
2. Beccari, C.V., Casciola, G., Mazure, M.L.: Design or not design? A numerical characterisation for piecewise Chebyshevian splines. Numer. Algor. **81**, 1–31 (2019)
3. Bister, D.: Ein neuer Zugang für eine verallgemeinerte Klasse von Tschebyscheff-Splines. Ph.D. thesis, University of Karlsruhe (1996)
4. Bister, D., Prautzsch, H.: A new approach to Tchebycheffian B-splines. In: Le Méhauté, A., Rabut, C., Schumaker, L.L. (eds.) Curves and Surfaces with Applications in CAGD, pp. 387–394. Vanderbilt University Press, Nashville (1997)

5. Bracco, C., Lyche, T., Manni, C., Roman, F., Speleers, H.: Generalized spline spaces over T-meshes: dimension formula and locally refined generalized B-splines. Appl. Math. Comput. **272**, 187–198 (2016)
6. Bracco, C., Lyche, T., Manni, C., Roman, F., Speleers, H.: On the dimension of Tchebycheffian spline spaces over planar T-meshes. Comput. Aided Geom. Design **45**, 151–173 (2016)
7. Bracco, C., Lyche, T., Manni, C., Speleers, H.: Tchebycheffian spline spaces over planar T-meshes: dimension bounds and dimension instabilities. J. Comput. Appl. Math. **349**, 265–278 (2019)
8. Buchwald, B., Mühlbach, G.: Construction of B-splines for generalized spline spaces generated from local ECT-systems. J. Comput. Appl. Math. **159**, 249–267 (2003)
9. Carnicer, J.M., Mainar, E., Peña, J.M.: On the critical lengths of cycloidal spaces. Constr. Approx. **39**, 573–583 (2014)
10. Coppel, W.A.: Disconjugacy. Springer, Berlin (1971)
11. Dyn, N., Ron, A.: Recurrence relation for Tchebycheffian B-splines. J. Anal. Math. **51**, 118–138 (1988)
12. Ince, E.L.: Ordinary Differential Equations. Dover Publications, New York (1956)
13. Karlin, S.: Total Positivity. Stanford University Press, Stanford (1968)
14. Karlin, S., Studden, W.J.: Tchebycheff Systems: With Applications in Analysis and Statistics. Interscience Publishers, New York (1966)
15. Koch, P.E., Lyche, T.: Exponential B-splines in tension. In: Chui, C.K., Schumaker, L.L., Ward, J.D. (eds.) Approximation Theory VI, vol. 2, pp. 361–364. Academic Press, Boston (1989)
16. Koch, P.E., Lyche, T.: Construction of exponential tension B-splines of arbitrary order. In: Laurent, P.J., Le Méhauté, A., Schumaker, L.L. (eds.) Curves and Surfaces, pp. 255–258. Academic Press, Boston (1991)
17. Kvasov, B., Sattayatham, P.: GB-splines of arbitrary order. J. Comput. Appl. Math. **104**, 63–88 (1999)
18. Lyche, T.: A recurrence relation for Chebyshevian B-splines. Constr. Approx. **1**, 155–173 (1985)
19. Lyche, T., Manni, C., Speleers, H.: Foundations of spline theory: B-splines, spline approximation, and hierarchical refinement. In: Lyche, T., Manni, C., Speleers, H. (eds.) Splines and PDEs. From Approximation Theory to Numerical Linear Algebra. Lecture Notes in Mathematics, vol. 2219, pp. 1–76. Springer International Publishing AG, Berlin (2018)
20. Lyche, T., Schumaker, L.L.: A multiresolution tensor spline method for fitting functions on the sphere. SIAM J. Sci. Comput. **22**, 724–746 (2000)
21. Manni, C., Pelosi, F., Sampoli, M.L.: Generalized B-splines as a tool in isogeometric analysis. Comput. Methods Appl. Mech. Eng. **200**, 867–881 (2011)
22. Manni, C., Pelosi, F., Speleers, H.: Local hierarchical h-refinements in IgA based on generalized B-splines. In: Floater, M., Lyche, T., Mazure, M.L., Mørken, K., Schumaker, L.L. (eds.) Mathematical Methods for Curves and Surfaces 2012. Lecture Notes in Computer Science, vol. 8177, pp. 341–363. Springer, Heidelberg (2014)
23. Manni, C., Reali, A., Speleers, H.: Isogeometric collocation methods with generalized B-splines. Comput. Math. Appl. **70**, 1659–1675 (2015)
24. Mazure, M.L.: Extended Chebyshev piecewise spaces characterised via weight functions. J. Approx. Theory **145**, 33–54 (2007)
25. Mazure, M.L.: Finding all systems of weight functions associated with a given extended Chebyshev space. J. Approx. Theory **163**, 363–376 (2011)
26. Mazure, M.L.: How to build all Chebyshevian spline spaces good for geometric design? Numer. Math. **119**, 517–556 (2011)
27. Mazure, M.L.: Constructing totally positive piecewise Chebyshevian B-spline bases. J. Comput. Appl. Math. **342**, 550–586 (2018)
28. Nürnberger, G., Schumaker, L.L., Sommer, M., Strauss, H.: Generalized Chebyshevian splines. SIAM J. Math. Anal. **15**, 790–804 (1984)
29. Pottmann, H.: The geometry of Tchebycheffian splines. Comput. Aided Geom. Design **10**, 181–210 (1993)

30. Schumaker, L.L.: Spline Functions: Basic Theory, 3rd edn. Cambridge University Press, Cambridge (2007)
31. Wang, G., Fang, M.: Unified and extended form of three types of splines. J. Comput. Appl. Math. **216**, 498–508 (2008)

Template Mapping Using Adaptive Splines and Optimization of the Parameterization

Svajūnas Sajavičius, Bert Jüttler and Jaka Špeh

Abstract We consider the construction of a spline map (a volumetric deformation) that transforms a template, which is given in the domain, into a target shape. More precisely, the domain is equipped with a set of surface patches (the template skeleton) and target patches for some of them (which are called the constraining patches) are specified. The constructed spline map approximately transforms the constraining patches into the associated target patches. Possible applications include isogeometric segmentation and parameterization of the computational domain. In particular, the approach should be useful when performing isogeometric segmentation and parameterization for a large class of computational domains possessing similar shapes. We present a solution approach, which is based on least-squares fitting. In order to deal with the influence of the parameterization, the well-established approaches of point and tangent distance minimization are employed for the iterative solution of the resulting nonlinear optimization problems. Additionally, we enrich the approach with spline space refinement. The efficiency and performance of the approach are investigated experimentally. We demonstrate that the optimization of the parameterization, which is used in the point or tangent distance minimization, is an essential step of the procedure. In addition, we use adaptive spline refinement in order to save computational resources. The proposed template mapping approach is also applied to a case of industrial interest, as well as to a volumetric example.

S. Sajavičius · B. Jüttler (✉) · J. Špeh
Institute of Applied Geometry, Johannes Kepler University, Linz, Austria
e-mail: bert.juettler@jku.at

S. Sajavičius
e-mail: svajunas.sajavicius@gmail.com

J. Špeh
e-mail: jaka@jakascorner.com

C. Giannelli and H. Speleers (eds.), *Advanced Methods for Geometric Modeling and Numerical Simulation*, Springer INdAM Series 35,
https://doi.org/10.1007/978-3-030-27331-6_9

1 Introduction

In this paper, we focus on the template mapping problem. The problem consists of the construction of a spline map (a volumetric deformation) that transforms the template domain into the target domain. The template map is constructed taken into account restrictions on mapping certain surface patches from the template domain: the template map transforms these surfaces (the constraining patches) into other predefined surfaces (the target patches).

The template mapping possesses applications in isogeometric segmentation and parameterization (see [33, 47]). These two steps are essential as preprocessing steps for NURBS-based numerical simulation, i.e., for isogeometric analysis (IGA) [22]. Consequently, they have attracted substantial interest from the scientific community. We summarize some of the related literature in the following.

IGA-suitable spline parameterizations of swept volumes are described in paper [1]. The paper [7] analyzes aspects of parameterization quality for geometric modeling in IGA. Solid modeling and domain parameterization using trivariate T-splines is discussed in [8, 29, 43].

The papers [16, 44] describe optimization based techniques for planar and volumetric domain parameterization in IGA. High-quality constructions for multi-patch NURBS parameterizations are introduced in [48]. The paper [10] uses inverse harmonic mappings for planar domain parameterization using truncated hierarchical B-splines (THB-splines), while [37] employs Powell–Sabin spline representations and [31] uses the Teichmüller mapping.

Unstructured spline spaces for IGA on manifolds are introduced in [35] (see also [24] for G^1 smooth discretizations). The conversion of the boundary representation (B-Rep) models into domain parameterizations for IGA is addressed in [2], while the construction of multi-patch parameterizations is discussed in [6]. Scaled boundary parameterizations for IGA are considered in [3]. The recent paper [19] adopts elliptic grid generation principles for IGA applications. Another approach to the computation of IGA-suitable planar parameterizations via PolySquare-enhanced domain partition is investigated in [46]. An approach for constructing low-rank parameterizations of planar domains is proposed in [32]. The paper [45] focuses on generating high-quality high-order Bézier triangular and tetrahedral elements for IGA on triangulations. A template segmentation is exploited in order to provide a multi-patch parameterization of a planar multiply-connected domain in [9]. The recent paper [18] gives an overview of isogeometric segmentation and parameterization and provides additional references.

The present paper is devoted to the technique of template mapping, which should be a useful tool when solving the isogeometric domain segmentation and parameterization problem for a large class of computational domains possessing similar shapes and equivalent topologies. Such classes occur naturally in engineering, e.g., when one is trying to identify the optimal design for a specific application. In this setting it appears to be a promising approach to transfer a pre-defined segmentation and parameterization from a (simplified) master domain (the "template") to each particular instance of the domain.

We solve the template mapping problem by applying an iterative procedure, which is based on least-squares fitting. In particular, we minimize an objective function that involves several terms representing the geometric error, smoothing and regularization.

For computational purposes, we need to discretize the objective function. The standard method to discretize the geometric error term is the *point distance minimization* (PDM) error term [21]. The optimization procedure based on the PDM error term is equivalent to an alternating optimization method. Though this method is quite robust and leads to significant improvements of the initial results, it is also known to have a low rate of convergence. Another approach, which employs the *tangent distance minimization* (TDM) error terms [4], is a Gauss–Newton-type method, thus providing quadratic convergence for zero-residual problems.

The method of *squared distance minimization* (SDM), which was introduced in [34], provides an alternative to the above-mentioned discretization methods for the geometric error terms (see [5] and the references cited therein). The SDM error terms are curvature-dependent, thus requiring C^2 smoothness, and the method lacks clear theoretical advantages with respect to the TDM method. A detailed discussion of the SDM method is therefore beyond the scope of the present paper.

In this paper, we adapt the PDM/TDM methods to the template mapping problem. In addition, we enrich the iterative approach with spline space refinement. Besides various other observations that we obtain during the experimental study, we also demonstrate that the local (adaptive) spline space refinement used instead of the global (uniform) refinement helps to save computational resources. The local refinement allows achieving a similar accuracy of the results with significantly reduced computational effort.

The rest of the paper is organized as follows. Section 2 presents the basic notation and definitions, formulates the template mapping problem and shortly describes possible applications to isogeometric segmentation and parameterization. In Sect. 3, the template mapping problem is discretized. We give a detailed description of the presented iterative approach for the template mapping problem in Sect. 4. The results of the experimental study are presented in Sect. 5. Finally, Sect. 6 summarizes the paper by formulating conclusions and identifying directions for future work.

2 Template Mapping Problem

In this section, we introduce the basic notation and definitions. Before formulating the template mapping problem in dimension independent form, possible applications to isogeometric segmentation and parameterization are shortly described.

Our interest in the *template mapping* originated in its possible applications to isogeometric segmentation and parameterization. To clarify the practical meaning of the theoretical concepts, which will be introduced below, we first summarize our approach to isogeometric segmentation and parameterization.

Assume that we have a B-Rep model of the computational domain, which we need to parameterize for the purpose of applications of IGA. If the domain is too complicated to be parameterized as a single patch, it needs to be segmented into quadrilateral or cuboidal subdomains (patches) that can be parameterized separately (multi-patch parameterization) in subsequent steps.

More precisely, we consider a template domain, described by a B-Rep model, which is topologically equivalent to the original computational domain (target domain). It is assumed that a segmentation of the template domain can be defined in a natural way and subdomains can be parameterized easily using some standard technique (for example, Coons patches). If, in addition, we have a map that deforms the template domain into B-Rep of the original computational domain, we arrive at a multi-patch parameterization of the original computational domain.

We now introduce the *template skeleton* and the *target patches*. Note that in the rest of the paper we use the notion *surface* to denote curves ($d = 2$) or two-dimensional surfaces ($d = 3$). We also use the hat symbol $\widehat{\ }$ for everything related to the template domain and skeleton.

Let us assume that $\widehat{\Omega}$ is a domain with piecewise smooth boundary in $\mathbb{R}^d$ ($d = 2$ or $d = 3$). We call this domain the *template domain*. In the template domain $\widehat{\Omega}$, the surface patches $\widehat{\Gamma}^k$, $k \in \mathcal{K} = \{1, 2, \ldots, M\}$, are given. Each of these patches is parameterized as

$$\widehat{\gamma}^k : (0, 1)^{d-1} \to \widehat{\Gamma}^k.$$

The surfaces $\widehat{\Gamma}^k$, $k \in \mathcal{K}^* = \{1, 2, \ldots, N\}$, $N < M$, are called the *constraining patches*, and the remaining ones ($k \in \mathcal{K} \setminus \mathcal{K}^*$) are referred to as the *free patches*. The union of the constraining and free patches forms the *template skeleton* $\widehat{\Gamma}$:

$$\widehat{\Gamma} = \bigcup_{k \in \mathcal{K}} \widehat{\Gamma}^k.$$

Each *constraining* patch $\widehat{\Gamma}^k$ corresponds to the *target patch* $\Gamma^k \subset \mathbb{R}^d$ ($k \in \mathcal{K}^*$). The target patches Γ^k are parameterized as

$$\gamma^k : (0, 1)^{d-1} \to \Gamma^k.$$

The constraining patches form the part of the template that must be mapped to the corresponding target patches with high accuracy. In contrast to this, there are no constraints on the mapping of free patches. The sole purpose of introducing these patches is to define the segmentation into quadrilateral ($d = 2$) or cuboidal ($d = 3$) subdomains.

We assume that the topological structure of the intersection of the constraining patches $\widehat{\Gamma}^k \subset \widehat{\Gamma}$ matches that of the target patches Γ^k.

Our aim is to construct the *template map*

$$\mathbf{s} : \widehat{\Omega} \to \mathbb{R}^d$$

that satisfies the requirements

$$\mathbf{s} \circ \widehat{\gamma}^k \circ \varrho^k\ (\mathbf{t}) = \gamma^k\ (\mathbf{t}), \quad \mathbf{t} \in (0,1)^{d-1}, \tag{1}$$

for $k \in \mathscr{K}^*$. Here ϱ^k: $(0,1)^{d-1} \to (0,1)^{d-1}$ are the *reparameterizations* of the constraining patches $\widehat{\Gamma}^k$. In general, the reparameterization functions ϱ^k can be chosen in different ways but they should always be bijective and regular. The requirements (1) ensure that the template map $\mathbf{s}$ transforms each constraining patch $\widehat{\Gamma}^k$ to the corresponding target patch Γ^k, $k \in \mathscr{K}^*$. We emphasize that the requirements (1) are formulated for the constraining and target patches only. For application purposes, the template map $\mathbf{s}$ is also supposed to be injective and regular in $\widehat{\Omega}$.

The surfaces $\widetilde{\Gamma}^k$ parameterized as

$$\widetilde{\gamma}^k : (0,1)^{d-1} \to \widetilde{\Gamma}^k : \mathbf{t} \mapsto \mathbf{s} \circ \widehat{\gamma}^k \circ \varrho^k\ (\mathbf{t})$$

are called the *mapped* (*transformed*) *constraining patches* (for $k \in \mathscr{K}^*$) or the *mapped* (*transformed*) *free patches* (for $k \in \mathscr{K}$). We expect that (cf. (1))

$$\gamma^k\ (\mathbf{t}) \approx \widetilde{\gamma}^k(\mathbf{t}), \quad \mathbf{t} \in (0,1)^{d-1},$$

is valid for each target patch Γ^k and the corresponding mapped constraining patch $\widetilde{\Gamma}^k$ $(k \in \mathscr{K}^*)$.

A schematic description of the template mapping problem is presented in Fig. 1. Alternatively, it can also be summarized in the following diagram:

$$\begin{array}{ccc} [0,1]^{d-1} & \xrightarrow{\varrho^k} & [0,1]^{d-1} \\ \downarrow{\scriptstyle \gamma^k} & & \downarrow{\scriptstyle \widehat{\gamma}^k} \\ \Gamma^k & \xleftarrow[\mathbf{s}]{} & \widehat{\Gamma}^k \end{array}$$

Note that only *approximate* commutativity of this diagram is to be expected.

We now introduce a *variational formulation* of the template mapping problem. We construct the template map $\mathbf{s}$ satisfying the constraints (1) by solving a nonlinear optimization (the least-squares) problem [25]. Indeed, we define and minimize the least-squares objective function:

$$F = \sum_{k \in \mathscr{K}^*} \| \gamma^k - \widetilde{\gamma}^k \|^2_{L_2([0,1]^{d-1})} + Q(\mathbf{s}) \to \min. \tag{2}$$

The term Q in the objective function stands for the smoothing and regularization. This term involves functionals known as the *quality measures*. Various quality measures are known from the rich literature on mesh generation (cf. [12, 28]) and parameterization in isogeometric analysis [10, 16, 19]. We will use the *uniformity* (*simplified thin plate energy*) functional

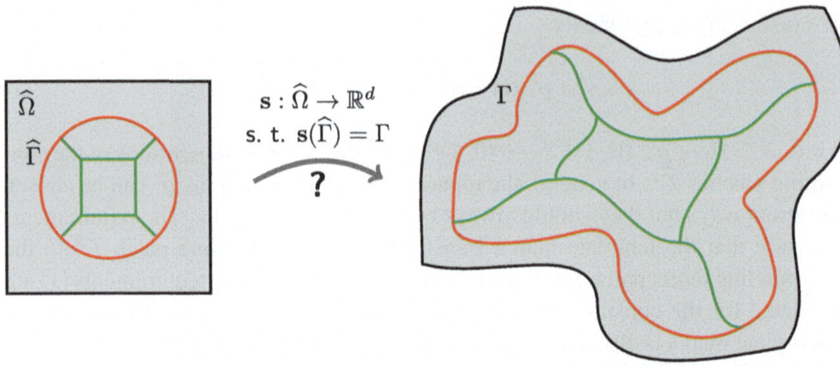

Fig. 1 The template mapping problem: find the map (a volumetric deformation) **s** that transfers the template skeleton (red and green, left) in such a way that the constraining patches (red, left) are transformed into the target patches (red, right). The mapping of the free patches (green, left) is not constrained

$$Q_u(\mathbf{s}) = \int_{\widehat{\Omega}} \left(\sum_{m,n=1}^{d} \|\mathbf{s}_{u_m u_n}\|^2 \right) \mathrm{d}\mathbf{u}.$$

were **s** with subscripts denote partial derivatives of the map **s**.

3 Discretization

In the previous section we formulated the template mapping problem in an abstract way. A proper template map is expected to satisfy certain requirements. In this section, the template mapping problem is discretized.

We will use a *spline approximation of the template map*. More precisely, we assume that the template map **s** is a spline function represented in a hierarchical spline space as a linear combination of the spline basis functions τ_j^ℓ defined by coefficients $\mathbf{c}_j^\ell \in \mathbb{R}^d$:

$$\mathbf{s}(\mathbf{u}) = \sum_{\ell=1}^{L} \sum_{j=1}^{J^\ell} \mathbf{c}_j^\ell \tau_j^\ell(\mathbf{u}), \quad \mathbf{u} \in \widehat{\Omega}. \tag{3}$$

Here the upper index ℓ stands for the hierarchical level, L denotes the total number of levels of refinement, and J^ℓ is the numbers of spline basis functions of level ℓ. For the simplicity of notation, all the coefficients $\mathbf{c}_j^\ell$ are collected in one matrix

$$\mathbf{c} = (\ldots, \mathbf{c}_j^\ell, \ldots) \in \mathbb{R}^{d \times n},$$

where $n = \sum_{\ell=1}^{L} J^\ell$ is the number of hierarchical spline basis functions in all levels.

Next we describe the *discretization of the geometric input data*. The target patches Γ^k are discretized by sampling points $\{\mathbf{p}_i^k\}_{i\in\mathscr{I}}$ corresponding to the initial parameter values $\{\mathbf{t}_i^k\}_{i\in\mathscr{I}}$:

$$\mathbf{p}_i^k = \gamma^k(\mathbf{t}_i^k),$$

where $\mathbf{t}_i^k \in (0,1)^{d-1}$ and $\mathscr{I}$ is an index set.

For the purpose of the closest point computation (see next section), we also discretize the constraining patches $\widehat{\Gamma}^k$, $k \in \mathscr{K}^*$. Each constraining patch is discretized by sampling points $\{\widehat{\mathbf{p}}_{\widehat{\imath}}^k\}_{\widehat{\imath}\in\widehat{\mathscr{I}}}$ corresponding to the parameter values $\{\widehat{\mathbf{t}}_{\widehat{\imath}}^k\}_{\widehat{\imath}\in\widehat{\mathscr{I}}}$, where $\widehat{\mathbf{t}}_{\widehat{\imath}}^k \in (0,1)^{d-1}$ and $\widehat{\mathscr{I}}$ is an index set. These sampled points are used for the initialization of the closest point computation. Consequently, their number is expected to be much larger then the number of points sampled on the target patches, $|\widehat{\mathscr{I}}| \gg |\mathscr{I}|$. In the rest of the paper we assume that $\widehat{\imath} \in \widehat{\mathscr{I}}$, $i \in \mathscr{I}$ and $k \in \mathscr{K}^*$.

Now we consider a *discrete representation of the reparameterization functions*. Assume that we have already the template map $\mathbf{s}$. For each target patch Γ^k and each sampled point $\mathbf{p}_i^k$, the reparameterization function ϱ^k defines the parameter value $\widehat{\mathbf{t}}_i^{k,*}$ such that the mapped point $\widetilde{\mathbf{p}}_i^{k,*} = \mathbf{s} \circ \widehat{\gamma}(\widehat{\mathbf{t}}_i^{k,*})$ is the closest point to the point $\mathbf{p}_i^k$ from all the points on the mapped constraining patch $\widetilde{\Gamma}^k$:

$$\widehat{\mathbf{t}}_i^{k,*} := \arg \min_{\widehat{\mathbf{t}}\in(0,1)^{d-1}} \|\mathbf{p}_i^k - \mathbf{s} \circ \widehat{\gamma}^k(\widehat{\mathbf{t}})\|.$$

In this way, the reparameterization functions ϱ^k are represented discretely by the *optimal parameter* values $\{\widehat{\mathbf{t}}_i^{k,*}\}$:

$$\varrho^k\colon (0,1)^{d-1} \to (0,1)^{d-1} : \mathbf{t}_i^k \mapsto \widehat{\mathbf{t}}_i^{k,*}.$$

The optimal parameter values are expected to ensure that the difference vectors

$$\mathbf{p}_i^k - \mathbf{s} \circ \widehat{\gamma}^k(\widehat{\mathbf{t}}_i^{k,*})$$

are parallel to the unit normal vectors[1] at the points $\mathbf{p}_i^k$ of the target patch Γ^k. The points $\widetilde{\mathbf{p}}_i^{k,*} = \widetilde{\gamma}^k(\widehat{\mathbf{t}}_i^{k,*})$ are called the *closest points* (or *foot points*) of the points $\mathbf{p}_i^k$ on the mapped constraining patch $\widetilde{\Gamma}^k$.

Finally we introduce the *discretized optimization problem*. Let $\{\mathbf{p}_i^k\}$ be a discrete set of points sampled on each target patch Γ^k. The objective function defined by (2) is discretized as

$$\widetilde{F} = \sum_k \sum_i e_i^k + \widetilde{Q}, \tag{4}$$

where $e_i^k = \min_{\widehat{\mathbf{t}}\in(0,1)^{d-1}} \|\mathbf{p}_i^k - \widetilde{\gamma}^k(\widehat{\mathbf{t}})\|^2$ are the *squared orthogonal distances* between the points $\mathbf{p}_i^k$ and the mapped target patch $\widetilde{\Gamma}^k$, and $\widetilde{Q}$ is a discretized smoothing and

[1] Different constraints are present at patch boundaries.

regularization term. The smoothing and regularization term is added to the objective function in order to avoid situations where the resulting linear system becomes singular or self-overlappings appear in the template map.

We minimize the objective function $\widetilde{F}$ with respect to the reparameterization functions ϱ^k and the coefficients $\mathbf{c}$:

$$\widetilde{F} \to \min_{\varrho^k, \mathbf{c}} .$$

This nonlinear least-squares problem is a *separable* and *constrained* optimization problem. Indeed, the reparameterization functions ϱ^k and the coefficients $\mathbf{c}$ can be treated as two separate groups of optimization variables, and the optimization problem can be formulated as the minimization of the objective function (4) with respect to the coefficients $\mathbf{c}$, i.e.

$$\widetilde{F} \to \min_{\mathbf{c}},$$

subject to the parameter optimization constraints. For the solution of this optimization problem we will apply an iterative procedure presented in the next section.

4 Iterative Solution Procedure

Our aim is to minimize the objective function (4) with respect to the coefficients of the discretized template map, and the reparameterization functions. We apply an iterative procedure for the solution of this optimization problem. In this section, we give a general outline of the procedure as well as describe its steps in detail.

As it was mentioned in the previous section, the optimization problem (3) is separable. Therefore, we treat the unknown coefficients $\mathbf{c}$ of the map $\mathbf{s}$ and the reparameterization functions ϱ^k in separate steps. The outline of the iterative procedure for the template mapping problem solution is the following:

Step 1: Discretization and initialization
Step 2: Control point computation
Step 3: Closest point computation
Step 4: Checking termination and refinement criteria:

- (4a) Termination criterion
- (4b) Refinement criterion

Step 5: Spline space refinement

In the next sections we will describe all the steps in detail.

Step 1: Discretization and Initialization

We assume the target patches Γ^k are discretized by sampled points $\{\mathbf{p}_i^k\}$ (see above). In addition, we also construct the initial version of the template map $\mathbf{s}$. The initial template map is defined by initial tensor-product spline basis and initial values of the coefficients **c**.

As the initial map for the iterative procedure, we use the identity map. This choice is suitable in cases when the constraining patches are quite similar to the target patches. Another possibility is to execute one iteration of the control point computation step (Step 2) with a simplified objective function (in order to obtain a simpler quadratic optimization problem).

Step 2: Control Point Computation

Suppose that the template map **s** defined by (3) is an initial map, or the current map generated in the previous iteration of the procedure. In this step, we minimize the objective function $\widetilde{F}$ with respect to the coefficients **c**. Clearly, the new values of the coefficients **c** define an updated map **s**.

For the discretization of the squared orthogonal distances (error terms) e_i^k in the discretized objective function (4) we implemented and examined the *point distance minimization* (PDM) and the *tangent distance minimization* (TDM) procedures (see e.g. [42] and references therein). The error terms e_i^k in the PDM are expressed as

$$e_{\mathrm{PDM},i}^k = \|\mathbf{p}_i^k - \widetilde{\gamma}^k(\widehat{\mathbf{t}}_i^k)\|^2,$$

while in the TDM procedure they are approximated as

$$e_{\mathrm{TDM},i}^k = \left[\left(\mathbf{p}_i^k - \widetilde{\gamma}^k(\widehat{\mathbf{t}}_i^k)\right)^\top \cdot \mathbf{N}_i^k\right]^2,$$

where $\mathbf{N}_i^k$ are the unit normal vectors at the points $\mathbf{p}_i^k$ on the target patch Γ^k. Both PDM and TDM procedures can be combined together. Then the squared orthogonal distances are expressed as

$$e_i^k \approx \omega_{\mathrm{PDM}}\, e_{\mathrm{PDM},i}^k + \omega_{\mathrm{TDM}}\, e_{\mathrm{TDM},i}^k,$$

where ω_{PDM} and ω_{TDM} are the weights controlling the influence of PDM and TDM errors terms.

In our implementation, the smoothing and regularization term is assumed to be

$$\widetilde{Q} = \omega_r\, \widetilde{Q}_r + \omega_u\, \widetilde{Q}_u,$$

where ω_r and ω_u are user-defined non-negative weights for the Tikhonov regularization term

$$\widetilde{Q}_r(\mathbf{c}) = \sum_{\ell=1}^{L} \sum_{j=1}^{J^\ell} \|\mathbf{c}_j^\ell - \overline{\mathbf{c}}_j^\ell\|^2$$

and the discretized uniformity functional $\widetilde{Q}_u$ (see [10]), while $\overline{\mathbf{c}}_j^\ell$ are the coefficients from the previous iteration.

The template map coefficients $\mathbf{c}$ are obtained by solving the linear system arising after the differentiation of the objective function (4) with respect to unknown coefficients $\mathbf{c}$,

$$\frac{\partial \widetilde{F}}{\partial \mathbf{c}} = 0.$$

We obtain a linear system since we consider only the quadratic uniformity functional as quality measure in our implementation. Other quality measures could be considered in addition to this simple one, but the implementation would then require suitable techniques from non-linear optimization, e.g., *Gauss–Newton-type* methods [25].

Step 3: Closest Point Computation

In this step, we perform an update of the parameter values $\{\widehat{\mathbf{t}}_i^{k,*}\}$ in such a way that the mapped points $\widetilde{\mathbf{p}}_i^k = (\mathbf{s} \circ \widehat{\gamma}^k)(\widehat{\mathbf{t}}_i^{k,*})$ are the closest points to the corresponding points $\mathbf{p}_i^k$ on the target patch Γ^k. Taking into account the discretization of the constraining patches, the initial parameter values are computed as

$$\widehat{\mathbf{t}}_i^{k,*} = \widehat{\mathbf{t}}_{\widehat{\imath}_0}^k \quad \text{with} \quad \widehat{\imath}_0 = \arg\min_{\widehat{\imath}} \|\mathbf{p}_i^k - \widetilde{\mathbf{p}}_{\widehat{\imath}}^k\|,$$

where $\widetilde{\mathbf{p}}_{\widehat{\imath}}^k = \mathbf{s} \circ \widehat{\gamma}^k(\widehat{\mathbf{t}}_{\widehat{\imath}}^k)$.

The initial parameter values are improved by executing a few Newton steps:

$$\widehat{\mathbf{t}}_i^{k,*} \leftarrow \widehat{\mathbf{t}}_i^{k,*} + \Delta\widehat{\mathbf{t}}_i^{k,*},$$

where

$$\Delta\widehat{\mathbf{t}}_i^{k,*} = \frac{\left(\mathbf{p}_i^k - \mathbf{s} \circ \widehat{\gamma}^k(\widehat{\mathbf{t}}_i^k)\right)^\top \cdot \left(\mathbf{s}_{u_1} \circ \widehat{\gamma}^k(\widehat{\mathbf{t}}_i^k)\right)}{\|\mathbf{s}_{u_1} \circ \widehat{\gamma}^k(\widehat{\mathbf{t}}_i^k)\|^2}$$

in the planar ($d = 2$) case, and

$$\Delta\widehat{\mathbf{t}}_i^{k,*} = \begin{pmatrix} \|\mathbf{s}_{u_1} \circ \widehat{\gamma}^k(\widehat{\mathbf{t}}_i^k)\| & (\mathbf{s}_{u_1}\mathbf{s}_{u_2}) \circ \widehat{\gamma}^k(\widehat{\mathbf{t}}_i^k) \\ (\mathbf{s}_{u_1}\mathbf{s}_{u_2}) \circ \widehat{\gamma}^k(\widehat{\mathbf{t}}_i^k) & \|\mathbf{s}_{u_2} \circ \widehat{\gamma}^k(\widehat{\mathbf{t}}_i^k)\| \end{pmatrix}^{-1} \begin{pmatrix} \left(\mathbf{p}_i^k - \mathbf{s} \circ \widehat{\gamma}^k(\widehat{\mathbf{t}}_i^k)\right)^\top \cdot \left(\mathbf{s}_{u_1} \circ \widehat{\gamma}^k(\widehat{\mathbf{t}}_i^k)\right) \\ \left(\mathbf{p}_i^k - \mathbf{s} \circ \widehat{\gamma}^k(\widehat{\mathbf{t}}_i^k)\right)^\top \cdot \left(\mathbf{s}_{u_2} \circ \widehat{\gamma}^k(\widehat{\mathbf{t}}_i^k)\right) \end{pmatrix}$$

in the volumetric ($d = 3$) case, where s_{u_1} and s_{u_2} denote the partial derivatives of the map $\mathbf{s}$. These formulas can be derived by considering the linear Taylor approximation of $\mathbf{s}$. It should be noted that this procedure needs to be modified on and near patch boundaries, in order to obtain valid results.

Step 4: Checking Termination and Refinement Criteria

The iterative procedure is repeated until a certain termination criterion is satisfied, e.g., the prescribed accuracy is achieved or the maximal number of iterations is reached. The procedure also terminates if the template map $\mathbf{s}$ becomes unacceptably irregular. We monitor the determinant of the Jacobian at certain points in $\widehat{\Omega}$ and terminate the procedure if the percentage of points with negative values of Jacobian determinant exceeds a prescribed threshold.

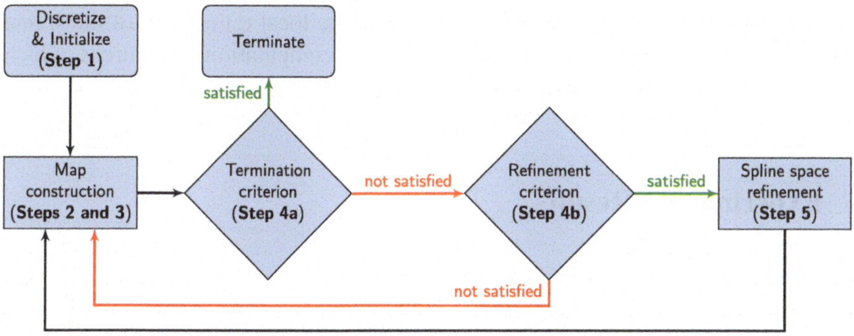

Fig. 2 The iterative approach for template mapping problem

In addition, the proposed approach for the template mapping problem can be enriched with the refinement of the spline space in which the template map **s** is represented. After several PDM/TDM iterations the accuracy of the results can be estimated and, if necessary, the spline space can be refined before executing another series of PDM/TDM iterations. As an indicator for the spline space refinement, the difference of the error in two successive iterations can be used: the spline space is refined whenever the error no longer changes significantly.

The flowchart representing the iterative approach is depicted in Fig. 2. The effectiveness of the approach combining PDM/TDM iterations and the spline space refinement will be demonstrated in Sect. 5. The details on the spline space refinement step will be given in the next section.

Step 5: Spline Space Refinement

The standard tensor-product constructions of the multivariate splines provide the possibility of the global (uniform) refinement only. This means that the insertion of a new knot into one of the knot vectors refines the entire column or row of cells in the mesh. In order to overcome this, various generalizations of tensor-product splines were proposed (see [13] and the references cited therein).

In our implementation of the approach, the local refinement is based on THB-splines [13, 14], which form another basis for the space of hierarchical splines [11, 17, 27]. In addition to the possibility of the local refinement, the THB-splines possess numerous nice mathematical properties. Compared to the hierarchical splines introduced in [27], THB-splines form a non-negative partition of unity and have the same or a smaller support. In addition, THB-splines are linearly independent and strongly stable with respect to the maximum norm [15].

In order to select hierarchical mesh cells that should be refined, various strategies can be applied. We can mention the *absolute threshold* and *relative threshold* approaches [26]. In the first strategy, the points where the error exceeds a user-defined threshold are marked for the refinement, while the latter approach marks a certain percentage of points with the largest errors.

The size of the refined area and, consequently, the size of the corresponding THB-spline basis can be affected by properly adjusting the *extension* parameter [26].

As we will see in computational experiments, the local refinement of the spline space allows to significantly reduce the amount of computational resources, which are required in the case of the global refinement.

5 Experimental Results

In this section we set up a test example and use it to analyze and compare different versions of the considered approach for the template mapping problem in terms of the accuracy and computational complexity (the amount of required computational resources). The experimental study consists of four experiments. In one of them, we apply the approach to a case of industrial interest. An experiment demonstrating the approach applicability to volumetric cases is also presented.

First we discuss briefly some *implementation details*. The template skeleton and target patches, which will be used in Experiments 1 and 2 of our experimental study, are presented in Fig. 3. For simplicity, in the first two experiments we analyze the template skeleton without free patches (all the patches are constraining). The template skeletons and the target patches of industrial and volumetric examples will be introduced below.

The iterative procedure is initialized by defining an initial map and sampling points $\mathbf{p}_i^k$ on each target patch Γ^k. In our experiments, as the initial map we use the identity map defined in a space of tensor-product B-splines of bi-degree $(3, 3)$ (in planar cases) or tri-degree $(3, 3, 3)$ (in volumetric case). The B-spline basis is defined using uniform knot vectors with 11 inner knots in each direction.

On each target patch in planar examples (Experiments 1–3), we sample 200 points corresponding to the parameters uniformly distributed on the unit interval. For the closest point computation, we discretize each constraining patch $\widehat{\Gamma}^k$ by sampling 10^4 points. Then, we map these points using the template map $\mathbf{s}$. From all these mapped points we find an initial closest point on the target patch Γ^k for each point $\mathbf{p}_i^k$. An initial closest point then is improved by executing two Newton steps (Step 3).

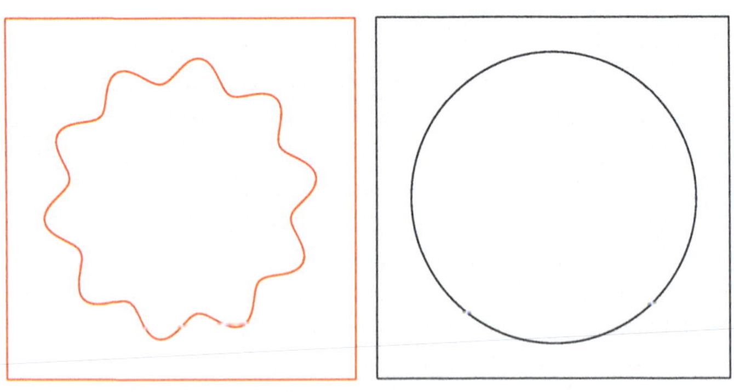

Fig. 3 Target patches (left) and constraining patches (right) used in Experiments 1 and 2

The selection of appropriate weights ω_r and ω_u of the Tikhonov regularization term and the uniformity functional in the discretized objective function (4) is one of the main challenges of the proposed approach. These weights affect both the accuracy and quality of the final result (see e.g. [26] for the detailed discussion) and, therefore, they should be selected very carefully. In Experiments 1, 2 and 4 we will only apply the Tikhonov regularization with the weight $\omega_r = 2 \times 10^{-2}$. In Experiment 3, we will also use the uniformity functional with weight $\omega_u = 2 \times 10^{-2}$. This was needed to obtain an acceptable accuracy of the results.

In our experiments, the hierarchical mesh cells for the local refinement are marked using the absolute threshold strategy with the threshold of the pointwise maximal L_2-error being constant and equal to 10^{-5}. The marked cells then are refined using the dyadic cell refinement.

The accuracy of the results is measured using the squared l_2-error:

$$E_{l_2} = \sum_k \sum_i \|\mathbf{p}_i^k - \widetilde{\mathbf{p}}_i^k\|^2.$$

Before the error estimation, the closest point computation (parameter optimization) step is executed.

All the steps of the examined approach have been implemented in G+SMO (Geometry + Simulation modules) C++ library for IGA [23]. This library includes an efficient implementation of the THB-splines.

Experiment 1: Comparison of PDM and TDM

In the first experiment, we are going to demonstrate that the closest point computation (parameter optimization) step (Step 3) is an essential component of the approach. Moreover, we compare convergence rates of the procedures and the regularity of the resulting template maps.

In Fig. 4 we demonstrate the template mapping results obtained using procedure with the fixed parameterization, as well as the approach based on PDM or TDM with the parameter optimization step. In this experiment, the template map **s** is constructed in the initial tensor-product spline space (the spline space refinement is not applied).

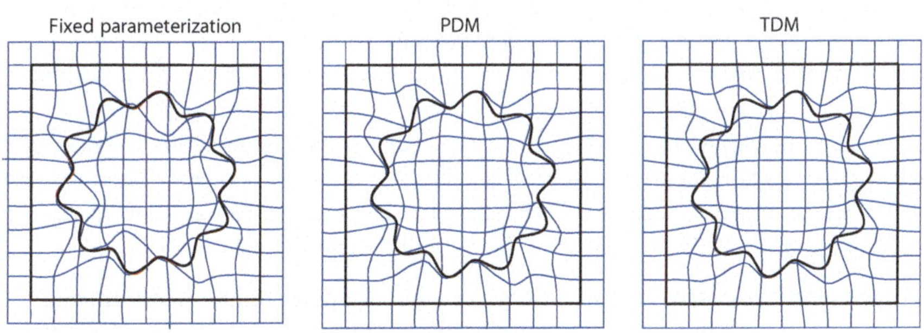

Fig. 4 The template mapping obtained after 20 iterations of different approaches: fixed parameterization (left), PDM (center) and TDM (right) (Experiment 1)

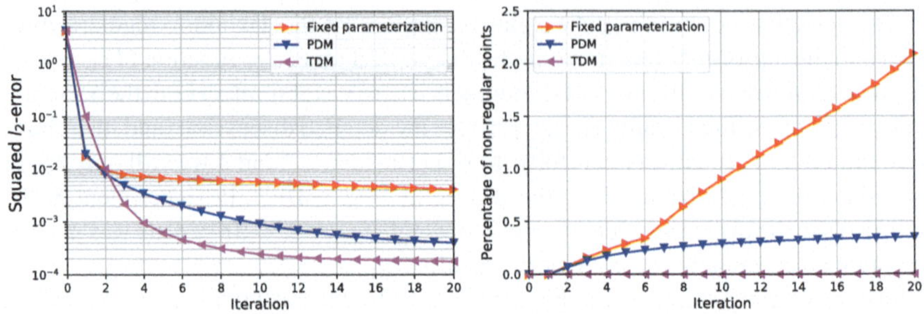

Fig. 5 Squared l_2-errors (left) and the percentage of points with negative values of the Jacobian determinant (right) (Experiment 1)

Table 1 Squared l_2-errors and the percentage of points with negative values of the corresponding map Jacobian determinant obtained after 20 iterations of the procedure with fixed parameterization and PDM/TDM procedures with the closet point computation step (Experiment 1)

	l_2-error	Non-regular points (%)
Fixed parameterization	4.16585×10^{-3}	2.0969
PDM	4.03419×10^{-4}	0.3552
TDM	1.76909×10^{-4}	0.0000

From Fig. 4 we can see that the resulting map **s** obtained using the fixed parameterization has self-overlappings. To estimate the regularity of the template map **s**, we computed the determinant of the map Jacobian at 10^6 points sampled in the parametric domain. The squared l_2-errors obtained during 20 iterations of the procedure as well as the percentage of points with negative values of the Jacobian determinant are presented in Fig. 5 (see also Table 1 for the corresponding numerical values after the last of 20 iterations). The most accurate results were obtained using the approach based on TDM. The PDM with the closest point computation step produced slightly less accurate results.

In the case of the procedure with fixed parameterization, the percentage of non-regular points (points with negative values of Jacobian determinant) exceeds 2%. The closest point computation in PDM procedure reduces this percentage to around 0.36%. In the map produced by the procedure based on TDM, we do not identify non-regular points at all.

Clearly, the approach benefits from the closest point computation (parameter optimization) step. Therefore, we will use the closest point computation in the rest of this study. Moreover, since TDM seems to have clear advances with respect to PDM, we limit ourselves to TDM in the remaining experiments.

Experiment 2: Comparison of Global and Local Spline Space Refinement

The goal of the second experiment is to compare the global and local spline space refinement strategies (Step 5) in terms of the accuracy of the results and the amount of the required computational resources.

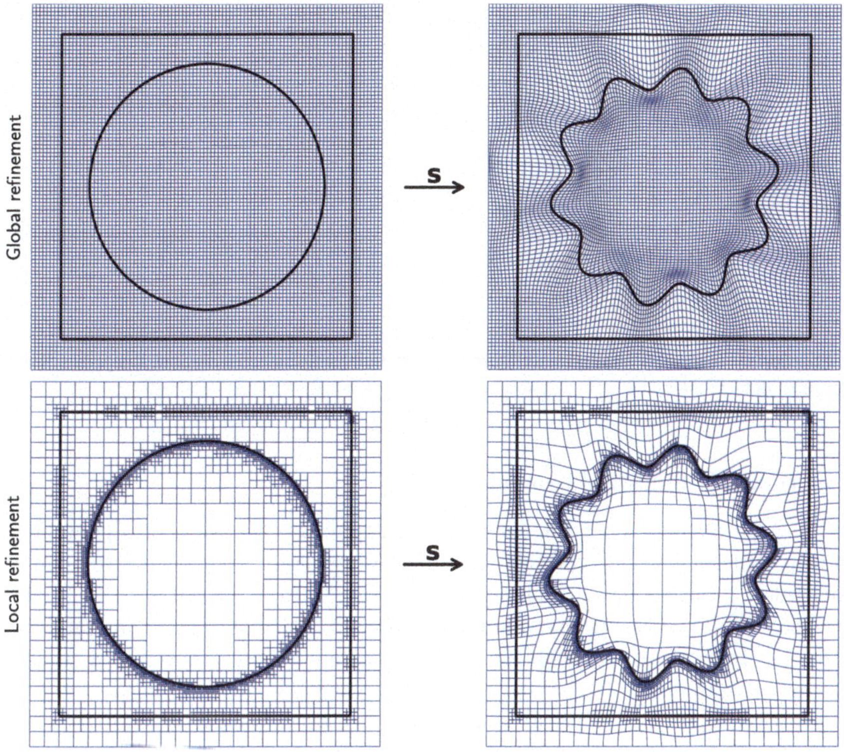

Fig. 6 The template mapping obtained after three iterations of global refinement (top) or seven iterations of local refinement (bottom) of the initial spline space (Experiment 2)

In this experiment, the spline space refinement is done after each single TDM iteration. The template mapping results are presented in Fig. 6. By using the same map regularity testing procedure as in Experiment 1, we do not identify any points with negative values of the Jacobian determinant. From Fig. 7 (left) we see that the number of degrees of freedom (the size of the spline space) grows exponentially if the spline space is refined globally. In case of the local spline space refinement, this number grows only linearly. This observation is expected and complies with the results obtained in [26].

From Fig. 7 (right) we also see that the local refinement leads to a similar accuracy of the results when using significantly coarser spline spaces (with much fewer degrees of freedoms in the spline map representation). Consequently, in order to achieve a similar accuracy, the local refinement strategy requires much less additional computational resources (computational time and memory) in comparison with the global one. Note that the procedure for adaptive refinement was stopped when the termination criterion was satisfied. This explains why the final error value exceeds the error obtained by using global refinement, where much fewer refinement steps were executed.

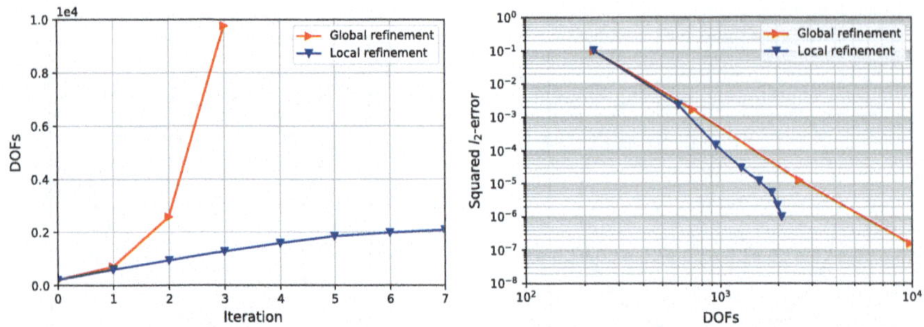

Fig. 7 The number of degrees of freedom (DOFs) (left), and the squared l_2-errors versus the number of degrees of freedom (right) obtained during the global and local refinement of the spline space (Experiment 2)

Experiment 3: Industrial Example

In this experiment, the presented approach is applied to two cases corresponding to the target patches, which represent the profiles of twin screw compressor rotors with four (male rotor) and six (female rotor) lobes (Fig. 8, top), see [20, 30, 36, 38–41] for additional details. The corresponding template skeletons are depicted in Fig. 8 (bottom). In this case, the template skeletons contain not only constraining patches but free ones too.

By this experiment we aim to demonstrate the approach applicability to the cases with industrial input and, in addition, investigate the possibility to combine series of TDM iterations and the local spline space refinement. We combine TDM procedure with local refinement of the spline space, i.e., the spline space refinement iteration is executed after each series of five TDM iterations. In addition to Tikhonov regularization ($\omega_r = 2 \times 10^{-2}$), the uniformity functional with the weight $\omega_u = 2 \times 10^{-2}$ is also used in this experiment. The identity map is used for the initialization of the iterative procedure.

The knot configurations in parametric and physical domains are exhibited in Figs. 9 and 10. The maps presented in these figures have no non-regular points. In addition, Fig. 11 demonstrates how the errors change after each iteration of TDM procedure. We see that after several iterations error decay slows down. The refinement of the spline space helps to speed up the convergence and obtain more accurate final results. We have already demonstrated in Experiment 2 that the local refinement strategy allows saving computational resources in comparison with the global one.

Experiment 4: Volumetric Example

So far, we applied the template mapping approach for the planar ($d = 2$) cases only. In the last experiment, we demonstrate the approach applicability to volumetric ($d = 3$) cases. The input data and the results are presented in Fig. 12.

The set of the target patches consists of four faces from a axis-aligned box and a patch made by extruding and rotating a curve from Experiments 1 and 2 (Fig. 3).

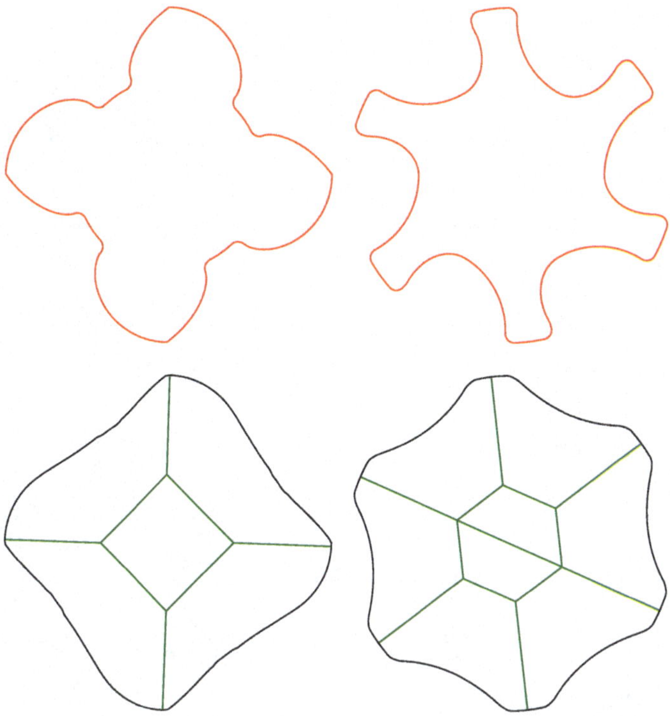

Fig. 8 Target patches (top) and template skeletons (bottom) used in Experiment 3. The target patches represent the profiles of a male rotor (left) and a female rotor (right) in a twin screw compressor. The template skeletons contain both the constraining patches (black) and free patches (green)

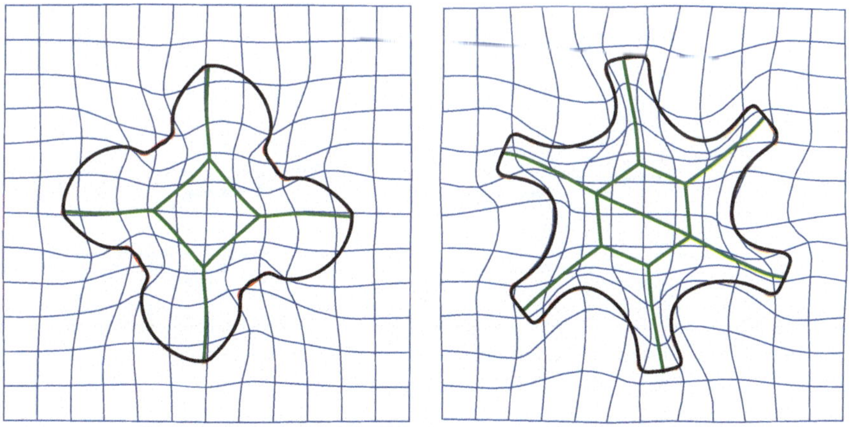

Fig. 9 The template mapping obtained after 20 iterations using TDM (Experiment 3)

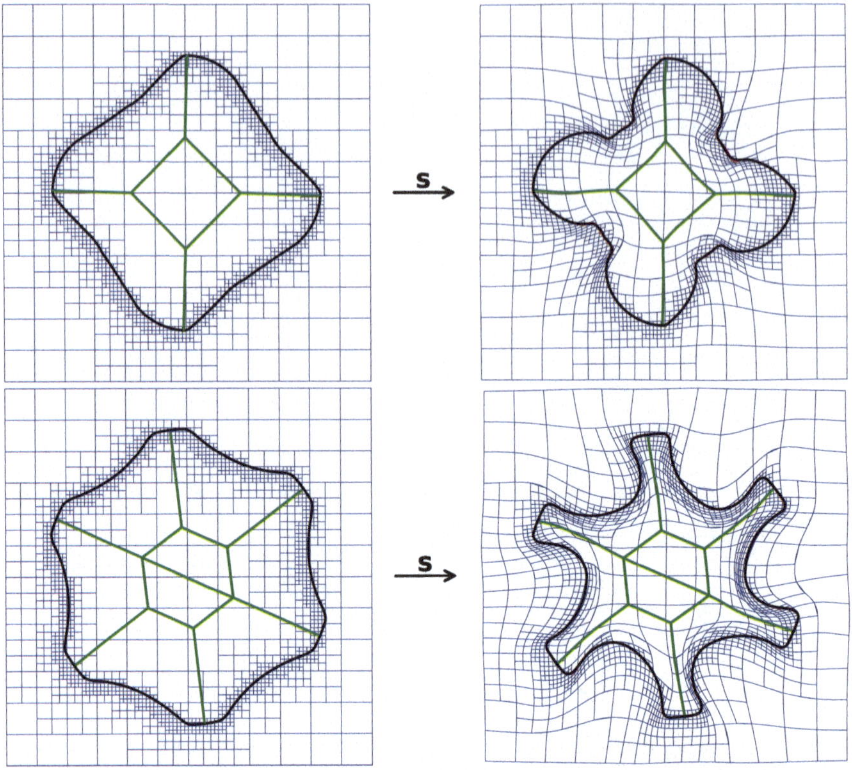

Fig. 10 The template mapping obtained after 20 iterations using TDM combined with three iterations of local refinement of the spline space (Experiment 3)

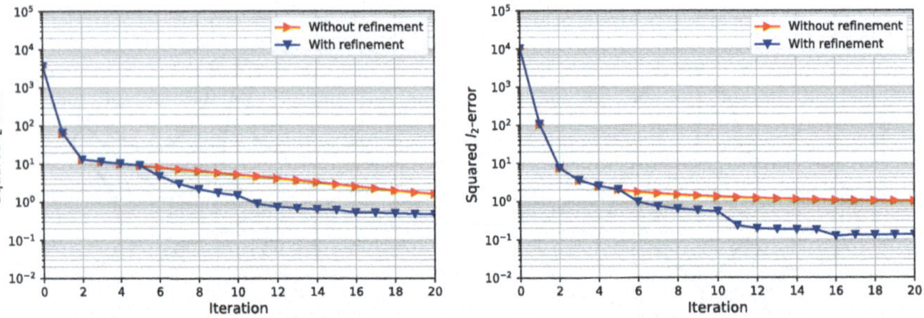

Fig. 11 Squared l_2-errors obtained by combining TDM and spline space refinement (Experiment 3): the profiles of twin screw compressor male rotor (left) and female rotor (right)

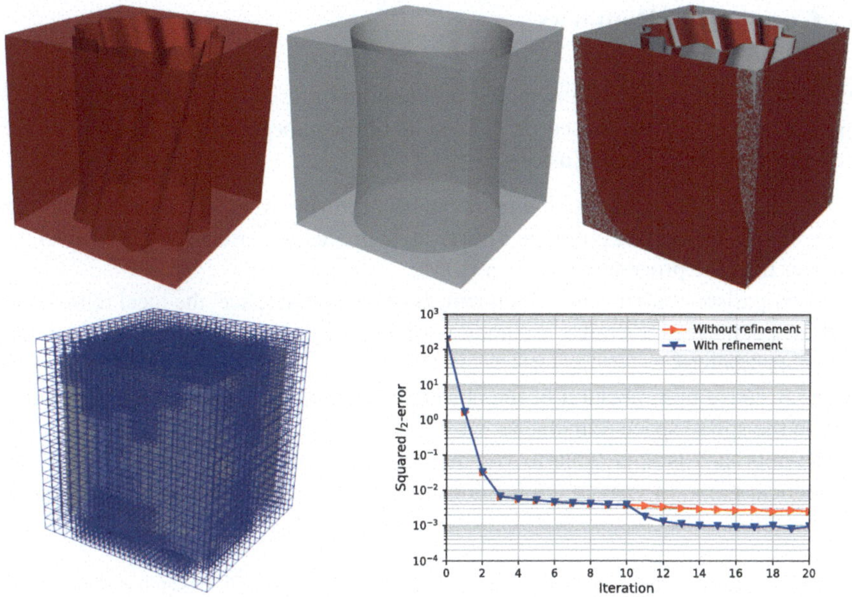

Fig. 12 Volumetric example of the template mapping (Experiment 4). Top row: target patches (left), template patches (middle), target patches and mapped template patches (right). Bottom row: hierarchical mesh after one iteration of adaptive spline space refinement (left) and squared l_2-errors (right)

Correspondingly, the set of template patches consists four faces from a box and a one-sheeted hyperboloid (cylindrical patch made by sweeping with varying radius).

In Fig. 12 we demonstrate how the mapped template patches match the target patches. The target patches were mapped using a volumetric template map obtained after two series of ten TDM iterations combined with one iteration of the adaptive spline space refinement. The initial template map is represented by 6,859 degrees of freedom, while after one iteration of the local refinement this number increased to 12,248. In case of the global spline space refinement, we would need 42,875 degrees of freedom. The map produced by the iterative procedure has no non-regular points.

We also compared the convergence speed of the procedures with and without spline space refinement. From Fig. 12 (bottom, right) we see that the local spline space refinement iteration, executed when the decay of the error already is slow, can slightly speed up the convergence. Although in this case the speedup is not extremely high, already the first TDM iteration after spline space refinement gives the results, which are more accurate in comparison to those obtained after the last iteration of the TDM procedure without spline space refinement. Therefore, this experiment once again confirms that the adaptive spline space refinement allows reducing the number of required iterations and, consequently, the amount of computational resources.

6 Concluding Remarks

We introduced the template mapping problem and presented an iterative adaptive approach for solving it. Based on the results of the experimental investigation, we arrive at the following conclusions.

- The closest point computation (parameter optimization) is beneficial for solving the template mapping problem. The iterative procedure based on TDM converges faster than the procedure based on PDM.
- In comparison with the global refinement of the spline space, the local refinement allows achieving a similar accuracy with significantly less amount of computational resources.
- The iterative minimization of the objective function can be efficiently combined with the spline space refinement.
- The approach is applicable to volumetric data. This is very important from the practical point, in view of applications to isogeometric segmentation and parameterization.

Acknowledgements Supported by the European Union's Horizon 2020 research and innovation programme under grant agreement No. 678727 (the MOTOR project), as well as by the ERC advanced grant CHANGE (GA no. 694515) and by the Austrian Science Fund (FWF NFN S117). We are grateful to our project partners from Dortmund for providing the data for the industrial example.

References

1. Aigner, M., Heinrich, C., Jüttler, B., Pilgerstorfer, E., Simeon, B., Vuong, A.V.: Swept volume parameterization for isogeometric analysis. In: Hancock, E.R., Martin, R.R., Sabin, M.A. (eds.) Mathematics of Surfaces XIII: 13th IMA International Conference. Lecture Notes in Computer Science, vol. 5654, pp. 19–44. Springer, Berlin (2009)
2. Al Akhras, H., Elguedj, T., Gravouil, A., Rochette, M.: Towards an automatic isogeometric analysis suitable trivariate models generation–application to geometric parametric analysis. Comput. Methods Appl. Mech. Eng. **316**, 623–645 (2017)
3. Arioli, C., Shamanskiy, A., Klinkel, S., Simeon, B.: Scaled boundary parameterizations in isogeometric analysis. Comput. Methods Appl. Mech. Eng. **349**, 576–594 (2019)
4. Blake, A., Isard, M.: Active Contours. Springer, London (1998)
5. Bo, P., Ling, R., Wang, W.: A revisit to fitting parametric surfaces to point clouds. Comput. Graph. **36**, 534–540 (2012)
6. Buchegger, F., Jüttler, B.: Planar multi-patch domain parameterization via patch adjacency graphs. Comput. Aided Des. **82**, 2–12 (2017)
7. Cohen, E., Martin, T., Kirby, R.M., Lyche, T., Riesenfeld, R.F.: Analysis-aware modeling: understanding quality considerations in modeling for isogeometric analysis. Comput. Methods Appl. Mech. Eng. **199**, 334–356 (2010)
8. Escobar, J.M., Cascón, J.M., Rodríguez, E., Montenegro, R.: A new approach to solid modeling with trivariate T-splines based on mesh optimization. Comput. Methods Appl. Mech. Eng. **200**, 3210–3222 (2011)
9. Falini, A., Jüttler, B.: THB-splines multi-patch parameterization for multiply-connected planar domains via template segmentation. J. Comput. Appl. Math. **349**, 390–402 (2019)

10. Falini, A., Špeh, J., Jüttler, B.: Planar domain parameterization with THB-splines. Comput. Aided Geom. Des. **35–36**, 95–108 (2015)
11. Forsey, D.R., Bartels, R.H.: Hierarchical B-spline refinement. ACM SIGGRAPH Comput. Graph. **22**, 205–212 (1988)
12. Frey, P.J., George, P.L.: Mesh Generation: Application to Finite Elements. Wiley, New York (2008)
13. Giannelli, C., Jüttler, B., Kleiss, S.K., Mantzaflaris, A., Simeon, B., Špeh, J.: THB-splines: an effective mathematical technology for adaptive refinement in geometric design and isogeometric analysis. Comput. Methods Appl. Mech. Eng. **299**, 337–365 (2016)
14. Giannelli, C., Jüttler, B., Speleers, H.: THB-splines: the truncated basis for hierarchical splines. Comput. Aided Geom. Des. **29**, 485–498 (2012)
15. Giannelli, C., Jüttler, B., Speleers, H.: Strongly stable bases for adaptively refined multilevel spline spaces. Adv. Comput. Math. **40**, 459–490 (2014)
16. Gravesen, J., Evgrafov, A., Nguyen, D.M., Nørtoft, P.: Planar parametrization in isogeometric analysis. In: Floater, M., Lyche, T., Mazure, M.L., Mørken, K., Schumaker, L.L. (eds.) Mathematical Methods for Curves and Surfaces: 8th International Conference. Lecture Notes in Computer Science, vol. 8177, pp. 189–212. Springer, Berlin (2014)
17. Greiner, G., Hormann, K.: Interpolating and approximating scattered 3D-data with hierarchical tensor product B-splines. In: Le Méhauté, A., Rabut, C., Schumaker, L.L. (eds.) Surface Fitting and Multiresolution Methods, pp. 163–172. Vanderbilt University Press, Nashville (1997)
18. Haberleitner, M., Jüttler, B., Scott, M.A., Thomas, D.C.: Isogeometric analysis: representation of geometry. Encyclopedia of Computational Mechanics. Wiley, New York (2017)
19. Hinz, J., Möller, M., Vuik, C.: Elliptic grid generation techniques in the framework of isogeometric analysis applications. Comput. Aided Geom. Des. **65**, 48–75 (2018)
20. Hinz, J., Möller, M., Vuik, C.: Spline-based parameterization techniques for twin-screw machine geometries. IOP Conf. Ser. Mater. Sci. Eng. **425**(012030) (2018)
21. Hoschek, J.: Intrinsic parametrization for approximation. Comput. Aided Geom. Des. **5**, 27–31 (1988)
22. Hughes, T.J.R., Cottrell, J.A., Bazilevs, Y.: Isogeometric analysis: CAD, finite elements, NURBS, exact geometry and mesh refinement. Comput. Methods Appl. Mech. Eng. **194**(39-41), 4135–4195 (2005)
23. Jüttler, B., Langer, U., Mantzaflaris, A., Moore, S., Zulehner, W.: Geometry + simulation modules: implementing isogeometric analysis. PAMM **14**, 961–962 (2014)
24. Kapl, M., Sangalli, G., Takacs, T.: Construction of analysis-suitable G^1 planar multi-patch parameterizations. Comput. Aided Des. **97**, 41–55 (2018)
25. Kelley, C.T.: Iterative Methods for Optimization. Frontiers in Applied Mathematics, vol. 18. Society for Industrial and Applied Mathematics (1999)
26. Kiss, G., Giannelli, C., Zore, U., Jüttler, B., Großmann, D., Barner, J.: Adaptive CAD model (re-)construction with THB-splines. Graph. Models **76**, 273–288 (2014)
27. Kraft, R.: Adaptive and linearly independent multilevel B-splines. In: Le Méhauté, A., Rabut, C., Schumaker, L.L. (eds.) Surface Fitting and Multiresolution Methods, pp. 209–218. Vanderbilt University Press, Nashville (1997)
28. Liseikin, D.V.: Grid Generation Methods. Springer, Netherlands (2010)
29. Liu, L., Zhang, Y., Hughes, T.J.R., Scott, M.A., Sederberg, T.W.: Volumetric T-spline construction using Boolean operations. Eng. Comput. **30**, 425–439 (2014)
30. Möller, M., Hinz, J.: Isogeometric analysis framework for the numerical simulation of rotary screw machines. I. General concept and early applications. IOP Conf. Ser. Mater. Sci. Eng. **425**(012032) (2018)
31. Nian, X., Chen, F.: Planar domain parameterization for isogeometric analysis based on Teichmüller mapping. Comput. Methods Appl. Mech. Eng. **311**, 41–55 (2016)
32. Pan, M., Chen, F., Tong, W.: Low-rank parameterization of planar domains for isogeometric analysis. Comput. Aided Geom. Des. **63**, 1–16 (2018)
33. Pauley, M., Nguyen, D.M., Mayer, D., Špeh, J., Weeger, O., Jüttler, B.: The isogeometric segmentation pipeline. In: Jüttler, B., Simeon, B. (eds.) Isogeometric Analysis and Applications

2014. Lecture Notes in Computational Science and Engineering, vol. 107. Springer, Cham (2015)
34. Pottmann, H., Leopoldseder, S.: A concept for parametric surface fitting which avoids the parametrization problem. Comput. Aided Geom. Des. **20**, 343–362 (2003)
35. Sangalli, G., Takacs, T., Vázquez, R.: Unstructured spline spaces for isogeometric analysis based on spline manifolds. Comput. Aided Geom. Des. **47**, 61–82 (2016)
36. Shamanskiy, A., Simeon, B.: Isogeometric simulation of thermal expansion for twin screw compressors. IOP Conf. Ser. Mater. Sci. Eng. **425**(012031) (2018)
37. Speleers, H., Manni, C.: Optimizing domain parameterization in isogeometric analysis based on Powell-Sabin splines. J. Comput. Appl. Math. **289**, 68–86 (2015)
38. Utri, M., Brümmer, A.: Improvement of the efficiency of twin-screw refrigeration compressors by means of dual lead rotors. In: International Compressor Engineering Conference, Paper 1428 (2016)
39. Utri, M., Brümmer, A.: Energy potential of dual lead rotors for twin screw compressors. IOP Conf. Ser. Mater. Sci. Eng. **232**(012018) (2017)
40. Utri, M., Brümmer, A., Hauser, J.: Comparison of thermodynamic efficiency between constant, dual and multiple lead rotors for an industrial air screw compressor. IOP Conf. Ser. Mater. Sci. Eng. **425**(012025) (2018)
41. Utri, M., Höckenkamp, S., Brümmer, A.: Fluid flow through male rotor housing clearances of dry running screw machines using dimensionless numbers. IOP Conf. Ser. Mater. Sci. Eng. **425**(012033) (2018)
42. Wang, W., Pottmann, H., Liu, Y.: Fitting B-spline curves to point clouds by curvature based squared distance minimization. ACM Trans. Graph. **25**, 214–238 (2006)
43. Wang, W., Zhang, Y., Liu, L., Hughes, T.J.R.: Trivariate solid T-spline construction from boundary triangulations with arbitrary genus topology. Comput. Aided Des. **45**, 351–360 (2013)
44. Wang, X., Qian, X.: An optimization approach for constructing trivariate B-spline solids. Comput. Aided Des. **46**, 179–191 (2014)
45. Xia, S., Qian, X.: Generating high-quality high-order parameterization for isogeometric analysis on triangulations. Comput. Methods Appl. Mech. Eng. **338**, 1–26 (2018)
46. Xiao, S., Kang, H., Fu, X.M., Chen, F.: Computing IGA-suitable planar parameterizations by PolySquare-enhanced domain partition. Comput. Aided Geom. Design **62**, 29–43 (2018)
47. Xu, G., Li, M., Mourrain, B., Rabzuck, T., Xu, J., Bordas, S.P.A.: Constructing IGA-suitable planar parameterization from complex CAD boundary by domain partition and global/local optimization. Comput. Methods Appl. Mech. Eng. **328**, 175–200 (2018)
48. Xu, G., Mourrain, B., Galligo, A., Rabczuk, T.: High-quality construction of analysis-suitable trivariate NURBS solids by reparameterization methods. Comput. Mech. **54**, 1303–1313 (2014)

The manufacturer's authorised representative in the EU is Springer Nature Customer Service Centre GmbH, Europaplatz 3, 69115 Heidelberg, Germany. If you have any concerns regarding our products, please contact ProductSafety@springernature.com

Printed and bound by CPI Group (UK) Ltd, Croydon, CR0 4YY
24/07/2026
02175924-0002